T0396496

Plasmonics and Super-Resolution Imaging

Plasmonics and Super-Resolution Imaging

edited by

Zhaowei Liu

Published by

Pan Stanford Publishing Pte. Ltd.
Penthouse Level, Suntec Tower 3
8 Temasek Boulevard
Singapore 038988

Email: editorial@panstanford.com
Web: www.panstanford.com

British Library Cataloguing-in-Publication Data
A catalogue record for this book is available from the British Library.

Plasmonics and Super-Resolution Imaging

ISBN 978-981-4669-91-7 (Hardcover)
ISBN 978-1-315-20653-0 (eBook)

Printed in the UK by Ashford Colour Press Ltd., Gosport, Hampshire.

Contents

Preface

Surface plasmons are collective electron oscillations at the interface between a metal and a dielectric. They were first predicted in 1957 by Rufus Ritchie, and then followed two decades of extensive exploration. Stemming from this first wave of research, the surface plasmon–based biosensor was commercialized in the early 1990s, and it is probably the most important application that can be identified now.

With advances in nanomanufacturing, the field of plasmonics was "rediscovered" from the late 1990s to the early 2000s, and a variety of contexts, including nanoscale light guiding in metal waveguides, surface plasmon–mediated anomalous light transmission, and the discovery of the perfect lens and the superlens, were predicted and experimentally demonstrated. Plasmonics has now become a major and perhaps the most fascinating part of nanophotonics.

Imaging is an important field of optical science and technology. Extending the resolution of a microscope into the nanoscale and breaking the diffraction limit have been long considered the holy grail in optics. The perfect-lens concept proposed by John Pendry in 2000 brought microscopy into a new era. A negative-refractive-index lens can not only refract light negatively but also recover the "lost treasure" carried by the evanescent waves, thus forming images with perfect resolution. This stimulated the later flourishing of the field of metamaterials, artificial materials with extraordinary material properties that do not readily exist in nature, including a negative refractive index.

Although negative-refractive-index materials at visible frequencies have been demonstrated in laboratories around the mid-2000s, the quality of those materials is simply not high enough for any practical imaging applications. Meanwhile, negative-permittivity materials, that is, plasmonic materials, have proved to be the most practical solution for super-resolution imaging. The unique dispersive property and resonant nature of surface plasmons lead to enhanced optical near-field and subwavelength confinement in space, forming the basic foundations for plasmonic-enhanced super-resolution and high-contrast microscopy technologies.

This book, therefore, was written to cover some major developments in the field of applying plasmonics for various imaging technologies and potential applications. It is a comprehensive and valuable reference for both students with an elementary knowledge of electromagnetism and applied optics, as well as researchers. The chapters were selected from a relatively large pool of work in order to provide readers with a balanced view of the developments in the field. The book starts with a few chapters describing different schemes to incorporate plasmonic principles for super-resolution microscopy. These include the far-field superlens, the metalens, metasurface-based lenses, surface plasmon microscopy, and the hyperlens. Subsequent chapters describe technologies that combine plasmonics with other established imaging methods, such as structure illumination microscopy (SIM), stimulated emission depletion (STED), and single-molecule spectroscopy. The book closes with a few chapters discussing specific applications in nanolithography, plasmonic coloring, and bioimaging. I hope readers find this book useful.

Zhaowei Liu

Chapter 1

The Far-Field Superlens

J. L. Ponsetto and Zhaowei Liu

Department of Electrical and Computer Engineering, University of California, San Diego, 9500 Gilman Drive, La Jolla, CA 92093, USA

zhaowei@ucsd.edu

1.1 Introduction

In this chapter we will discuss a powerful tool for deep subwavelength imaging known as the far-field superlens (FSL). The FSL takes the concept shown by previous examples of "perfect" lenses and near-field superlenses and adds the capability for far-field image detection. To start, we will review the motivating factors for developing this tool and some essential background concepts. Next we will discuss the underlying physical mechanisms behind FSL functionality, followed by experimental examples of the FSL in action. Finally, we will explore several variations and modifications of the FSL design, showing its exciting potential in a wide range of applications. At the end of this chapter, the key ideas will be summarized, placing the FSL and its performance in context amongst other lens-based imaging methods.

Plasmonics and Super-Resolution Imaging

Edited by Zhaowei Liu

ISBN 978-981-4669-91-7 (Hardcover), 978-1-315-20653-0 (eBook)

www.panstanford.com

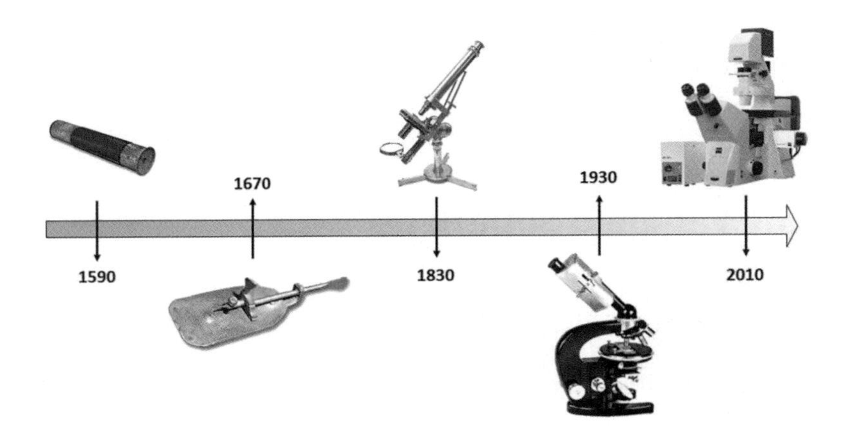

Figure 1.1 Advances in microscopy over the years. 1590: The first compound microscope. 1670: Lens making improves, enabling >250× magnification. 1830: Chromatic and spherical aberrations are minimized. 1930: Phase contrast microscopy invented. 2010: Modern microscopes have additional useful capabilities such as digital image capture, polarization control, and fluorescence imaging.

1.1.1 Background

Optical magnification and microscopy has revolutionized the scientific world since its invention in the late sixteenth century. Over the years, improvements to optical microscopy have vastly expanded its capabilities, some of which are briefly summarized in Fig. 1.1. In recent decades, however, microscopy has found itself increasingly constrained by a fundamental barrier: the diffraction limit of light. It is the wave-like quality of light that leads to diffraction and limits resolution [1]. Heisenberg's uncertainty principle insists that there is a limit to how well we can know the position and momentum of a photon simultaneously, $\Delta p \cdot \Delta x = \hbar \Delta k_x \cdot \Delta x \geq \hbar/2$. To get precise knowledge of the position x, we want to maximize our spatial frequency component Δk_x, but we are limited to k-values within the propagating regime, $k_x \leq 2\pi n/\lambda_0$. Here, p represents momentum, $\hbar$ is the reduced Planck constant, n is the index of refraction of the surrounding medium, and λ_0 is the illuminating wavelength in vacuum. In practice, the resolution may be further limited by factors such as point spread functions (PSFs) and experimental noise, dependent on the optical components in use. In the optical microscopy community, a PSF-based calculation originally used by Ernst Abbe in 1873 is generally adopted to predict the minimum resolvable point-to-point distance d in the object plane as approximately

$$d = \frac{\lambda_0}{2 \cdot \text{NA}}, \tag{1.1}$$

where λ_0 is the free-space wavelength and $\text{NA} = n \cdot \sin(\theta)$ is the numerical aperture of the system, with θ representing the maximum collection angle of the microscope objective. For visible wavelengths in free space, this limits imaging resolution to a few hundred nanometers. No information from finer features beyond the spatial frequency detection bandwidth will be collected. One solution to improve the absolute resolution of a microscope is to use shorter wavelengths for illumination, but in many cases these higher-energy waves can damage the object one wishes to examine. The question then, for a myriad of applications such as biological imaging, lithography, and optical data storage, is as follows: How can we bypass the diffraction limit and achieve super-resolution? In recent years, a number of methods for far-field super-resolution imaging have been proposed and demonstrated [2–6], such as the FSL, the hyperlens, superoscillation, and fluorescence-assisted probing. Compared to other far-field super-resolution imaging techniques, the FSL is desirable due to its wide field of view and its potential for real-time imaging capability.

In this initial section we will discuss some of the key advances preceding the FSL. Specifically, we will examine the perfect lens and the near-field superlens. In doing this, we will uncover important physical mechanisms that are essential to a whole range of imaging tools, including the FSL.

1.1.2 Negative Refraction and the Perfect Lens

For an isotropic medium, the dispersion equation for propagation of electromagnetic waves through matter is

$$\mathbf{k}^2 = \frac{\omega^2}{c^2} n^2, \tag{1.2}$$

where $\mathbf{k}$ represents the wavevector, ω the angular frequency, and c the speed of light. The refractive index n is determined by the complex, frequency-dependent material properties permittivity ε and permeability μ.

$$n^2 = \varepsilon\mu \tag{1.3}$$

For the sake of analytical simplicity, let us assume negligible material loss and take n, μ, and ε as real numbers. Starting with Maxwell's equations for a monochromatic plane wave, we then can write,

$$\vec{\mathbf{k}} \times \vec{\mathbf{E}} = \frac{\omega}{c}\mu\vec{\mathbf{H}} \tag{1.4}$$

$$\vec{\mathbf{k}} \times \vec{\mathbf{H}} = -\frac{\omega}{c}\varepsilon\vec{\mathbf{E}}, \tag{1.5}$$

where $\vec{\mathbf{E}}$ and $\vec{\mathbf{H}}$ represent the electric and the magnetic field, respectively. We can see that when $\varepsilon, \mu > 0$, the resulting $\vec{\mathbf{E}}$, $\vec{\mathbf{H}}$, and $\vec{\mathbf{k}}$ form a right-handed triplet of vectors, as seen in natural materials. However, if ε and μ are both negative, the vectors form a left-handed triplet. Such a condition leads to a negative index of refraction [7]. This change in sign turns out to have a number of fascinating implications, especially in imaging. Perhaps the most striking change in a material of negative refractive index is the refraction direction. Refraction at an interface will be directed at an opposite angle compared to positive-index materials. This effect has been widely exploited to create metamaterials with special properties [8, 9].

In negative-index materials, divergent propagating waves will be refocused, as the spreading ray angles are reversed [7]. More recently, it was posited [10–12] that in negative-index materials, evanescent nonpropagating waves may also be enhanced and contribute to a focused image. Therefore, a slab of negative-index material may act as a perfect lens, perfect in the sense that it may in theory perfectly resolve an image by including the evanescent components of the object's field.

The perfect lens design makes use of a slab of negative-index material with $\varepsilon = -1$ and $\mu = -1$, which has an impedance $Z = \mu/\varepsilon = 1$. Since the impedance is matched with free space, there will be no reflections at the slab interface. A negative-index medium can amplify the evanescent field of an object such as a dipole emitter. It should be noted that energy conservation is not violated, because evanescent waves do not transport energy. Starting from Maxwell's equations and the relevant dispersion relations with z as the lens axis,

$$k_z = \sqrt{\frac{\omega^2}{c^2} - k_x^2 - k_y^2}; \quad \frac{\omega^2}{c^2} > k_x^2 + k_y^2 \tag{1.6}$$

$$k_z = i\sqrt{k_x^2 + k_y^2 - \frac{\omega^2}{c^2}}; \quad \frac{\omega^2}{c^2} < k_x^2 + k_y^2 \tag{1.7}$$

Note that Eq. 1.6 represents propagating waves, while Eq. 1.7 represents evanescent waves that have a purely imaginary k_z. For a slab with permittivity, permeability, and refractive index all equal to -1, with the appropriate boundary conditions for the surrounding free space, we can solve for the transmission T [10],

$$T = e^{-ik_z d_s} \tag{1.8}$$

for both perpendicular (s) and parallel (p) polarizations, where d_s is the slab thickness. The above equation is remarkable because it implies amplification of evanescent waves. However, in addition to requiring material properties that do not exist in nature, this result is only obtained under the conditions of negligible loss and good impedance matching, and thus it has not yet been realized experimentally.

Making isotropic negative-index materials at optical frequencies has proven challenging thus far. Some metals, such as silver and aluminum, exhibit large negative real permittivities with low loss. However, the magnetic properties of such metals, governed by μ, are not as helpful. There are no known natural materials with negative permeability. Fortunately, in the quasi-electrostatic $\omega << c_0\sqrt{k_x^2 + k_y^2}$ approximation at optical frequencies, we can ignore the spatial variation of incident $\vec{\mathbf{E}}$ within a small subwavelength geometry, such as a thin silver slab. This is advantageous because it effectively decouples the electric and magnetic fields, meaning that for p-polarized light, only the permittivity will be relevant for transmission through the slab. So, for an operating wavelength where the silver permittivity is –1, we can expect behavior similar to that of a perfect lens.

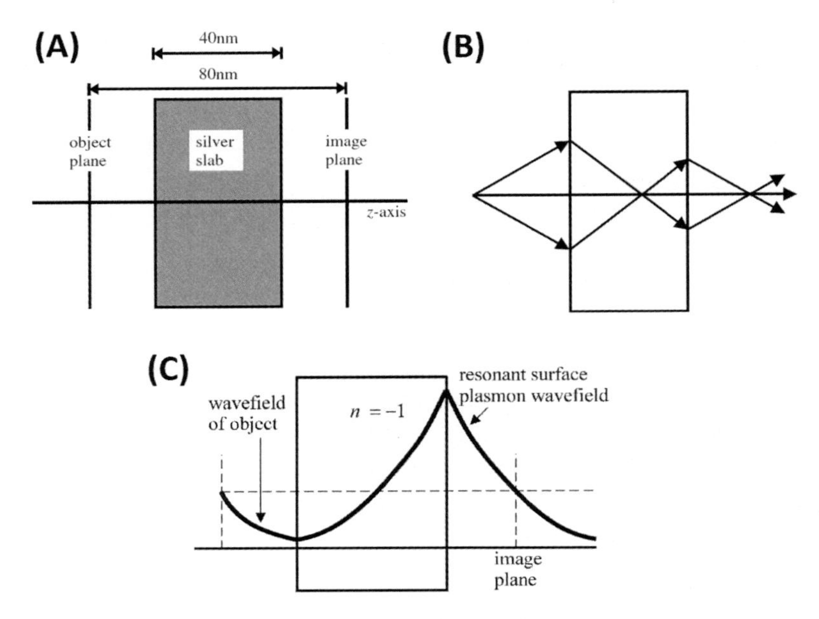

Figure 1.2 (A) Basic schematic of a silver slab superlens, which in certain cases approximates the perfect lens concept. (B) Propagating wavevectors from the object are refocused due to the negative index within the slab. (C) Enhancement of evanescent fields is theoretically predicted to occur within the negative-index medium.

Figure 1.2 shows an example of a thin silver slab serving as a superlens at optical frequencies. A ray-tracing diagram illustrates how propagating components will be refocused twice, once within the slab and again a short distance away from the back surface, at which point the image can be captured. At the same time, evanescent components are enhanced in the form of resonant plasmons on the back surface of the slab.

The intuitive reason for the enhancement of evanescent fields can be looked at from multiple perspectives. Above, we analyze the dispersion relation and boundary conditions to find an amplification in the transmission of evanescent components [10]. Additional understanding can be gained from looking at the concurrent plasmonic mode behavior, as will be discussed in the next section.

1.1.3 The Near-Field Superlens

The perfect lens promises perfect resolution under perfect conditions. Without an ideal negative-index material, this may not be realized experimentally for quite some time. However, a variant called the near-field superlens has proven immensely useful [13–18]. The design and scheme is similar to that of the perfect lens: a thin slab of silver is placed in the near field of the object, and the evanescent fields are captured, amplified, and refocused on the other side of the slab. The superlens design anticipates the imperfect material properties and is optimized to achieve the best experimental resolution possible. The operating wavelength is chosen such that the frequency-dependent permittivity of silver is equal to -1 in order to match equally and oppositely with the surrounding free space. In such a situation, the transmission of monochromatic p-polarized light can be calculated by Fresnel equations to be

$$T_{\mathrm{p}}\left(k_x\right)=\frac{4\varepsilon_2\varepsilon_3 e^{-k_x d_s}}{\left(\varepsilon_1+\varepsilon_2\right)\left(\varepsilon_2+\varepsilon_3\right)-\left(\varepsilon_2-\varepsilon_1\right)\left(\varepsilon_2-\varepsilon_3\right)e^{-2k_x d_s}}, \quad (1.9)$$

where ε_1 is permittivity in the region containing the source, ε_2 in the silver slab, and ε_3 in the region opposite the source. We can see that if the adjacent permittivities are equal and opposite, resonant amplification will occur. Note that this result also relies on the quasi-static approximation. The impedance-matched low-loss silver slab focuses and shifts the phase of propagating components and amplifies the evanescent fields to assemble a focused, super-resolved image on the other side of the slab.

The resonant enhancement of evanescent wavevectors from a source near a superlens can alternatively be explained by plasmonics. We have spoken at length about what happens when we select a frequency such that $\varepsilon = -1$. This is also exactly the condition for a surface plasmon (SP) resonance [18]. Light at this frequency drives resonant plasmonic fields at the interface when adjacent permittivities are equal and opposite. The governing equations for plasmons at an interface are [19]

$$k_x^2=\frac{\varepsilon_1\varepsilon_2}{\varepsilon_1+\varepsilon_2}\frac{\omega^2}{c^2} \quad (1.10)$$

$$k_{j,z}^2 = \frac{\varepsilon_j^2}{\varepsilon_1 + \varepsilon_2} \frac{\omega^2}{c^2}; \quad j = 1, 2 \tag{1.11}$$

$$\varepsilon_1 (\omega) \cdot \varepsilon_2 (\omega) < 0 \tag{1.12}$$

$$\varepsilon_1 (\omega) + \varepsilon_2 (\omega) < 0 \tag{1.13}$$

Examining the above equations, we note that k_x must be real in order to propagate along the interface. This necessitates that the denominator (sum) and numerator (product) of Eq. 1.10 must have the same sign. Additionally, for a bound solution, k_z must be imaginary in both media, so based on Eq. 1.11, the sum $\varepsilon_1 + \varepsilon_2$ must be negative. When $\varepsilon_2 (\omega)$ approaches $-\varepsilon_1 (\omega)$ the excitable k_x band is broadened. At the right frequency, evanescent wavevectors from a subwavelength object couple resonantly to plasmon modes on the opposite surface of a silver slab placed in the near field, thus enabling enhancement and super-resolution imaging at a focal point close to the slab.

Figure 1.3 shows the layered design of a silver near-field optical superlens for experimental demonstration. A patterned chromium film served as the object. The polymethylmethacrylate (PMMA) layer was a spacer, and photoresist (PR) was placed on the other side of the silver superlens to record the image in the near field. The wavelength used was a common lithographic wavelength. We can clearly see, both in the 2D "NANO" image and in the intensity profile, that the superlens provides significantly higher resolution than the diffraction-limited case.

All of our discussion so far has been limited to resonant plasmon-enhanced near-field phenomena. In practice, many free-space imaging systems use far-field detection techniques, which will not directly benefit from the near-field superlens discussed above. In the following section, we will focus on the key advancement made to bring the super-resolution imaging power of the superlens into the propagating far-field regime.

1.2 One-Dimensional Far-Field Superlens Theory

For most optical setups, the simplest and most cost-effective way to detect an image is in the far field with a sensor such as a

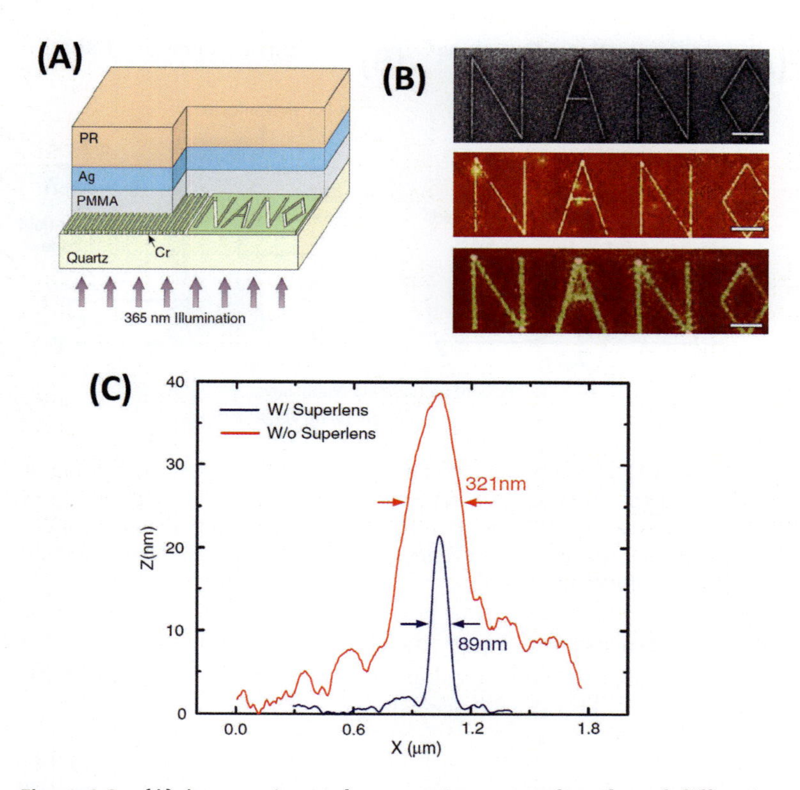

Figure 1.3 (A) An experimental setup using a superlens for subdiffraction lithography. (B) From top to bottom: Focused ion beam (FIB) image of the object. Atomic force microscope (AFM) recording of superlens-developed image. AFM recording of diffraction-limited developed image. The scale bars correspond to 2 microns. (C) Spatial line width in this case was improved from 321 nm down to 89 nm with the superlens.

charge-coupled device (CCD). The near-field superlens improves resolution by enhancing evanescent waves, but a detector placed in the far field of an object can only collect propagating waves. An FSL, therefore, has to bring the super-resolving information carried by the evanescent fields into the propagating regime [20, 21]. This is done by introducing an additional diffractive grating on top of a superlens to reduce the **k**-vector magnitude.

Generally, the conversion from a large **k**-vector evanescent wave to a small **k**-vector propagating wave can be accomplished via a

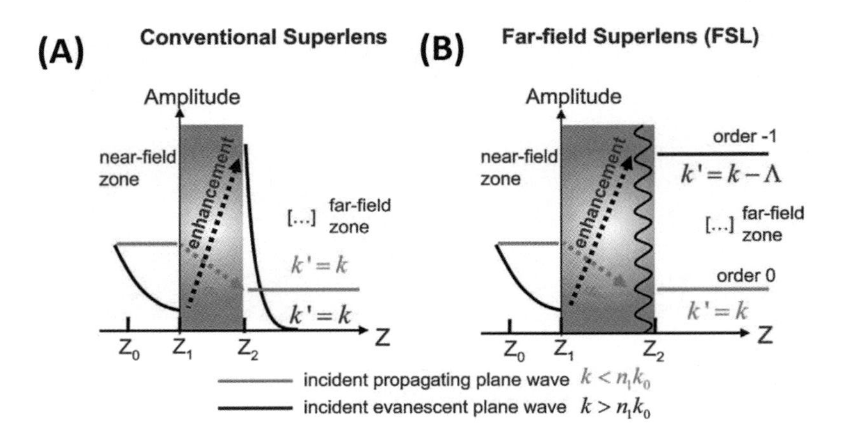

Figure 1.4 (A) The superlens enhances evanescent waves originating at point Z_0 so that they can be captured in the near field of point Z_2 to form a super-resolved image. (B) The FSL gives a similar enhancement but additionally translates the evanescent information into the propagating regime via a diffraction grating.

diffraction process from a periodic grating. For instance, waves with incident transverse wavenumbers k_i diffracted by a grating with periodicity d will be altered according to

$$k_{\mathrm{m}} = k_i + m\frac{2\pi}{\Lambda}, \tag{1.14}$$

where m is an integer representing diffraction order and can be positive, zero, or negative. When m takes a negative value, such as -1, the evanescent waves may be converted to propagating waves by selecting an appropriate grating period Λ. Unfortunately, diffraction from a grating always comes with multiple orders, which means that a far-field detectable wave with **k**-vector $\mathbf{k}_{\mathrm{m}}$ may originate from different $\mathbf{k}_i$ through different diffraction orders. This one-to-multiple diffraction process, also referred to as the *wavevector-mixing issue*, is not desirable in imaging. For exactly this reason, a grating on its own is not commonly treated as an imaging device. Therefore, the grating-assisted FSL design needs to solve the wavevector-mixing problem.

An overview of the FSL principle of operation in comparison to the near-field superlens is shown in Fig. 1.4. No far-field detector can directly collect evanescent information. Therefore, a well-designed

FSL should provide a known, one-to-one transformation of high-k evanescent information into the far-field propagating regime. This unambiguous transformation is essential for final reconstruction of a super-resolved image. The optical transfer function (OTF) of an FSL should give efficient first-order diffraction over a desired band of **k**-vectors and suppress all other orders. This can be done with careful engineering of the geometric and material properties of a relatively simple diffractive grating. Fine-tuning of key FSL parameters is critical in supressing the unwanted diffraction orders. These parameters can be tested and optimized with rigorous coupled-wave analysis (RCWA) prior to fabrication.

To make sense of the diffracted signal exiting the FSL, the original k-space location of the evanescent information has to be identified. In other words, we need to know which diffractive order we are collecting from the grating. In the design discussed in this chapter, diffractive orders higher than $+/-1$ are supressed because SP modes on the metal superlens surface are inherently lossy for very large wavevectors. Design of an OTF, as shown in Fig. 1.5A, allows us to differentiate between the $+1$ and -1 orders, achieving a one-to-one transformation suitable for reconstruction and super-resolution imaging. Under p-polarization, evanescent waves within the wavevector bands $(nk_0 < |k| < nk_0 + k_\Lambda)$ are enhanced by the superlens. All other wavevectors are suppressed. The grating wavevector $\mathbf{k}_\Lambda$ determines the spatial frequency shift imparted by the first-order diffraction process.

By intensity measurements in the far field, a super-resolution image can be reconstructed on the basis of the known OTF. For a well-designed FSL, the raw image obtained can be assumed to primarily contain waves originating from first-order diffraction. Thus, the first step is to take the Fourier transform of the raw image and shift the collected signals by $\mathbf{k}_\Lambda$, according to the one-to-one OTF scheme. The standard, diffraction-limited wavevector information can be obtained by taking the Fourier transform of a diffraction-limited image collected separately. Then, an inverse Fourier transform of the assembled extended spectrum can reconstruct the full super-resolved information in real space.

With a real structure, an ideally rectangular OTF becomes rounded, but the functionality is the same, resulting in a clear

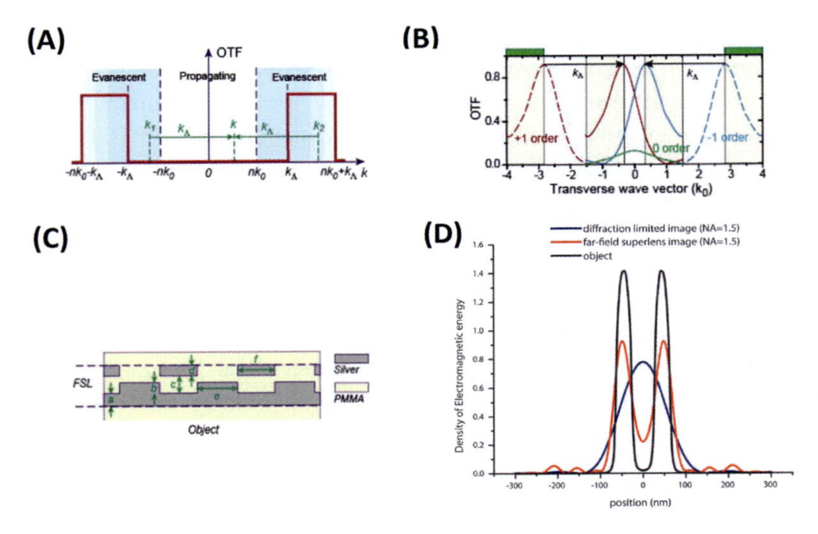

Figure 1.5 (A) Idealized FSL OTF shown as a red line. A known one-to-one transformation from the evanescent band (blue shading) allows for reconstruction of a super-resolution image. (B) RCWA calculated k-space OTF of a silver FSL with $\mathbf{k}_\Lambda = 2.5\mathbf{k}_0$. Dashed curves represent the original evanescent waves, and solid curves show the shifted waves after the grating, now in the propagating regime. The green curve shows the suppressed zero-order diffraction. (C) Key parameters for the superlens and grating components, which can be tuned to achieve the desired OTF. Note that the topmost layer of silver serves primarily to attenuate zero-order diffraction. (D) The far-field superlens image (shown in red) is able to resolve an object (black) that is normally unresolved in a diffraction-limited image (blue).

resolution improvement. In Fig. 1.5B, we draw the OTF passbands (green bars) analogous to those in Fig. 1.5A, with the cutoff determined by a requirement that the ratio of $+1$ to -1 should be larger than e to avoid wavevector mixing. Note that this specially designed FSL only serves to gather information from the band passed by the OTF. For our calculated structure, as shown in Fig. 1.5B, with $\mathrm{NA} = 1.5$, $\mathbf{k}_\Lambda = 2.5\mathbf{k}_0$, this means that only wavevectors in the band $2.8\mathbf{k}_0 < |\mathbf{k}| < 4\mathbf{k}_0$ are unambiguosly retrieved. A diffraction-limited image can capture the lower-k components $|\mathbf{k}| < 1.5\mathbf{k}_0$. In fact, with a clever setup described in a later section of this chapter, these bands can be detected rapidly in a single experimental setup. The remaining gap from 1.5 to $2.8\mathbf{k}_0$ will usually not ruin the imaging result, but it can also be filled in via additional imaging techniques such as structured illumination [23], finally yielding continuous detection over a large band of wavevectors $|\mathbf{k}| < 4\mathbf{k}_0$, clearly indicating that super-resolution is achieved.

The structural geometry shown in Fig. 1.5C was optimized with $a = 35$ nm, $b = d = 55$ nm, $c = 100$ nm, $e = 45$ nm, and $f = 105$ nm. The silver dispersion parameters were based on previous experimental data [22], and the PMMA had a refractive index of 1.52. This structure yielded clear resolution enhancement as shown in Fig. 1.5D.

A number of things can be addressed to further optimize this design. As previously discussed, reducing all unwanted diffraction orders to zero transmission will aid in a cleaner reconstruction. The operational bandwidth could be enhanced via structured illumination or a grazing angle illumination. Also, as we will discuss later in this chapter, design considerations can tune the working wavelength to meet specific application requirements. A thin metal slab with a grating works in the quasi-static approximation, but a true, isotropic negative-index material would give more flexibility. Despite the advances of the FSL, the lens must still be placed in the near field of the object, which introduces experimental difficulty.

1.3 One-Dimensional Experimental Demonstration

An experimental realization of the FSL was achieved [24]. The structure of this device consists of a subwavelength grating added to the back surface of a thin silver slab. The two main functional pieces of this design are as follows: One, the slab selectively enhances the evanescent waves from the object, and two, the grating converts the enhanced evanescent waves into propagating waves for far-field detection.

This FSL is inserted between the object specimen and a normal microscope objective. The FSL OTF must first be measured experimentally for different lateral wavevectors. The collected calibration data allows for reconstruction of an accurate final image.

1.3.1 Verifying the Transfer Function

As an illustrative example, let us take a look at a special situation with a simple object that is itself a grating. For reasons that will be shown here, grating objects are useful for initial calibration measurements of the FSL in order to determine the experimental OTF. For a grating object, in the presence of a grating-based FSL, the wavevectors produced by diffraction are as follows:

$$\alpha_{l,m} = k_i + l\frac{2\pi}{\Lambda_O} + m\frac{2\pi}{\Lambda_{FSL}}, \qquad \alpha_{l,m} < n\mathbf{k}_0 \tag{1.15}$$

Here, l and m are integers representing the diffractive orders from the object and the FSL, respectively. In addition, $\mathbf{k}_0$ is the wavevector of the light in a vacuum, and Λ_O and Λ_{FSL} are the periods of the object and FSL, respectively. Note that both Λ_O and Λ_{FSL} are subwavelength, so the first-order diffraction from either the FSL or the grating object alone will be evanescent.

Assuming a well-designed FSL and object, such that the only nonnegligibly propagating diffracted orders are $\alpha_{0,0}$, $\alpha_{-1,+1}$, $\alpha_{+1,-1}$, we can write the transmitted magnetic far field for normally incident monochromatic light as

$$H_t(x, z) = a_0 e^{i\alpha_0 x + i\gamma_0 z} + a_{-1} e^{i\alpha_{-1} x + i\gamma_{-1} z} + a_1 e^{i\alpha_1 x + i\gamma_1 z} \tag{1.16}$$

with complex $a_p = |a_p| e^{\varphi_p}$ and

$$\alpha_p = p2\pi \frac{\Lambda_O - \Lambda_{FSL}}{\Lambda_O \Lambda_{FSL}} \quad \text{and} \quad \gamma_p = \sqrt{nk_0^2 - \alpha_p^2} \tag{1.17}$$

For a symmetric grating, we find some simplifying equalities:

$$a_{-1} = a_1^*, \qquad \alpha_{-1} = -\alpha_1, \qquad \gamma_{-1} = \gamma_1 \tag{1.18}$$

$$H_t(x, z) = a_0 e^{i\gamma_0 z} + 2|a_1| cos(\alpha_1 x + \varphi_1) e^{i\gamma_1 z} \tag{1.19}$$

The intensity distribution is the squared magnitude of H_t

$$I_t(x, z) = |H_t(x, z)|^2 \tag{1.20}$$

$$\begin{aligned} I_t(x, z) = {} & |a_0|^2 + 4|a_1|^2 \cos^2(\alpha_1 x + \varphi_1) \\ & +4|a_0||a_1| \cos((\gamma_0 - \gamma_1) z + \varphi_0) \cos(\alpha_1 x + \varphi_1) \end{aligned} \tag{1.21}$$

The intensity distribution is dependent on z, and for the sake of reconstruction we can scan for the optimal position in the z plane of the microscope to ensure that $\cos\left(\left(\gamma_0 - \gamma_1\right) z + \varphi_0\right) = 1$, and thus the intensity contrast is maximized along the x direction. Now we have further simplified the intensity equation

$$I_t\left(x, z\right) = \left|a_0\right|^2 + 4\left|a_1\right|^2 cos^2\left(\alpha_1 x + \varphi_1\right) + 4\left|a_0\right|\left|a_1\right| cos\left(\alpha_1 x + \varphi_1\right), \quad (1.22)$$

where $|a_0|$ and $|a_1|$ can be retrieved from analysis of the measured DC and AC spatial frequency components, respectively. Note that some error is introduced because the intensity from the microscope is collected from a finite depth of field (DOF) rather than a single z depth. Fortunately, the interference pattern along z is usually significantly longer than the DOF, so we can safely ignore this source of distortion.

The negative first-order OTF of the FSL is

$$t_{-1}\left(k\right) = \frac{\left|a_1\left(k - \Lambda_{\mathrm{FSL}}\right)\right|}{\left|b_1\left(k\right)\right|} \quad (1.23)$$

Here $|b_1(k)|$ is the field amplitude incident on the FSL. This is difficult to measure in experiment, so we calculate it for the various experimental gratings using RCWA, starting from Maxwell's equations. Incorporating these results with the $|a_1(k - \Lambda_{\mathrm{FSL}})|$ retrieved from intensity measurements, we can calculate the transfer function for different object periodicities one at a time using the above procedure. Piecing these data points together, we can obtain the full experimental OTF of our FSL.

1.3.2 Building a Far-Field Superlens

In this section we will look into the details of how to fabricate an FSL [25]. An FSL should be held at a constant z distance and a laterally fixed position in the near field of the object or specimen. For proof-of-concept purposes, an object and an FSL can be fabricated together as a single structure to facilitate a simpler, verifiable imaging demonstration. We will presently discuss how to make such a structure with common lithography techniques.

First, a 40 nm layer of chromium (Cr) was deposited via e-beam evaporation on a quartz wafer substrate, as shown in Fig. 1.6. The desired object geometry was then inscribed into the Cr film with a focused ion beam (FIB) using accelerated gallium atoms to

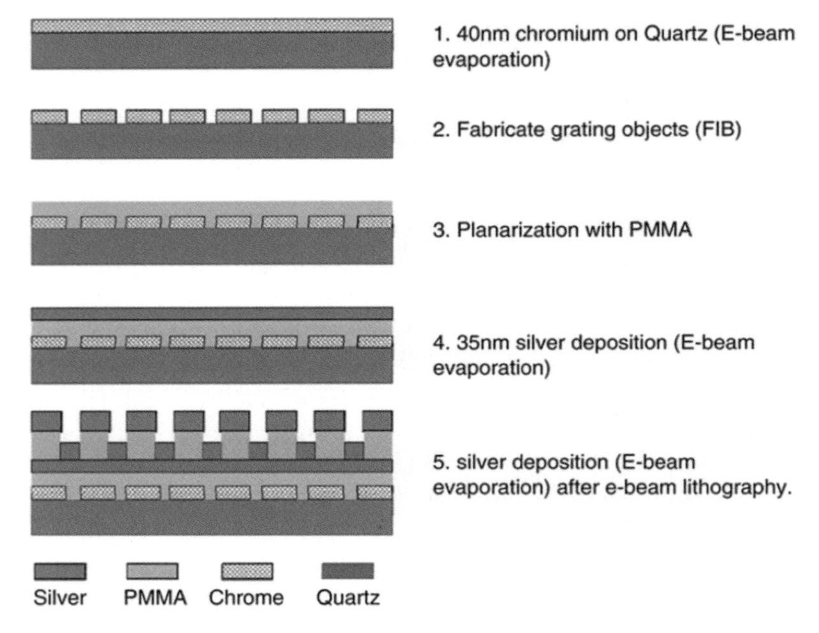

Figure 1.6 Fabrication steps for making an FSL with an example object fabricated directly underneath. Arbitrary objects patterned into the chrome layer can be resolved beyond the diffraction limit by the FSL.

mill 40 nm wide object grooves. Next, a spacing layer of PMMA was introduced with four spin coatings, totaling a 700 nm initial thickness. This planarizes the surface since the coating process is a semiconformal deposition.

Once planarized, the PMMA was then etched down to just 35 nm with a Tegal plasma etcher. When the desired thickness was reached, the sample was baked above the glass transition temperature of PMMA (105°C) for 30 minutes. This reflowing process reduces the surface roughness from plasma etching. This process was also tested for deep metal grooves of 100 nm, and a planar result was still achieved with PMMA coating. Finally, the FSL slab structure was fabricated on top of the PMMA. A 35 nm uniform silver film was deposited via e-beam evaporation. Surface roughness was kept to a minimum to reduce scattering loss. The FSL grating structure was patterned using e-beam lithography (EBL) and an additional layer of

PMMA with 100 nm thickness. After exposure, a 55 nm thick silver layer was deposited with e-beam evaporation, with the highest layer providing further attenuation to the zero-order transmission. The grating period was 150 nm by design. This fabrication procedure was used to create the samples for the experimental imaging results shown in the following section.

1.3.3 Experimental Imaging

The experimental setup for demonstration of FSL super-resolution is schemetically illustrated in Fig. 1.7A. The object is placed in the near field of the FSL, which was fabricated by the method described in the previous section. Far-field images were obtained with a Zeiss Axiovert microscope using an oil immersion objective (100X magnification, NA = 1.4). The sample was illuminated at normal incidence with a laser at a wavelength of 377 nm. The images were recorded with a Princeton Instruments VersArray 1300F UV-CCD.

Imaging measurements were first calibrated by an experimental verification of the OTF using several grating objects with different spatial periods, as discussed in detail in Section 1.3.1. The experimental OTF comparison with the RCWA calculation is shown in Fig. 1.7B, demonstrating good agreement. The confirmed OTF data is treated as known information to assist the image reconstruction at a later stage.

Imaging results will now be shown for two nanowires that served as line sources. By comparing a scanning electron microscope (SEM) image, a diffraction-limited image, and a reconstructed FSL image, super-resolution can be demonstrated (see Fig. 1.7C). The wire width in this example was 50 nm and the gap between the wires was 70 nm. Compared to the operational wavelength of 377 nm, the object geometry was deeply subwavelength.

As would be theoretically expected, the p-polarized waves undergo the superlensing process, while the s-polarized waves do not. This difference is due to the plasmonic coupling present in the case of p-polarization, which enables the enhancement of the evanescent wavevectors. Since s-polarized waves, in the quasi-static approximation, are governed only by the metallic permeability (which is positive), these waves do not couple high-k information

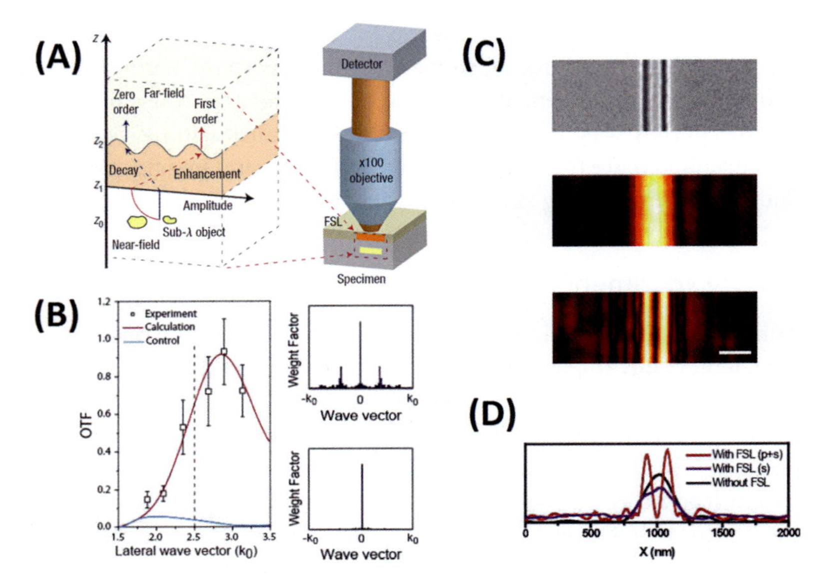

Figure 1.7 (A) Schematic illustration of the experimental setup and the sample comprising both the FSL and a subwavelength object. (B) OTF measurement and calculation. The blue control curve was obtained by replacing the silver slab with a PMMA layer. On the right, the k-space spectra of collected object information from p- (upper right) and s- (lower right) polarized light are shown. (C) From top to bottom: The SEM image, diffraction-limited image, and super-resolved FSL image of two wire objects. The white scale bar represents 200 nm. (D) Imaged intensity profiles of the two parallel nanowires, showing the resolution improvement gained with the FSL, combining both polarizations.

to the far field. We see that the purple curve in Fig. 1.7D does not resolve the object, nor does the diffraction-limited image taken in the absence of the FSL. By comparison with an SEM image, we can verify the accurately super-resolved image captured with the FSL. By combining the recorded data from s- and p-polarizations, the reconstructed intensity profile from the FSL shows the two clearly resolved lines with subwavelength spacing.

The image processing and reconstruction procedure, as specified in previous sections, is straightforward and not computationally intensive, allowing for quick production of the final image. In the case of this particular demonstration, the line pair was purposely

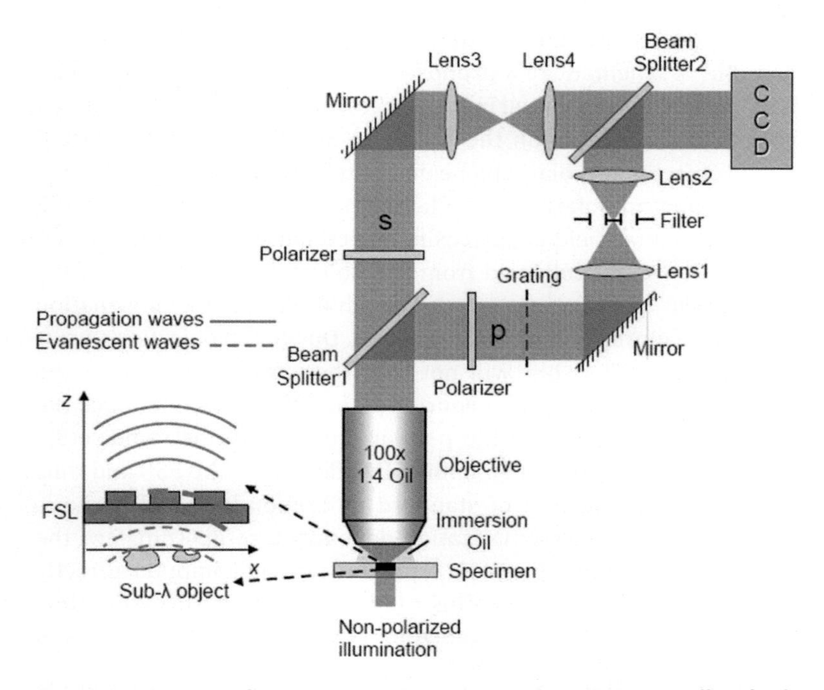

Figure 1.8 A setup for imaging with a 1D superlens that can collect both standard and a super-resolved wavevector bands simultaneously using two different polarizations (s and p). This enables real-time reconstruction performed directly by the optical elements.

oriented at a small angle with respect to the FSL grating direction, so the signal from different diffracted orders would be easier to separate in Fourier space. The spectrum obtained from p-polarized illumination will be shifted back to the original, higher **k**-vector region according to the design of the FSL. In this specific case, the shift in **k** is equal to $2.51\mathbf{k}_0$. The extended spectrum from p-polarized illumination will be combined with the low **k**-vector spectrum obtained from s-polarized illumination and reconstructed via inverse Fourier transform to generate the super-resolution image. These results constitute an initial proof of concept for the FSL in experiment, but many advances are still to be discussed.

Aside from the aforementioned numerical image reconstruction method, the super-resolved image may be directly reconstructed with optical hardware so that real-time FSL imaging becomes possible. A schematic of a proposed optical setup for far-field real-time FSL imaging is shown in Fig. 1.8. Normally incident unpolarized laser light is incident on the specimen and the FSL and collected by the objective. A polarizing beamsplitter creates two optical paths for the different polarizations. The p-polarized light component will experience a far-field superlensing effect and will allow for high-k information to be collected from the object. An additional grating in this arm of the setup is used to shift the high-k information back to its appropriate k-space position. Due to the objective's prior magnification, this shift will not take the wavevectors out of the propagating regime. Next, a Fourier-plane filter is used to block the zero-order component of the p-polarized light. In the same setup, the s-polarized light will be minimally affected by the FSL and thus allow for the collection of standard diffraction-limited $\mathbf{k}$-vectors. In total, this distinct polarization-dependent setup transfers the burden of reconstruction from a postprocessing computer directly to the optical components. This removes any postprocessing time limitation, enabling rapid acquisition of super-resolution images in real time.

1.4 Tuning the Operational Wavelength

It has been shown [26] that FSLs can be cleverly designed to operate at arbitrary, user-selected illumination wavelengths. This is done by tuning the geometry as well as the material properties of both the FSL and the surrounding dielectric material. Two example results of wavelength tuning by adjusting the permittivity values and structural geometry are discussed and numerically demonstrated in this section. This flexibility of the FSL is useful for optimized super-resolution imaging of phenomena at specific operating frequencies.

A working FSL wavelength of 457.9 nm (a typical Argon laser wavelength) will be sought in this example. From the silver slab in an optimized FSL, we want a high evanescent enhancement factor over a broad wavevector band. First, the slab thickness and surrounding dielectric material permittivity are selected such that the working wavelength for the slab superlens λ_{NSL} is close to our

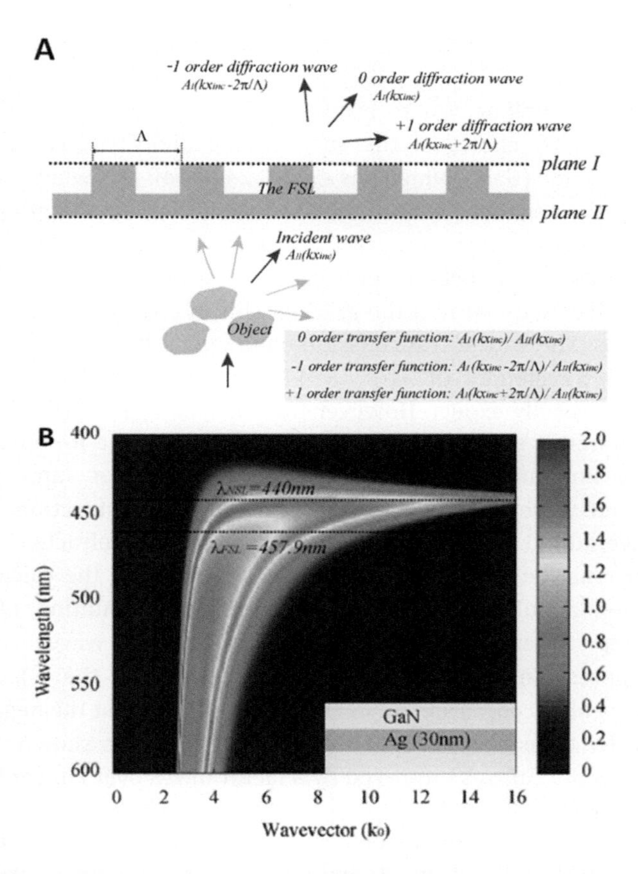

Figure 1.9 (A) The FSL grating is designed so that a one-to-one transformation from the evanescent to the propagating regime can be used for super-resolution reconstruction. (B) The colorbar shows the calculated transmission (log scale) of a silver slab surrounded by GaN, showing the importance of operational wavelength choice, and the relation with the device's spatial bandwidth.

desired FSL working wavelength, λ_{FSL}. As shown in Fig. 1.9B, this is done by calculating the transmission for targeted wavelengths and wavevectors. Second, the FSL grating geometry and metal permittivity are optimized to achieve the desired OTF for the diffracted orders. As shown in Fig. 1.9A, the design of the grating is such that for a given evanescent wavevector within the targeted band, there is a one-to-one transformation into the propagating regime.

First, we look at how to bring an FSL to the desired operating wavelength by choice of the surrounding dielectric. We choose gallium nitride (GaN), which has $\varepsilon_{\mathrm{d}} = 6.24$ at 440 nm, which closely matches the negative permittivity of silver at that wavelength ($\varepsilon_{\mathrm{m}} = -6.5 + 0.19i$).

The transmission coefficients can be first calculated for just the 30 nm silver slab without the grating. This of course is a near-field superlens. The widest transmission band opens at the permittivity-matched wavelength of $\lambda_{\mathrm{NSL}} = 440$ nm (see Fig. 1.9B), as we would physically expect. However for our FSL, $\lambda_{\mathrm{FSL}} = 457.9$ nm is an even better wavelength to operate at. In fact, the **k**-dependent transfer function at 457.9 nm will be favorable for suppressing unwanted wavevectors in the FSL imaging reconstruction. Once we have added the grating to our structure, the refractive index of GaN must be taken into account when analyzing the calculated diffracted angular spectrum. Successful implementation of this multistep strategy is confirmed for a given incident wavevector by the suppression of all diffractive orders except for the -1 order, which is enhanced, as shown in Fig. 1.10A. We see that the negative-first-order transmission is dominant, as required. As shown in Fig. 1.10C, resolution was improved by a factor of 2.5 over conventional microscopy with this FSL.

Another method for wavelength tuning is to adjust the effective metal permittivity. Using a metal–dielectric composite, we can decrease the composite permittivity to match the surrounding dielectric at visible frequencies, as shown by effective medium theory (EMT),

$$\varepsilon_{\mathrm{e}} = \frac{1}{2}\left[(2p-1)(\varepsilon_{\mathrm{m}} - \varepsilon_{\mathrm{d}}) \pm \sqrt{(2p-1)^2(\varepsilon_{\mathrm{m}} - \varepsilon_{\mathrm{d}})^2 + 4\varepsilon_{\mathrm{m}}\varepsilon_{\mathrm{d}}}\right] \quad (1.24)$$

Here, ε_{e} is the effective permittivity and ε_{m} and ε_{d} are the permittivities of the metal and dielectric, respectively. The volume filling ratio is p, and the sign is chosen so that the imaginary part of ε_{e} is positive. The accuracy of EMT depends on subwavelength feature geometry; otherwise the approximation is inaccurate. So, for extremely small object wavevectors approaching the composite

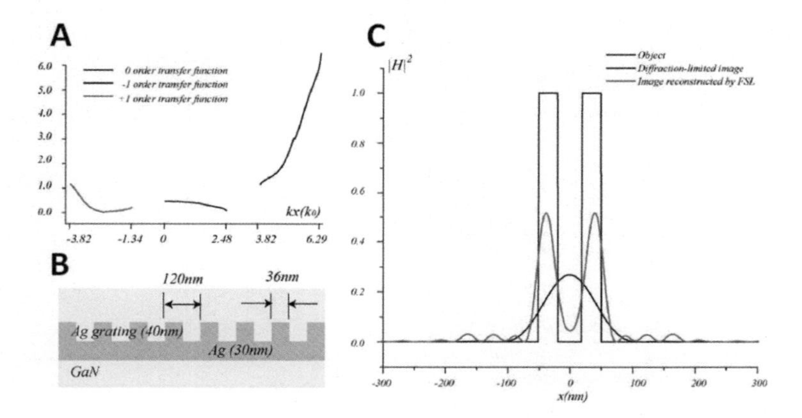

Figure 1.10 (A) A k-space plot of the transfer function for different diffractive orders shows the desired dominance of a the -1 order. (B) Structural parameters of the optimized silver FSL. (C) Real-space imaging results of the structure, with careful choice of the surrounding dielectric. Super-resolution is achieved.

structure size, this could distort our physical picture. However, for 20 nm composite features, the EMT approximation will hold for up to $\lambda/10$ resolution. Here, an Ag–air composite (shown in Fig. 1.11) is used as the metal slab, and Al_2O_3 is used as the surrounding dielectric. For a composite with $p = 0.85$, the effective real part of ε_e is -3.27, almost oppositely matched to that of Al_2O_3 at $\varepsilon_d = 3.17$. In a similar manner to the last example, we see that the -1 order of diffraction dominates with an appropriate choice of grating geometry, and impressive super-resolution is attained.

Comparing the two methods shown in this section, the first is simpler from a fabrication standpoint. However, the second method is more flexible in terms of tunability, since the composite filling ratio can be arbitrarily chosen. These two design methods can always be combined, if necessary.

1.5 The Two-Dimensional Far-Field Superlens

Thus far, we have only discussed 1D imaging with the FSL. However, this is by no means the limit of what can be done. In this section

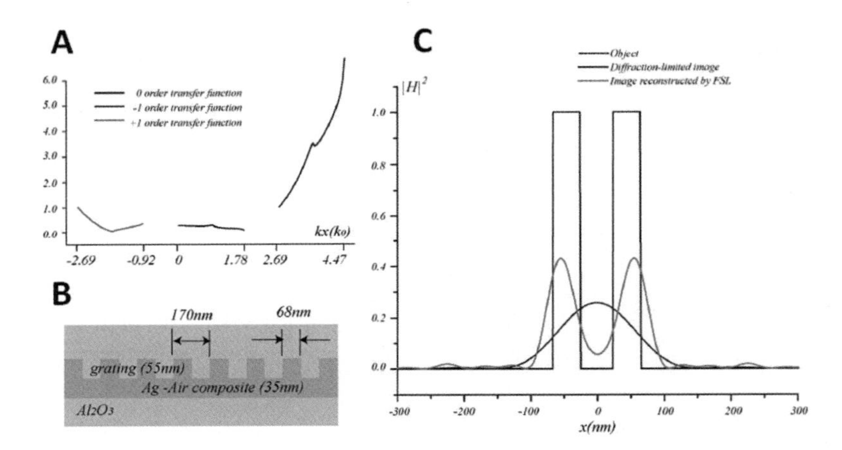

Figure 1.11 (A) The negative first-order diffraction transfer function shows significant enhancement, while other orders are suppressed, as desired. (B) The optimized geometry of the Ag–air composite FSL. (C) Subdiffraction imaging results for a structure tuned by adjusting the effective metal permittivity.

we will discuss the theory of 2D imaging with FSLs [27]. Two-dimensional images may be acquired with resolution beyond the diffraction limit by rotating the FSL relative to the object and combining several subimages in a reconstruction step. By rotating the FSL and recording the angular spectrum at various orientations, one can fill in a very large 2D area in k-space.

This chapter section also introduces an alternate multilayer FSL design that can intrinsically work over a broad range of visible wavelengths. The metamaterial FSL is a metal–dielectric multilayer with a 1D subwavelength grating on the back surface. This design improves performance in two important ways. One, the negative first-order diffraction shows increased dominance, with transmission 3 orders of magnitude larger than other orders. Second, the EMT properties of the multilayer structure allow for a wavelength-tunable design, provided the layer thicknesses are much smaller than the operating wavelength.

Macroscopically, evanescent enhancement in the 2D FSL relative to propagating components can be understood via EMT. For transverse magnetic (TM) polarization, the dispersion relation of the multilayer is

$$\frac{k_x^2}{\varepsilon_{\mathrm{e},z}} + \frac{k_z^2}{\varepsilon_{\mathrm{e},x}} = k_0^2, \tag{1.25}$$

where $\varepsilon_{\mathrm{e},z} > 0$ and $\varepsilon_{\mathrm{e},x} < 0$ are the effective permittivities of the structure in the z and x directions, respectively. Looking at the hyperbolic blue isofrequency curve in Fig. 1.12C we see that only wavevectors $\mathbf{k} > \mathbf{k}_\mathrm{m}$ are allowed to propagate in the multilayer medium. Microscopically, as we have discussed before, SPs can explain the physical picture. Surface plasmon polariton (SPP) modes split into symmetric and antisymmetric modes on a thin film, due to coupling between modes on the front and back of the film. In the presence of many very thin films, a multitude of hybridized modes may exist and be coupled, and some can be highly compact. In this case, the transmission for high **k**-vector waves can be quite efficient over a wide, continuous range (Fig. 1.12D), since many frequencies can be supported by the various SPP modes. For the propagating waves, however, adding layers simply means a thicker metal structure overall and less transmission. This helps allow for better dominance of the -1 diffracted order.

Similar to the 1D case, the transfer function is calculated with the aid of RCWA for reconstruction purposes. The data is collected by six rotated orientations of the FSL along with a single reflection-mode diffraction-limited measurement. A Fourier space numerical algorithm is used to reconstruct the extended angular spectrum and a real-space image. It is shown in Fig. 1.13 that super-resolution is achieved with a working wavelength of 405 nm and a final resolution of 100 nm. For this demonstration, a subdiffraction distribution of objects is examined. In the diffraction-limited case, we see the standard OTF based on regularly propagating wavevectors. In the case of the 2D FSL, an additional ring of k-space information is captured and used to form a super-resolved image. In future work, the field of view could potentially be much larger than what was shown here as a proof of concept. Care must be taken to avoid large gaps in the collected k-space region, otherwise artifacts will be introduced when incorporating very high-k information into the final image. For experimental realization of this concept, one needs the ability to rotate the FSL relative to the sample in the near field, which could introduce complications and slow imaging speed considerably.

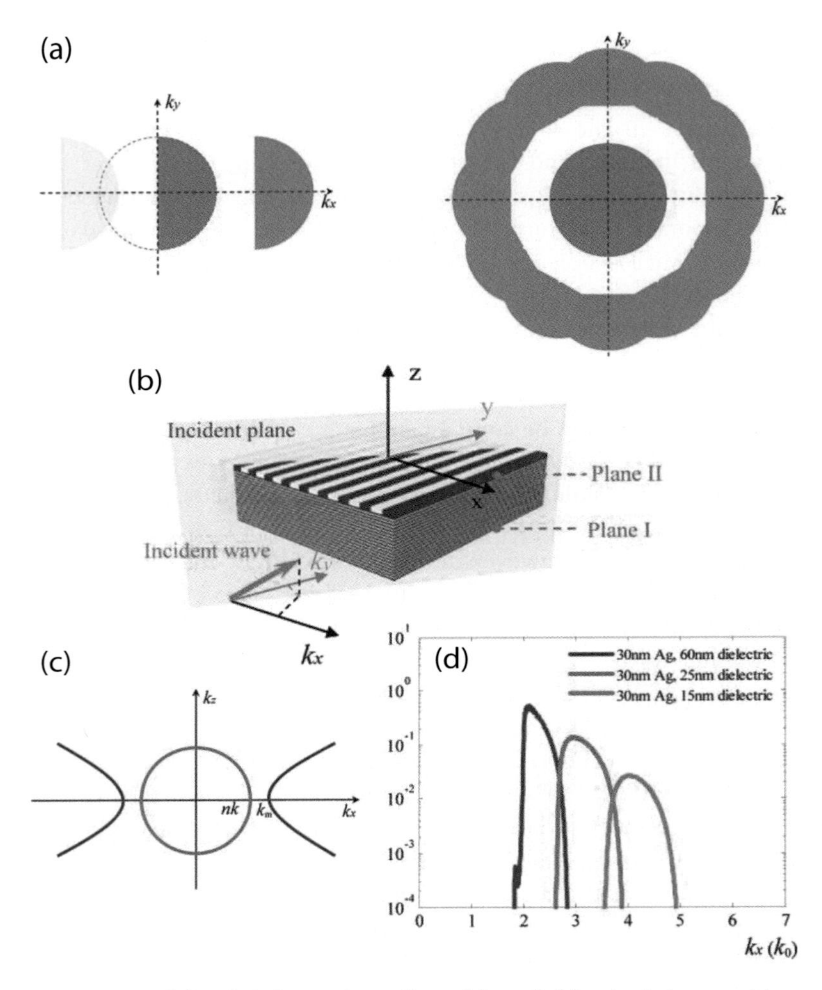

Figure 1.12 (A) Left: Information collected from (left) a single image with a 2D FSL, originating from the +1, 0, and −1 diffracted orders. Note that only the right halves of the collection regions are shown. The requirement of a one-to-one reconstruction means we have to discard some information with ambiguous origin. Right: Negative first-order information collected from measurement at six orientations (green regions) and one simple diffraction-limited reflection image (red region). (B) 2D FSL design, showing multilayer structure and a grating on top. (C) Dispersion relation of the multilayer 2D FSL. (D) Broadband transmission performance (for different Ag filling fractions).

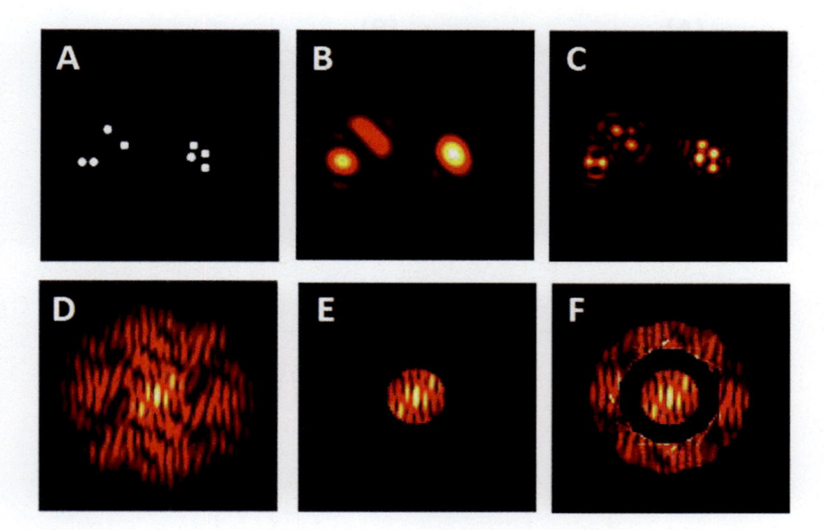

Figure 1.13 (A–C) Real and (D–F) *k*-space depictions of (left to right) the actual object, the conventional diffraction-limited image, and the calculated 2D FSL image. Super-resolution of the original objects is readily apparent with use of the 2D FSL. The real-space field of view is 2 μm × 2 μm.

1.6 Summary

Discussions of the past, present, and future of the FSL are plentiful [28–30]. In recent years, many exciting developments have taken place in the field of super-resolution, especially surrounding the idea of the FSL [31–35]. As technology, fabrication, and theories are improved, new opportunities will open up to researchers.

The value of the FSL is best understood in the context of other available imaging tools, as summarized in Fig. 1.14. In a standard diffraction-limited microscopy setup such as epifluorescence microscopy, light scattered or emitted from an object falls into two categories, propagating and evanescent. Propagating light will be collected by a lens and be focused into the imaging plane. Evanescent wavevectors will decay exponentially away from the object surface and will be lost in the far field.

In a near-field superlens setup, these evanescent components can be amplified by a superlens placed in the near field of the

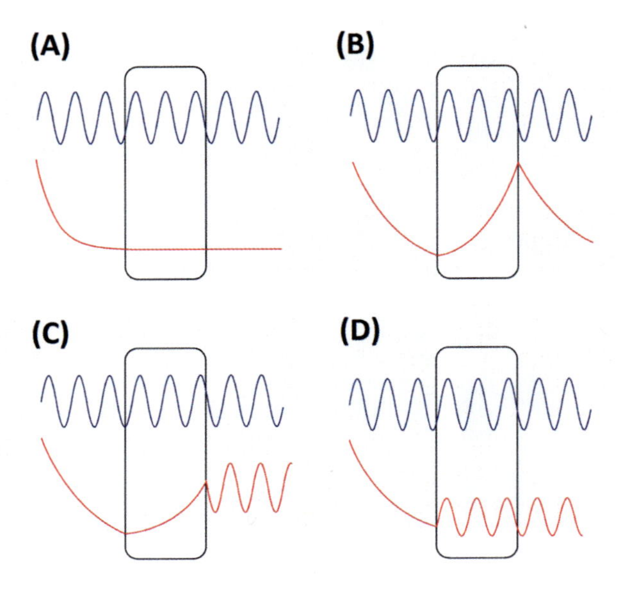

Figure 1.14 Comparison of imaging performance for propagating (blue) and evanescent (red) wavevectors. (A) The conventional lens. (B) The near-field superlens. (C) The FSL. (D) The hyperlens.

object. Thus, the image can be refocused again in the near field on the opposite side of the lens. The image will be super-resolved because both propagating and evanescent components contribute to the focused image. The FSL behaves similarly to a near-field superlens, with the added feature of a grating on the back side to reduce the evanescent wavevectors so that they may propagate in free space after enhancement in the lens. This allows for a much more convenient far-field detection setup. By comparison, a hyperlens [2] is also placed in the near field of the object and, by design, allows for propagation and magnification of evanescent waves within the lens itself, thus shrinking the wavevectors and preparing the waves for free-space propagation upon exiting the back surface of the hyperlens.

In summary, the FSL is an exciting development in the field of super-resolution optics. It has the ability to amplify evanescent object wavevectors and convert them via a one-to-one diffraction grating transfer function to propagating wavelengths without loss of information. This is done by taking advantage of the short wavelengths of resonant plasmons accessible in thin metal films at optical frequencies. The FSL is a valuable tool for surpassing the diffraction limit without the constraints and experimental difficulties of near-field detection setups or slow scanning processes. Some challenges for FSL operation are bandwidth limitations, reconstruction complexity and sensitivity to noise, and experimental difficulty associated with placing the FSL in the near field of the object plane. Variations of the FSL have shown promise for 2D imaging, white light imaging, and all-dielectric imaging. In the future, possibilities for improving the FSL include development of more flexible, simple, and robust designs and implementations. Potential applications for the FSL include bioimaging of subcellular organelles, defect detection, data storage, and lithography, amongst others. This chapter has discussed the current state of affairs in the world of FSLs, but the door remains open for exciting future innovations to improve the practicality and performance of these tools.

References

1. Born, M., Wolf, E., and Bhatia, A. (2002). *Principles of Optics: Electromagnetic Theory of Propagation, Interference and Diffraction of Light* (Cambridge University Press).
2. Liu, Z., Lee, H., Xiong, Y., Sun, C., and Zhang, X. (2007). Far-field optical hyperlens magnifying sub-diffraction-limited objects, *Science*, **315**, p. 1686.
3. Jiang, W. X., Qiu, C. W., Han, T. C., Cheng, Q., Ma, H. F., Zhang, S., and Cui, T. J. (2013). Broadband all-dielectric magnifying lens for far-field high-resolution imaging, *Adv. Mater.*, **25**, pp. 6963–6968.
4. Hell, S. W. (2010). Far-field optical nanoscopy, in A. Gräslund, R. Rigler and J. Widengren (eds.), *Single Molecule Spectroscopy in Chemistry, Physics and Biology, Springer Series in Chemical Physics*, Vol. 96 (Springer Berlin Heidelberg), pp. 365–398.
5. Jacobs, Z., Alekseyev, L. V., and Narimanov, E. (2006). Optical hyperlens: far-field imaging beyond the diffraction limit, *Opt. Express*, **14**, pp. 8247–8256.
6. Huang, F. M., and Zheludev, N. I. (2009). Super-resolution without evanescent waves, *Nano Lett.*, **9**, pp. 1249–1254.

7. Veselago, V. G. (1968). The electrodynamics of substances with simultaneously negative values of ε and μ, *Sov. Phys. Uspekhi*, **10**, pp. 509–514.
8. Shalaev, V. M. (2007). Optical negative-index metamaterials, *Nat. Photonics*, **1**, pp. 41–48.
9. Yen, T. J., Padilla, W. J., Fang, N., Vier, D. C., Smith, D. R., Pendry, J. B., Basov, D. N., and Zhang, X. (2004). Terahertz magnetic response from artificial materials, *Science*, **303**, pp. 1494–1496.
10. Pendry, J. B. (2000). Negative refraction makes a perfect lens, *Phys. Rev. Lett.*, **85**, pp. 3966–3969.
11. Pendry, J. B., and Ramakrishna, S. A. (2003). Refining the perfect lens, *Physica B*, **338**, pp. 329–332.
12. Smith, D. R., Schurig, D., Rosenbluth, M., Schultz, M., Ramakrishna, S. A., and Pendry, J. B. (2003). Limitations on subdiffraction imaging with a negative refractive index slab, *Appl. Phys. Lett.*, **82**, pp. 1506–1508.
13. Liu, Z., Fang, N., Yen, T.-J., and Zhang, X. (2003). Rapid growth of evanescent wave by a silver superlens, *Appl. Phys. Lett.*, **83**, pp. 5184–5186.
14. Fang, N., Liu, Z., Yen, T.-J., and Zhang, X. (2003). Regenerating evanescent waves from a silver superlens, *Opt. Express*, **11**, pp. 682–687.
15. Cubukcu, E., Aydin, K., Ozbay, E., Foteinopoulou, S., and Soukoulis, C. M. (2003). Subwavelength resolution in a two-dimensional photonic-crystal-based superlens, *Phys. Rev. Lett.*, **91**, p. 207401.
16. Fang, N., Lee, H., Sun, C., and Zhang, X. (2005). Sub-diffraction-limited optical imaging with a silver superlens, *Science*, **308**, pp. 534–537.
17. Fang, N., and Zhang, X. (2003). Imaging properties of a metamaterial superlens, *Appl. Phys. Lett.*, **82**, pp. 161–163.
18. Ramakrishna, S. A., Pendry, J. B., Schurig, D., Smith, D. R., and Schultz, S. (2002). The asymmetric lossy near-perfect lens, *J. Mod. Opt.*, **49**, pp. 1747–1762.
19. Novotny, L., and Hecht, B. (2012). *Principles of Nano-Optics* (Cambridge University Press).
20. Durant, S., Liu, Z., Steele, J. M., and Zhang, X. (2006). Theory of the transmission properties of an optical far-field superlens for imaging beyond the diffraction limit, *J. Opt. Soc. Am. B*, **23**, pp. 2383–2392.
21. Liu, Z., Durant, S., Lee, H., Pikus, Y., Fang, N., Xiong, Y., Sun, C., and Zhang, X. (2007). Far-field optical superlens, *Nano Lett.*, **7**, pp. 403–408.

22. Johnson, P. B., and Christy, R. W. (1972). Optical constants of noble metals, *Phys. Rev. B*, **6**, p. 4370.
23. Gustafsson, M. G. L. (2000). Surpassing the lateral resolution limit by a factor of two using structured illumination microscopy, *J. Microsc.*, **198**, pp. 82–87.
24. Liu, Z., Durant, S., Lee, H., Pikus, Y., Xiong, Y., Sun, C., and Zhang, X. (2007). Experimental studies of far-field superlens for sub-diffractional optical imaging, *Opt. Express*, **15**, pp. 6947–6954.
25. Lee, H., Liu, Z., Xiong, Y., Sun, C., and Zhang, X. (2008). Design, fabrication and characterization of a far-field superlens, *Solid State Commun.*, **146**, pp. 202–207.
26. Xiong, Y., Liu, Z., Durant, S., Lee, H., Sun, C., and Zhang, X. (2007b). Tuning the far-field superlens: from UV to visible, *Opt. Express*, **15**, pp. 7095–7102.
27. Xiong, Y., Liu, Z., Sun, C., and Zhang, X. (2007a). Two-dimensional imaging by far-field superlens at visible wavelengths, *Nano Lett.*, **7**, pp. 3360–3365.
28. Zhang, X., and Liu, Z. (2008). Superlenses to overcome the diffraction limit, *Nat. Mater.*, **7**, pp. 435–441.
29. Narimanov, E. E. (2007). Far-field superlens: optical nanoscope, *Nat. Photonics*, **1**, pp. 260–261.
30. Verma, R., and Lee, H. (2011). Far-field superlensing, US Patent App. 12/658,342.
31. Qiu, C., Zhang, X., and Liu, Z. (2005). Far-field imaging of acoustic waves by a two-dimensional sonic crystal, *Phys. Rev. B*, **71**, p. 054302.
32. Shen, N. H., Foteinopoulou, S., Kafesaki, M., Koschny, T., Ozbay, E., Economou, E. N., and Soukoulis, C. M. (2009). Compact planar far-field superlens based on anisotropic left-handed metamaterials, *Phys. Rev. B*, **80**, p. 115123.
33. Wang, Z., Guo, W., Li, L., Luk'yanchuk, B., Khan, A., Liu, Z., Chen, Z., and Hong, M. (2011). Optical virtual imaging at 50 nm lateral resolution with a white-light nanoscope, *Nat. Commun.*, **2**, p. 218.
34. Lemoult, F., Fink, M., and Lerosey, G. (2012). A polychromatic approach to far-field superlensing at visible wavelengths, *Nat. Commun.*, **3**, p. 889.
35. Regan, C. J., Dominguez, D., Peralta, L. G., and Bernussi, A. A. (2013). Far-field optical superlenses without metal, *J. Appl. Phys.*, **113**, p. 183105.

Chapter 2

Beating the Diffraction Limit with Positive Refraction: The Resonant Metalens Approach

Geoffroy Lerosey, Fabrice Lemoult, and Mathias Fink
Institut Langevin, ESPCI ParisTech and CNRS, Paris, France
geoffroy.lerosey@espci.fr, fabrice.lemoult@espci.fr

In this chapter, we introduce the resonant metalens, a super-lens that uses the positively refracting band of locally resonant metamaterials to achieve super-resolution imaging from the far field. We first describe its principles very qualitatively, detail its principal applications, and underline its limitations. Then we propose initial experimental demonstrations of the concept in the field of microwaves and audible acoustics, proving that it can be used to focus waves well below the diffraction limit in these domains or to image far below the Abbe limit and from the far field. We finally devote the last section of the chapter to a numerical demonstration of the potentialities of the approach in the visible range of the electromagnetic spectrum using a resonant metalens made out of plasmonic nanoresonators. We give a few clues as to how these numerical results could be obtained on a real optical table with

Plasmonics and Super-Resolution Imaging
Edited by Zhaowei Liu

ISBN 978-981-4669-91-7 (Hardcover), 978-1-315-20653-0 (eBook)
www.panstanford.com

spatial light modulators by drastically simplifying the requirements of the imaging and focusing procedure.

2.1 Introduction

Waves, whatever their nature, acoustic or electromagnetic, for instance, are subject while propagating to phenomena such as refraction and diffraction. Diffraction, in particular, imposes a fundamental limit to any imaging system based on waves, such as a microscope: two objects can be distinguished from a remote distance if and only if they are separated by a distance at least larger than half a wavelength. This limitation is known as the diffraction limit or diffraction barrier. It stands also for any system aimed at focusing waves: the thinner focal spot that can be obtained using the best-available lens presents a width that is comparable to half a wavelength [53].

To understand this limit, the example of a transmission microscope used to image a living organism such as a virus may be used. When light impinges from the far field on such a medium, it is scattered into many waves due to the local refractive index changes. The resulting wave field, on the object, presents spatial variations that can be as small as these refractive index fluctuations, which can be much smaller than the wavelength of the light used in the experiment. Looking at this wave field from the Fourier domain, that is, in the momentum space, the latter is clearly composed of waves whose wavenumber can lie above and below that of the waves propagating in free space. The latter, which carry the spatial information of the object down to a value of about half a wavelength, can propagate without trouble to the collecting apparatus and participate to the image. The former on the other hand, which contain the very fine—or subwavelength—information about the object, present an imaginary wavevector in the direction of the imaging apparatus and therefore cannot reach it. They are named evanescent waves because their amplitude decreases out very fast from the object, on distances smaller than the wavelength, and the region to which they are confined is known as the near field of the sample. To obtain a super-resolved image of the virus mentioned here, one must have access to these evanescent waves in order to obtain the subwavelength information of the object. Similarly, focusing waves at scales smaller than the wavelength requires us to be able to generate these evanescent waves [25].

Collecting the evanescent part of the spectrum can be done using near-field scanning optical microscopes, which are the optical analogues of stethoscopes. They capture the evanescent waves in the near field of a sample using thin tips [2, 47, 65, 78]. Yet these microscopes require point-by-point mechanical and hence time-consuming scanning, which precludes any real-time imaging application with super-resolution. This explains why there have been recently many proposals to design superlenses based on the concept of metamaterials [16, 18, 60, 61], that is, lenses that permit one to create at once the image of an object with subwavelength resolution or to focus waves well below the diffraction limit.

The first and most famous example of these lenses is without any doubt the superlens imagined by John Pendry about 15 years ago. It consists in using a slab of negative-index metamaterial in order to drastically amplify the evanescent waves present on a source or an object so as to collect them in the far field [63]. But this theoretical concept has proved to be very limited by losses inherent to any material in the visible part of the spectrum, and therefore scientists have worked toward other solutions. In particular, concepts such as the far-field superlens [17, 69] or the hyperlens have been proposed [31, 52], both of which try to convert the evanescent waves present in the vicinity of a sample into a propagating one that can be acquired using regular collecting components. These techniques have all one thing in common, apart from being based on metamaterials, that they use monochromatic waves only.

In this chapter, we will introduce the concept of resonant metalens that we have developed during the last few years and that permits one to beat the diffraction limit using positive refraction in locally resonant metamaterials associated to a broadband approach. We will explain in the first part the basis of the concept, and in particular we will underline how the finite size of the locally resonant metamaterial lens permits one to code the subwavelength information of a polychromatic source placed in its near field into propagating waves that can efficiently reach the far field and hence be collected. We will explain how this allows super-resolution imaging of complex objects from the far field, as well as subwavelength focusing of waves from the far field. Then in the second part we will propose the results obtained with two experimental realizations of the resonant metalens. The first demonstration concerns electromagnetic waves in the microwave part of the spectrum and a lens made out of resonant metallic wires assembled on a subwavelength scale, while the second one deals with acoustic waves propagating in an array of soda cans, that is, Helmholtz resonators, which are very good subwavelength resonators for sound. Finally, in the last part of this chapter, we will propose an extension of this idea to the field of optics through numerical calculations. We will show that a plasmonic nanorod constitutes a good subwavelength resonant unit cell to design a resonant metalens in the visible part of the spectrum, and underline the peculiarities of this design as compared to the previously exposed ones. We will expose numerical proofs of subwavelength imaging and focusing obtained from the far field using this plasmonic resonant metalens. To conclude we will discuss potential simplifications which could lead to experimental realization.

2.2 Principles of the Resonant Metalens

To explain the physical mechanisms underlying the concept of the resonant metalens, we first briefly introduce the notion of meta-material and more precisely that of locally resonant metamaterials. Then we explain how using a finite-size slab of such media alongside a polychromatic—or broadband—approach constitutes a resonant metalens. We describe its first key point, which is that this allows one to literally code the subwavelength information of a source placed in the near field of the metalens into the spectrum of the field generated inside it. We then review the second key point of the resonant metalens, which is that the resonant nature of the metalens permits a very efficient conversion of the evanescent waves carrying this subwavelength spatial information into propagating waves that can be measured in the far field. We finally discuss potential applications of the resonant metalens and underline its limitations.

2.2.1 Locally Resonant Metamaterials

Metamaterials have now reached a scientific maturity and have obtained a good public exposure, yet defining them is still important. Indeed, there is still a confusion about what is and what is not a metamaterial. A general definition of a metamaterial is a composite material made out of inclusions of matter in an otherwise homogeneous matrix, these inclusions being randomly or periodically organized at a spatial scale for which the waves in the homogeneous matrix cannot resolve [18, 60, 61]. As a consequence, it is generally stated that these composite media must present a typical spatial scale that is small compared to the wavelength, and as a rule of thumb, it is widely accepted that an upper limit is $\lambda/3$, where λ is the wavelength of the waves propagating in the host matrix [75]. The main property of these media is that they behave for an incident wave as a homogeneous medium which has effective properties than can be calculated analytically in the best case or extracted experimentally otherwise. Metamaterials have been proposed first in electromagnetism, where they can present negative effective permittivity [60] and permeability [61], and later in acoustics, where they possess effective bulk or Young's modulus (or equivalently effective compressibility) and density [20, 48, 51].

Consequently any artificial media composed of a mixture of two or more materials can be termed "metamaterial," provided that the constituents are finely enough chopped. They need to be distinguished from photonic crystals, another class of man-made composite media, which are almost mandatorily periodic, present a typical spatial scale of about half a wavelength, and are governed by Bragg scattering [32, 33, 86]. The concept of metamaterials permits one to obtain materials presenting properties not available in natural ones, yet there is a class of metamaterials that exhibit properties truly unavailable from nature. These are the locally resonant metamaterials.

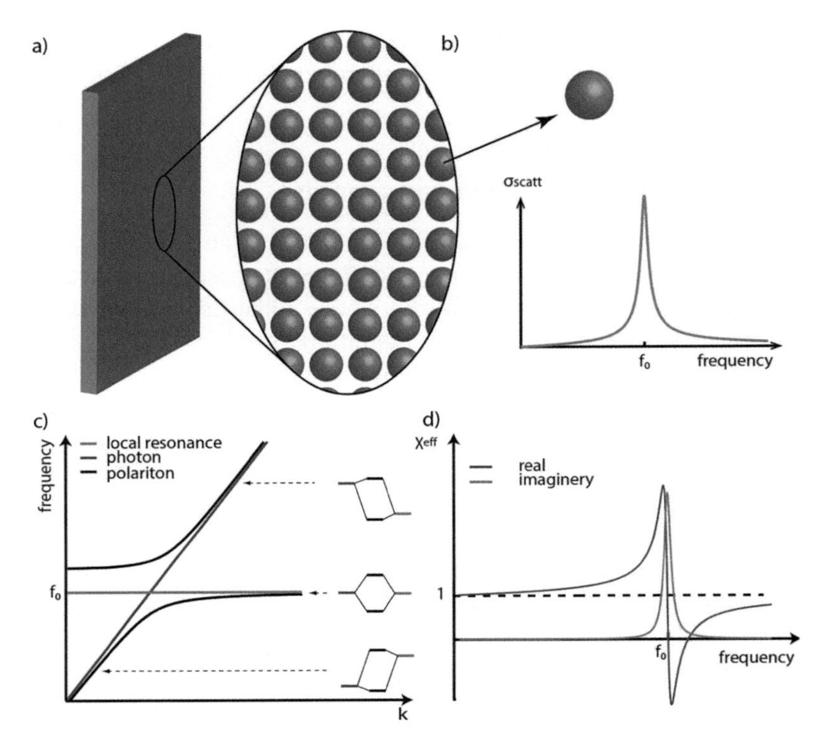

Figure 2.1 A locally resonant metamaterial (a) is an artificial material whose unit cell (b) exhibits a resonance behaviour when excited by waves. Such a medium can then be described by (c) a dispersion relation that shows a polariton behaviour or by (d) an effective property χ (which can be ε, μ, or n).

Figure 2.1 schematizes what can be called a locally resonant metamaterial. Just as any metamaterial, it is composed of unit cells that are organized at a scale that is much smaller than the wavelength (Fig. 2.1a). But the particularity of this type of metamaterial lies in the fact that its unit cell is a resonator of subwavelength dimension, as depicted in Fig. 2.1b. The scattering cross section of each unit cell that is embedded in the host matrix is therefore resonant and can be very high around its resonance frequency. This composite medium was proposed initially by John Pendry in electromagnetism [60, 61], and almost at the same time

Ping Sheng made a very similar proposal in acoustics, a concept he called *locally resonant sonic materials* [51].

Locally resonant metamaterials are the exact equivalent of dielectrics for light. In the latter, atoms, which are deep-subwavelength resonators, are excited by the incoming light waves and their relaxation participates in the total optical field. This gives rise to variations of the optical index of refraction, which can present values larger or lower than that of vacuum. Stated otherwise, the interaction of light with microscopic resonant scatterers creates at the macroscopic scale an effective index of refraction. Locally resonant metamaterials behave just the same way: the interaction of waves with the mesoscopic scatterers gives birth at the macroscopic scale to effective properties that are linked to the resonant nature of the unit cells. There is, however, a big difference between atoms in dielectrics and the unit cells of metamaterials. While the former present an albedo that is close to zero, meaning that their scattering cross section is much smaller than their absorption cross section, the latter can present albedos close to unity [36]. In other words, if most of the incoming interacting light is absorbed by atoms in the form of inelastic scattering, the subwavelength resonators that compose such a metamaterial can reradiate most of it.

Nevertheless the physics of dielectrics is very similar to that of locally resonant metamaterials. We make the approximation that the metamaterial unit cells are not strongly coupled by any near-field interaction. This approximation is not valid for any unit cell [77], as we will see in the last part of this chapter, but we use it since it permits us to capture the essence of locally resonant metamaterials. Under this approximation, the interaction of waves and subwavelength resonators in such a metamaterial creates a polariton, just as light interacting with atoms in a dielectric. Namely, there is an avoided crossing between the local resonance of the unit cell and the plane-wave dispersion line (Fig. 2.1c), which gives rise to a binding branch of subwavelength modes below the resonant frequency of the unit cell, a band gap above it, and above this band gap an antibinding branch of suprawavelength modes. We have recently interpreted this behavior in terms of Fano interference [21] between the continuum of plane waves propagating in the matrix of the metamaterial and the local resonance [44].

A more common way of describing the physics of locally resonant metamaterials consists in using the idea of effective properties. For instance, a given unit cell in electromagnetism will act on the effective permittivity of the metamaterial or on its permeability, depending on if it is electrically or magnetically resonant. In our formal description of Fig. 2.1, it can be either, and the description remains the same. The same principles are also valid for the case of acoustic waves, for instance, in which case the effective properties can be the bulk modulus or the density of the metamaterial, depending on the subwavelength acoustic resonator. As schemed in Fig. 2.1d the resonant unit cells create below their resonant frequency a band of very high effective property equivalent to the branch of subwavelength modes, then a band of negative effective properties the band gap, and finally a band of low effective property the suprawavelength modes branch.

Much attention has been paid in the past 10 years to using the negative effective property band of such metamaterials. Indeed, if one can realize a metamaterial presenting two almost co-localized subwavelength unit cells which are resonant at the same frequency and which act on both properties of the medium—the permittivity and permeability in electromagnetism and the bulk modulus and density in acoustics—one can obtain a metamaterial that has both its effective properties negative. This results, as was pointed out more than 40 years ago by Veselago [81], in a medium whose effective index is negative. Pendry proposed in a seminal paper in 2000 [63] that a slab of such a metamaterial should behave as a perfect lens for imaging and focusing purposes since it amplifies infinitely the evanescent waves coming from a source or object, making them measureable in the far field with conventional optical components. Again, this approach has been shown to be largely hampered by losses of materials, especially for applications in optics where materials are relatively dissipative.

Yet, quite surprisingly, the high-effective-property band offered by locally resonant metamaterials has not really been considered up to now for superlensing applications. But, this band supports wave that are evanescent waves in the host medium since the dispersion relation is placed below (in frequency) the dispersion relation of the homogeneous matrix. In the following section, we will explain how and under which conditions they can indeed be used for such a purpose, before proposing examples of applications.

2.2.2 Coding the Subwavelength Information of a Source into the Complex Spectrum of a Polychromatic Wave Field

Of course, it may seem trivial to use a slab of high-effective-property metamaterial to beat the diffraction limit, since the index of the latter can be much higher than that of air, meaning that the resolution of an imaging system based on such material should be much better than the one of the same system in air, akin to solid immersion microscopes in optics. Yet we want here to remind the reader that beating the diffraction limit means being able to generate (or measure) wave fields that oscillate on a scale smaller than this diffraction scale, namely the matrix's wavelength. We will now show how this can be achieved using a locally resonant metamaterial lens, starting by demonstrating the interest of using a broadband or polychromatic approach rather than a monochromatic one.

To start with one should notice that contrary to what is usually modeled theoretically or simulated using numerical tools, when dealing experimentally with the concept of metamaterials, infinitely extended media are not and cannot be used. Rather, experiments are always realized using finite-size metamaterials similar to the formal one that is schemed in Fig. 2.1a. This fact has very profound consequences that are most of the time wrongly neglected. To understand this point we will consider the case in which a source is placed in the near field of such a metamaterial, as depicted in Fig. 2.1a.

Here we remind the reader that we are interested in the band of high effective properties of the locally resonant metamaterial, that is, the frequency range between DC and f_0. Again, this band consists of a continuum of evanescent modes whose dispersion relation lies below the line of that of the plane waves propagating in the homogeneous matrix. Stated otherwise, waves excited by a source placed in the near field of the finite-size metamaterial slab cannot leave the latter, and are reflected by the metamaterial/air interfaces. Namely, the metamaterial slab now behaves as a cavity whose eigenmodes and eigenfrequencies can, in principle, be calculated using the effective property of the metamaterial (which can also be spatially dispersive) and the boundary conditions at the interfaces of the finite-size composite medium, which can be quite complex. This results in a very complicated mathematical problem which has been solved for some metamaterials only [73], yet we will show later on that it is not necessary to solve such problem in order to use a slab of locally resonant metamaterials as a resonant metalens.

Another and maybe more natural way to envision the properties of a finite slab of locally resonant metamaterial that we will use as a resonant metalens is simply to consider what it is exactly. Actually, it is a bunch of closely spaced subwavelength resonators which all present the same resonant frequency f_0. Because they are placed relatively close one to another, these resonators are coupled, and hence, from a set of identical resonators, one obtains an ensemble of collective modes whose resonant frequencies are distinct, apart for geometrical degeneracy, and that are distributed on the frequency axis. Very similar and well known examples of the phenomenon include an array of spring/mass systems, which are all independently a natural resonator and display a collective behavior when connected together, or a set of electronic LC resonators which are cascaded and again support collective eigenmodes. Yet, there exists a major difference between these two very famous examples and a locally resonant metamaterial slab. Indeed, in the former each resonator is coupled to the other ones through its direct neighbors, and hence the ensemble supports a single band (if the system is infinite) or a set of collective eigenmodes (for a finite system) due to hybridization between the resonators. On the contrary, in the latter the resonant unit cells are coupled to each other through the continuum of plane waves propagating in the matrix, leading to a hybridization between the resonators and the plane waves, which gives rise to the polariton behavior mentioned previously, and hence two bands or sets of eigenmodes, the binding ones and the antibinding ones [30, 36].

Whether one chooses the first or the second approach to understand a finite-size sample of locally resonant metamaterial, one certainly conclude that it does not present a continuous effective property band but rather supports discrete eigenmodes that result

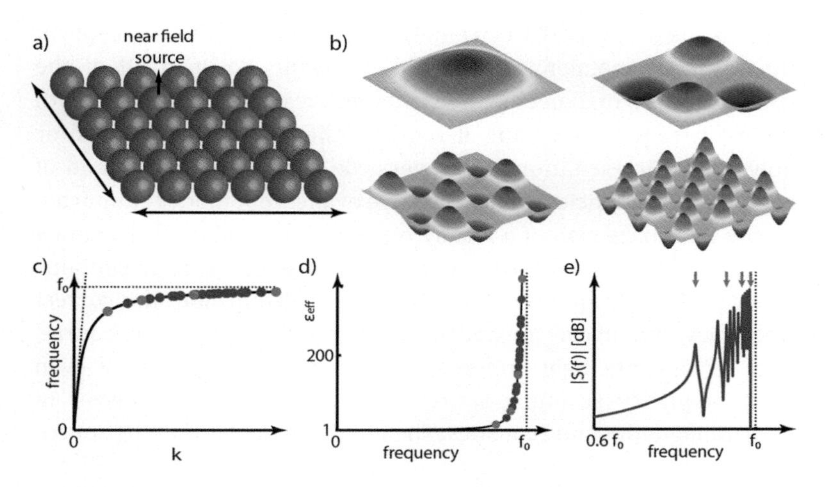

Figure 2.2 (a) A finite-size, locally resonant sample supports (b) stationary eigenmodes, which discretizes (c) the band structure and (d) the effective property. (e) On the frequency domain this manifests as a superposition of resonant peaks.

from the collective behavior of the ensemble of resonators and of which we plot a few examples in Fig. 2.2b. These eigenmodes, as the figure clearly shows, can display spatial variations that span from the size of the metamaterial slab down to as small as the period between the resonators. Namely, the wave field corresponding to these modes can be homogeneous on the entire slab or on the contrary can change sign every resonator. Very importantly, these collective modes are associated with discrete resonance frequencies that span an interval between DC and f_0. These frequencies are all distinct, except for some modes which are degenerated due to symmetry reasons, meaning that each mode can be generated using monochromatic excitations with chosen frequencies. In other words, each collective mode is associated with a resonance frequency which results from the discretization of the high-effective-property band of the infinite locally resonant metamaterial, which can be represented in the dispersion relation as in Fig. 2.2c or in the effective property curve as in Fig. 2.2d, where each mode of our formal finite-size metamaterial has been pointed by a blue disk, while the four represented in Fig. 2.2b have been noted with a red disk.

This last aspect is extremely important in the concept of the resonant metalens. Indeed, if one wants to span all of the spatial scales, one needs to excite the resonant metalens with different frequencies. Thus, it requires the use of a broadband or polychromatic excitation that covers the whole frequency band of resonance frequencies of the collective evanescent modes supported by the finite-size slab of a locally resonant metamaterial. The most natural choice for the latter is a pulse, in which case it contains at once all the frequency components centered around the correct frequency, but it can as well be a temporally incoherent source, such as a white light supercontinuum in optics, and may even be a frequency scanning source such as a network analyzer in electromagnetics and acoustics. The important fact is that the source of waves used with the resonant metalens must contain all the resonant frequencies of its collective modes.

The first principle of the resonant metalens can be explained quite easily. Suppose a source such as that described above is placed in the near field of the resonant metalens, as in Fig. 2.2a, this source being either a point source or a more complex object. Spatially, the source can be decomposed on the eigenmodes of the resonant metalens by modal decomposition, just as a pinch on a guitar cord decomposes on its harmonic. Now since the source contains all the resonant frequencies of the resonant metalens, it can excite all of its collective modes. As a consequence, the Fourier decomposition of the spatial information of the source is translated in the spectrum of the field that is excited inside the resonant metalens. Namely, the spatial information of the source, at a subwavelength scale as small as the period of the metamaterial, is coded in the spectrum of the wave field that it generates inside the resonant metalens or, equivalently, in the temporal evolution of the wave field if the latter is generated by a time-varying source. As an example, a homogeneous object the size of the resonant metalens will only excite its fundamental mode and hence its lowest resonance frequency, while a point source will excite all the eigenmodes of the resonant metalens, with a phase and an amplitude that are directly given by its spatial decomposition onto these eigenmodes, as we formally plot in Fig. 2.2e, where we again point the modes mapped in Fig. 2.2b, with red arrows.

This is a key component of the resonant metalens: Thanks to the dispersive and finite-size nature of the locally resonant metamaterial and thanks to a broadband/polychromatic excitation, it is capable of coding the spatial information of a source in the complex spectrum of the wave field it supports. Now one may reply that this information is still useless because it remains inside the resonant metalens and, being stored in evanescent waves, cannot escape it to reach the far field. We will see in the next section that this assertion is wrong and that the resonant metalens is an extremely efficient evanescent to propagating waves converter.

2.2.3 Efficient Conversion of Evanescent Waves to Propagating Ones, Thanks to Resonant Amplification

Again, to use the subwavelength or evanescent modes supported by a locally resonant metamaterial for true subdiffraction or super-resolution imaging and focusing purposes, it is necessary for the latter to be converted into propagating waves by some mechanism. This mechanism comes for free with a resonant metalens, and to understand it we need to define what an evanescent wave is.

An evanescent wave is a mathematical tool that describes a field presenting spatial variations that are smaller than the wavelength of the waves near an interface. There are not propagating solutions of the wave equation, but they only permit us to satisfy the fields' continuities near this interface. For example, when two materials with different indexes of refraction share a common interface it is well known that total internal reflection can occur. This comes from the fact that the medium of a larger index allows the propagation of waves whose momentum is too large for propagation in the lower-index one. In that case to match the fields at the interface one has to introduce an evanescent wave. This evanescent wave exponentially decays from the interface because its momentum tangential to the interface is higher than the lower-index medium's momentum or equivalently the spatial variation of the field tangential to the interface is smaller than the host wavelength. But, the concept of evanescent waves is not limited to planar interfaces and another phenomenon responsible for the generation of evanescent waves is scattering. When waves of a given wavelength impinge on an object that is smaller than the wavelength, to satisfy the continuity of the fields near this object one needs again to introduce mathematical solutions that vary on a scale comparable to this object, namely evanescent waves. In other words, the wave field near this small scatterer possesses momenta that can be higher than the free-space momentum due to the existence of these evanescent waves. In both cases, the evanescent waves are stick to the discontinuities that generated them [25], the reason why they are termed "evanescent." Yet the reason why it cannot survive more than a few portions of wavelength from its birthplace lies in the fact that an evanescent wave is inherently not a propagating solution of the wave equation in free space.

In the case of the resonant metalens, we have shown that the locally resonant medium can support propagating solutions that have a higher momentum than the one allowed in the host matrix. Viewed from the matrix these waves are therefore in the evanescent part of the momentum spectrum. To simplify the explanation, we adopt here a 2D formalism where such waves, three of which are pictured in Fig. 2.3a, propagate in an infinitely extended locally resonant medium along the horizontal direction. In the Fourier domain, with respect to the horizontal direction of our 2D space, these waves in the resonant medium are therefore Dirac delta functions (Fig. 2.3a). Interestingly, these delta functions are placed above the propagation window of the matrix, meaning that the horizontal component of their wavevector is larger than the momentum of the waves propagating in the surrounding medium. To satisfy the continuity of the waves at the interface, it means that evanescent waves exist along the vertical direction, and they necessarily decrease exponentially in the vertical direction.

Now, it is clear that the exponentially decaying nature of these evanescent waves comes from their infinite extension, and that they are in this respect mathematical objects. In real life, any wave field is of finite extension, and therefore purely evanescent waves do not exist per se, the reason why a point scatterer supports evanescent waves in its near field but also scatters propagating

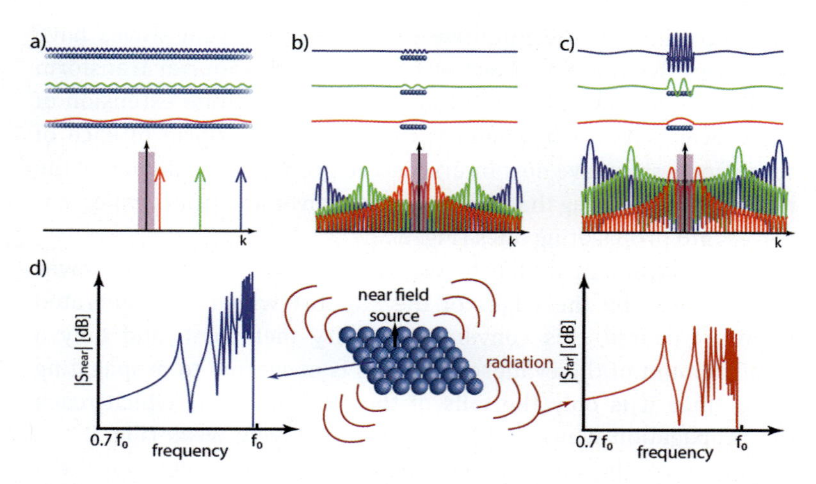

Figure 2.3 (a) Infinitely extended evanescent modes supported by an infinite locally resonant medium. The corresponding Fourier transforms are Dirac distribution in *k*-space (outside of the propagation filter purple rectangle). (b) Adding finiteness creates reflections: stationary waves now exist and the overlap with the propagation filter in *k*-space is no more null. (c) Depending on this overlap (the radiation leakage) the mode gains amplitude. (d) A near-field broadband source excites the modes. In the near field (left) this manifests by resonant peaks with growing amplitude, while modes radiate with the same efficiency (right).

waves toward the far field. This observation has very profound consequences, which may not seem drastic in the usual contexts of imaging or focusing waves, but which are for that of the resonant metalens. Keeping our very simplified 2D problem, we now bound the evanescent waves discussed above, as schemed in Fig. 2.3b. This has two consequences, first, since the evanescent waves are now standing waves total internal reflection occurs at both ends they must present even spatial Fourier spectra, but this is not the main conclusion. Much more important is the fact that the bounded waves, which have been multiplied by a rectangular function standing for the finiteness of the medium, are no longer Dirac delta functions. Indeed, multiplying a function by another one in the spatial domain amounts to convolve it by the Fourier transform of the latter in the reciprocal space. This is schematized again in Fig. 2.3b: the Dirac delta functions of the infinitely extended evanescent waves have been convolved by Sinc functions which are the Fourier transform of the rectangular functions responsible for the finite extension of these waves. As a consequence, the Fourier transform of each of these finite-size wave now intersects the propagation window of the host matrix, meaning that there is some conversion from evanescent waves into propagating ones (Fig. 2.3b).

This conversion is due to the fact that the evanescent waves are scattered by the edges of the support which has generated them. Of course, this conversion is very inefficient, and only a small amount of the evanescent waves is converted to propagating ones, since it is only the tails of the Sinc functions which reach the propagation window, and they are relatively weak (Fig. 2.3b). Furthermore, the more subwavelength the evanescent wave, the further from the propagation window in the Fourier domain, and hence the weaker its conversion to propagating waves that can reach the far field. Conversely, the larger the extension of the finite-size evanescent wave, the thinner the Sinc function in the Fourier domain and, again, the weaker its value when it reaches the propagation window, or, equivalently, the more inefficient its conversion to propagating waves. These two reasons explain why super-resolved images cannot usually be obtained from the far field, even when imaging small objects. The contribution of the converted evanescent waves to the propagating ones is quite negligible. Moreover there is naturally no coding of the subwavelength information in this case, contrary to that of the resonant metalens. It means that even with a relatively efficient conversion of the evanescent waves, meaning a low resolution and a small object, the subwavelength information is still lost since mixed with the diffraction limited one. This is why scattering-type near-field optical microscopes use heterodyning to get rid of the background of propagating waves and have access to the converted evanescent waves only [55].

As a matter of fact, the resonant metalens approach benefits from another mechanism that makes the conversion of evanescent to propagating waves efficient, and it stems from its resonant nature. Indeed, as mentioned before, each collective mode supported by the resonant metalens, and which oscillates at a different resonance frequency, is an evanescent wave of finite extension. It radiates at a given time a small amount of energy in the far field due to scattering of the evanescent modes off its edges, and the more subwavelength the mode, the smaller this amount. Consequently, any source placed in the near field of the resonant metalens feeds each collective mode, which is now a cavity and therefore accumulates energy over time. Hence, the energy of a given mode inside the resonant metalens is increased by the Purcell effect [67], or equivalently by a resonant amplification, which drastically modifies the behavior of the evanescent to propagating wave conversion of the finite-size lens. Indeed, at this point, the properties of the resonant metalens are ruled by energy conservation: the energy coming from the source and which cannot escape the resonant metalens builds up in the lens over time at each frequency. Finally, and neglecting the losses, because the waves accumulated in each resonant collective mode have no other way to exit the system than to be scattered in the far field, each mode is converted into propagating waves with the same efficiency. The most subwavelength modes simply take longer to escape than those that are closer from the matrix's dispersion relation. This fact is clearly observed in Fig. 2.3c, where we plot again the finite-size evanescent waves supported by the resonant metalens, but this time including their resonant enhancement. Clearly their amplitude increases in the metamaterial lens which translates into the fact that their Fourier transforms cross the propagation window with equal amplitude, whatever their subwavelength character.

Finally, one concludes from this simple explanation that owing to its resonant and finite-size nature, the resonant metalens permits one to convert all the evanescent waves it supports into propagating waves with the same efficiency, for collection and measurement with conventional components from the far field. This is the second key aspect of the resonant metalens. Overall, associated with the first key point discussed in the previous section, it means that a resonant metalens codes the subwavelength information of a source placed in its near field into the spectrum of the wave field the latter generates inside it, but also scatters efficiently this information to the far field in the form of propagating waves, as schemed in Fig. 2.3d. We will now discuss the potential applications of such a superlens, and mention its limitations.

2.2.4 Applications and Limits of a Resonant Metalens

We have just seen that the resonant metalens can code in time or frequency and send in the far field the spatial information of a source or object placed in the near field with a subwavelength resolution that can be as small as its period. Now we expose its applications regarding imaging and focusing beyond the diffraction limit, starting with super-resolution focusing. Of course, as we mentioned previously, one needs to use a broadband approach to take full benefit of the resonant metalens. It may seem rather complicated, due to the highly scattering and complex nature of the lens, to exploit it for focusing applications, notably if conventional beamforming is considered. In fact, there exists a very convenient and simple method that can take full benefit of the resonant metalens to beat the diffraction barrier from the far field for focusing and imaging applications: it is the concept of time reversal [11, 24, 45].

Time reversal is based on the temporal reversibility of the equations which rules the propagation of waves in media. It states that if a wave field is solution of the wave equation, then a dual wave field, which is the time-reversed version of the initial one, is also solution of the same equation. Consequently, when a source emits a pulse in a given medium it generates a given wave field $\Phi(t)$. The time-reversed wave field $\Phi(-t)$, that is, the reversed played version of $\Phi(t)$, is also a solution of the wave equation and it focuses waves back to the position of the original source. This concept, although it looks quite magical stated as it is in fact very easy to understand in the context of a modal approach, and hence for applications regarding the resonant metalens.

Suppose that like in Fig. 2.3d, a source emits a short pulse in the near field of a resonant metalens, the latter being frequency centered such that it excites all the collective subwavelength eigenmodes of the lens. We pointed it out that the spatial information of the source is coded in the spectrum or time dependence of the wave field generated in the resonant metalens and is scattered in the far field very efficiently. This behavior comes out very naturally as a consequence of time reversal, in a focusing experiment. Indeed, when the source emits the pulse, it excites the eigenmodes of the resonant metalens with a zero phase: all of the modes beat coherently at the time of the pulse emission and at the location of the source. Now when the wave field is stored by the lens and converted into far-field radiation, it gives rise to a long temporally varying wave field, a signature of the resonant nature of the metalens. At any position in space, the wave field is constituted of a set of frequencies which are the resonant frequencies of the resonant metalens, and the long signal measured on a sensor results from a temporally incoherent sum of all these frequency components. Indeed, the propagation of the waves in the lens and in free space has for consequence to phase-shift all the frequency components composing the wave field independently. A natural solution that comes to mind in order to focus waves in the near field of the resonant metalens at the position of the original source is then to phase-conjugate all the frequency components of the wave field generated by the source, such that the phase shift induced by wave propagation in and out of the lens is perfectly cancelled, putting back all the modes in phase at a given time and at the initial source location.

This is exactly what time reversal does. In a typical time reversal experiment, schematized in Fig. 2.4a, we record the field created by a source placed in the near field of the lens on a set of locations in the far field. Then these signals are time-reversed, that is, flipped in time, which corresponds in the Fourier domain to a phase conjugation at each frequency contained in the signal. Consequently, the modes excited in the lens by the time reversal experiment beat until they are all in phase at a given time and at the initial source location, creating an instantaneous focusing of the wave, or, stated otherwise, a spatiotemporally focused wave field. At all other locations on the resonant metalens, the modes do not beat in phase at this time since the phase shift that has been cancelled is uniquely defined for the initial location. We have used this concept in various domains to focus waves much below the diffraction limit with resonant metalenses and we will present later on a few examples of our obtained results. Obviously, the number of sensors and sources used to perform the time reversal experiments influences the quality and resolution of the focusing, a fact that we cannot develop here due to

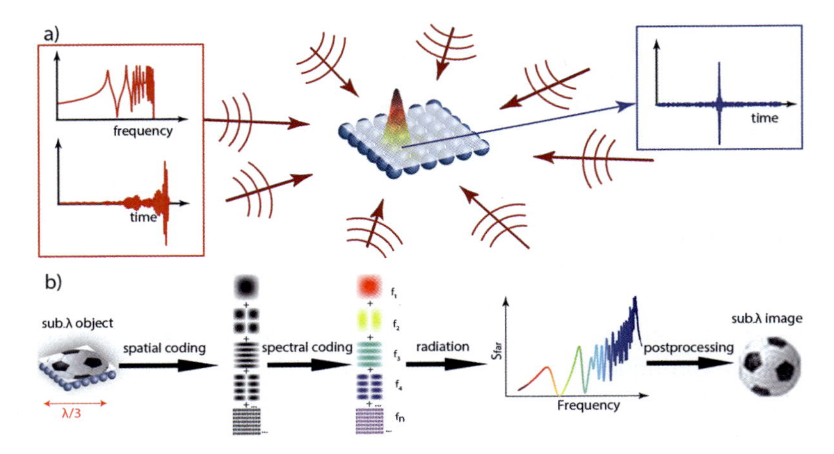

Figure 2.4 (a) Superfocusing scheme: We emit time-dependent wave fields from the far field. They sum up constructively both in time and in space to exhibit the subwavelength focus. (b) Subwavelength imaging: A subwavelength object is placed at the input of the resonant metalens. The broadband near field emitted by the object decomposes onto the subwavelength eigenmodes. Each mode experiments its own resonance and radiates the field toward the far field at its own resonance frequency: the resonant metalens has colored the emission. Then a super-resolved image can be built.

lack of space, but that is discussed in Ref. [40] and also in the next section.

Similarly to the case of subwavelength focusing beyond the diffraction limit, the resonant metalens can be used to perform subwavelength imaging from the far field, at the price of a rather thorough prior knowledge of the properties of the lens. As we scheme in Fig. 2.4b, the concept is very similar to that of the focusing. We place a contrast object, here formally a soccer ball, in the near field of a resonant metalens. The spatial information of the latter, whether it is itself a source or if it is illuminated from the far field, decomposes on the collective modes of the resonant metalens. Since every mode has a distinct resonance frequency, it amounts to "coloring" each spatial Fourier component of the object before efficient radiation of the information in the form of propagating waves to the far field. Evidently, if this information can be acquired in the far field, it means that it can be decoded in order to reconstruct the object with a resolution that can be much better than the diffraction limit. Here we do not want and do not have enough room to describe the various reconstruction methods that can be employed and that were published before [40], but we have to point out that each method necessitates quite a large a priori knowledge of the resonant metalens and most preferably the set of impulse responses or complex spectra acquired between the near field of each resonator of the lens and multiple directions in the far field. This knowledge in hand, matched filtering, inverse filtering, or some hybrid version between these solutions may be used, which permits one to extract the spatial information of the object from the far-field scattered waves. We will present in the last section of this chapter a numerical example of super-resolution imaging at visible wavelengths using a plasmonic resonant metalens. Again, the resolution and quality of image reconstruction are closely related to the number of sensors used for the imaging process. This fact, discussed in detail in Ref. [43], will be commented in the last part of this chapter.

Finally, before we conclude this section on the basic principles of a resonant metalens, we would like to point out very briefly its limitations. An obvious one comes from the fact that, even though locally resonant metamaterials are usually considered as homogeneous materials presenting effective properties, they are made out of resonant inclusions. Consequently, the fastest varying mode will always be limited to the period of the medium if a monopolar resonator is used. Hence the resolution of a resonant metalens cannot be better than its period, or average period if it is not ordered, even though the focal spot it can generate may be even smaller due to edge effects. Yet another much more drastic fact limits the resolution of a resonant metalens both for focusing and imaging applications: dissipation. Indeed, as we have seen it, the whole concept of the resonant metalens, the time/frequency coding it allows, and the efficient evanescent to propagating wave conversion, all result from the fact that waves are trapped for a long time within the lens: this ultimately limits its resolution. Indeed, the more subwavelength a mode, the longer its lifetime inside the resonant metalens. This is clearly explained by considering that the more evanescent a wave, the lower its conversion efficiency to propagating waves by the resonant metalens. Another way to understand it in the case of a purely polaritonic metamaterial is to notice that on the edges of the Brillouin zone, that is, at high wavenumbers, the subwavelength modes band is very flat. It means that their group velocity is very small, which in turn implies that they spend a longer time in the resonant metalens. Either way, the modes that stay longer in the lens are more sensitive to dissipation, and since they are the most subwavelength ones, this is the main resolution limit of this kind of superlens.

2.3 Experimental Demonstrations with Microwaves and Sound

In the previous section of this chapter, we have explained the principles of the resonant metalens and underlined its potential applications and limitations. Before we show how such a lens could be used in the context of subwavelength imaging and focusing in the visible range using plasmonic resonators, we first review the original demonstration of the resonant metalens in the microwave domain and a nice transposition to the acoustic one.

2.3.1 Original Demonstration: A Wire Medium–Based Resonant Metalens for Microwave Applications

The first demonstration of the concept was published in 2007 in *Science* [46], in which we proved that a random collection of metallic wires could be used alongside time reversal to focus microwaves beyond the diffraction limit. One point that was not obvious in this experiment, using the random collection of metallic scatterers, denoting the microstructured medium, was that the wires were all approximately of equal length. Considering this microstructured medium through this new prism, it can be visualized as an array of resonant wires, and this system can be interpreted at the light of the resonant metalens approach. Indeed, they are all around a quarter wavelength long and placed on top of a copper ground plane; hence they are resonant monopoles. Such an array of resonators should behave as a matrix of coupled oscillators, similar to a system of N

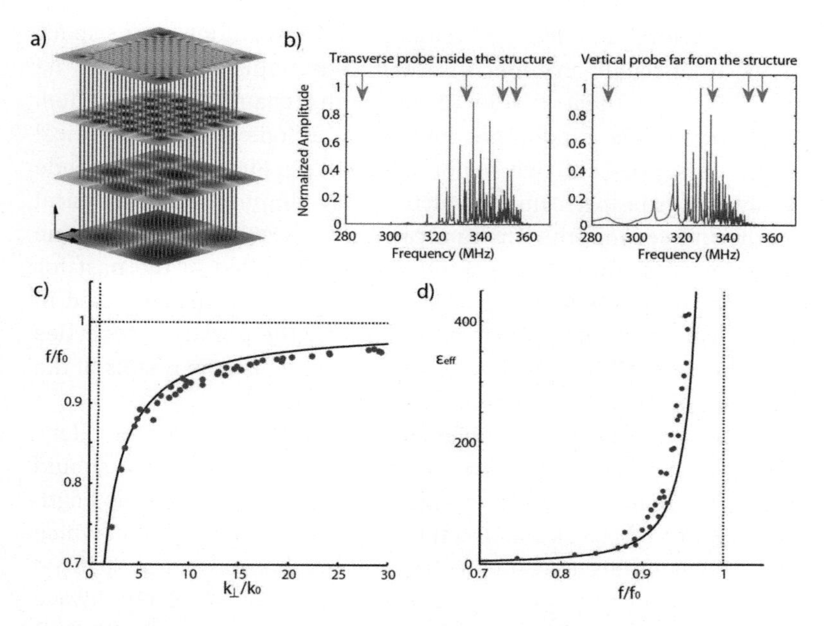

Figure 2.5 (a) 3D representation of a finite-size wire medium. Superimposed: amplitude of a few collective modes supported by the metalens. (b) Results of the transient 3D simulations: inside the structure and in the far field. Red arrows: resonance frequencies of the four modes mapped in (a). (c) Dispersion relation obtained from the 3D simulations and (d) its corresponding description in terms of effective permittivity.

coupled mass/spring systems. This is the idea that we have explored in order to grasp the physics of such a medium and infer the links with the results of time reversal focusing at a subwavelength scale.

To study such an array of resonant unit cells, we have simplified the system [39]; the medium (Fig. 2.5a) is composed of a periodic array of $N = 20 \times 20$ equal-length copper wires (length $L = 40$ cm; period $a = 1.2$ cm; diameter $d = 3$ mm). The wires are oriented along the z direction and periodically spaced in the xy plane. The first resonance f_0 of a wire is around 375 MHz, and they are therefore organized on a very deep subwavelength at the equivalent resonant wavelength $\lambda_0 = 80$ cm, that is, a period equals to $\lambda_0/70$. The studied array is therefore a very good resonant metalens: viewed from the top the resonant metallic wires are almost point-like atoms.

We perform all our measurements and simulations in free space or in an anechoic chamber in order to decouple the effect of the medium from those of the reverberating chamber used in Ref. [46]. Leaving the random medium for a periodic array of identical resonators presents two major advantages: First, a Bloch mode analysis can be performed, which greatly simplifies the analytical study, and second, the medium can be analyzed within the frame of the wire medium that has been deeply studied in the past for different purposes [8, 9, 74]. This is the approach that we used in the original papers to obtain quantitatively the physical properties of this system [39, 41], yet here we will simply use the results of the previous section to explain these qualitatively.

As we expect from our demonstration of the resonant metalens concepts, the studied array of resonant metallic wires should support eigenmodes which oscillate at a very deep-subwavelength scale, which can be as small as the period of the medium, and which all present different resonant frequencies. We have first realized simulations of such a resonant metalens using CST Microwave Studio, a transient simulation-based software. A short pulse (10 ns) centered at 375 MHz is emitted by a small dipole source placed in the near field of the studied resonant metalens. We superimpose on the latter in Fig. 2.5a the spatial profile of four modes extracted from this simulation via a Fourier analysis: it confirms that the lens supports collective modes that can oscillate spatially at scales as small as its period. In Fig. 2.5b, we show the spectrum of the electric field measured with a probe placed inside the resonant metalens, and as expected, the latter displays many resonant peaks. These peaks are the signature of all the resonant collective modes that are excited by the probe due to the finite size of the lens, its resonant nature and its dispersive behavior near f_0. Again, this confirms that an array of resonant metallic wires indeed behaves as the resonant metalens described in Section 2.1 of the chapter.

Using the full results of the simulation Fourier analysis, we can present the full set of subwavelength collective modes supported by the resonant metallic wires. We do so on a dispersion relation (Fig. 2.5c) and using the metamaterial approach, that is, using an effective property representation (Fig. 2.5d). Such a system of N periodic electric wires is analyzable within the frame of classical coupled dipoles: a set of N equations is written modeling the behavior of the current in each wire and including coupling coefficients linking a specific wire to its neighbors. This approach, however, requires the knowledge of the coupling matrix and some computation, and there is actually a more elegant way to obtain the desired results. In fact, we calculated the dispersion relation of the infinite medium of resonant wires similar to the finite-size resonant metalens that we study here, using three different approaches in Ref. [41].

The first approach is based on considering that each collective mode is constituted of transverse electromagnetic (TEM) modes trapped in a Fabry–Perot that has the length of the wires plus an evanescent tail which depends on the wavevector of the mode. We then analytically obtained the effective length of the Fabry–Perot for each mode, which gave its resonance frequency and thus the dispersion relation of the medium. In the second approach, we calculated analytically the fields inside the wire medium, and considered the reflections of the collective modes at the wire–air interfaces. Since all the modes are subwavelength, and hence evanescent if the medium is infinite, this led us to evaluate a phase shift due to the so-called Goos–Hänchen shift [26] at each reflection, which depends once again on the wavevector of the considered mode and, again, permits one to calculate the dispersion relation. Finally, another less physical but mathematically easier method is to write analytically the fields inside the resonant wire medium and to apply the continuity equations for the electric and magnetic fields at the air–wire interfaces. This approach, similar to that used by Pendry when dealing with surface waves on structured metals known as spoof plasmons [62], leads to the same dispersion relation as the last one, and both are very close to that obtained with the dispersive Fabry–Perot. This dispersion relation is plotted in Fig. 2.5c and it is in very good agreement with the eigenmodes extracted from the simulation.

We point here that this dispersion relation is formally equivalent to that obtained for spoof plasmons propagating on structured metals in 2D, that is, surface waves propagating on metals carved with $\lambda/4$ deep rectangular grooves. Indeed, in our case we create surface waves which are evanescent on both sides of the resonant wires; these are polaritons due to the hybridization between the plane-wave continuum and the resonance of the electric wires. Conversely, 2D spoof plasmons are polaritons which result from the hybridization between the plane-wave continuum and the groove resonance. Hence they are surface waves of the same kind, polaritonic surface waves, yet they are solely evanescent at one interface, since the other is metallic.

Using this dispersion relation, one can also extract the effective parameters of the infinite metamaterial, and superimpose it on the effective properties of the collective modes obtained from the simulations. Since the wires are sensitive to an electrical excitation it is admitted that such a medium can be modeled in terms of an effective permittivity and this is what we extracted in Fig. 2.5d. It is clear that the resonant wire medium can support modes with very high effective permittivity that one needs to achieve subwavelength imaging or focusing from the far field. It is also clear that since the resonant metalens, supporting true polaritons in this case, is very dispersive, its collective modes all display distinct resonant frequencies. Therefore, the first key principle of the resonant metalens is satisfied; namely, a source or object placed in its near field decomposes on the eigenmodes of the metalens, and its spatial information is coded in the spectrum of the wave field generated inside the lens. We now verify the second key principle of the resonant metalens, namely that it converts very efficiently the evanescent modes into propagating waves that can be collected in the far field. For that reason, we plot on Fig. 2.5b, next to that measured inside the resonant metalens, the spectrum of the field measured in its far field under the same conditions. Clearly resonant peaks are observable in the far-field spectrum at the same frequencies as in the near-field one, proving via this simulation that the resonant metalens not only can code the spatial information of a source at a subwavelength scale, but also can send it in the far field with a good efficiency.

We now describe experiments that prove that the studied resonant metalens can be used to focus waves using time reversal onto focal spots as small as 1/25th of a wavelength and to realize images from the far field with a resolution as fine as the period of the metalens, 1/80th of the wavelength.

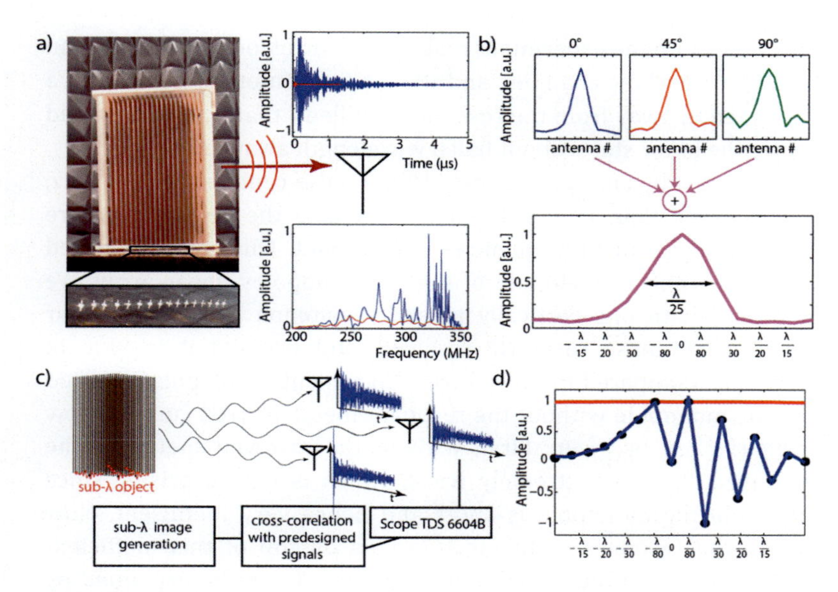

Figure 2.6 Experimental results. (a) The experimental resonant metalens on its ground copper plane. Signals and spectra received in the far field after emission of a short pulse with the lens (blue) and without as a control curve (red). (b) Focal spots obtained after one-channel time reversal when tilting the position of the far-field antenna: $\lambda/25$ widths are demonstrated. A three-channel time reversal experiment keeps the same resolution but decreases the side lobes (bottom). (c) Sketch of the imaging procedure. (d) The imaging experiment: 16 monopoles generated a subwavelength phase and amplitude profile in the near field of the lens (black points). We plot the result of the image reconstruction: a true $\lambda/80$ resolved image of the initial pattern is reconstructed (blue), while it remains impossible without the lens (red).

Experimentally, we reproduce the exact replica of our simulated resonant metalens except that it is supported by a Teflon structure (Fig. 2.6a), which slightly shifts the resonance frequencies of the collective modes. We place under the resonant metalens a copper ground plane that shields all the cables used in the experiment without perturbing too much the properties of the lens: indeed, since each mode it supports is evanescent, it only sees this ground plane through an evanescent tail, and its effect is thus rather limited. On this ground plane, we have soldered 16 very small and hence inefficient monopoles, which serve as emitters or receivers depending on the experiment realized. We do all our experiments in a 20 m^2 anechoic chamber, and the far-field antennas are placed a few meters away from the resonant metalens, that is, in its far field since the latter starts about half a wavelength away in air.

We start by emitting a short (10 ns) pulse centered at 300 MHz, a bit lower than in the simulation to take the Teflon structure shift of the resonant frequencies into account. This pulse is emitted with the central monopole of the 16 monopoles linear array. We measure with one vertically polarized antenna placed in the far field the impulse response received, and we plot it in blue in Fig. 2.6a. Compared to the control experiment of the pulse emitted by the monopole without the resonant metalens (red curve), a few remarks can be given. First, without the resonant metalens, the received signal is extremely weak, which is not surprising since the emitting monopole is short and hence very inefficient. Now when the resonant metalens is placed on top of this inefficient source, the received signal is much higher. It can be explained by the Purcell effect [67], or the resonant amplification of the field by the resonant metalens; the latter has impedance matched the unmatched monopole which now emits efficiently in the far field. Also, the received signal with the lens is much longer (5 μs) than the initially emitted pulse (10 ns), and the control pulse which is about the same duration since the experiment is realized in an anechoic room. This is a clear signature of the extremely resonant nature of the lens.

Looking now at the Fourier transform of these signals, the control spectrum and result spectrum are again very different. While the control one is almost flat since no reverberation occurred in the anechoic chamber, and because the emitting monopole is not resonant, the spectrum of the impulse response with the resonant metalens displays a large number of very resonant peaks, very similar to those observed in the simulation. These peaks are the signature of the resonant behavior of each collective mode. They prove at once that the resonant metalens supports a large number of very resonant subwavelength modes and that these modes, although of subwavelength nature, are efficiently converted to propagating waves that are acquired easily by the receiving antenna in the far field. We will finally show how these properties can be used to realize subwavelength imaging and focusing from the far field.

We start by repeating the time reversal experiments of [46] using our periodic system in an anechoic chamber. We have recorded the field produced by a small monopole placed in the near field of the medium using three antennas in the far field of the structure using the same emission pulse. These three antennas are placed at 0°, 45°, and 90° around the medium, and we record the impulse responses between these antennas and the monopole placed in the near field of the medium. We now perform time reversal, focusing using the three far-field antennas. To that aim, the impulse responses from each one of the three far-field antennas were time-reversed and sent back using a single antenna at a time. Measuring the temporally varying field received on the array of monopoles placed around the initial source position in the near field of the wire medium, we obtain the temporal signal resulting from time reversal focusing. Clearly, plotting the maximum over time of the signals received on each small monopoles, we prove that we have obtained a focal spot that is deeply subwavelength (around $\lambda/25$) using any of the three far field antennas, that is, using a single antenna and in free space (anechoic chamber). A multichannel time reversal experiment using simultaneously the three antennas gives a similar focal spot, with lower side lobes (Fig. 2.6b).

Finally, we perform a very simple proof of concept experiment of far-field imaging beyond the diffraction limit using the wire based resonant metalens. To that aim, an object is emulated using the 16 monopoles of the array: each one emits a 10 ns pulse centered around 300 MHz with various amplitudes and phases. We record using the same three antennas the far field emitted by the resonant metalens when this object is placed in the near field of the lens. The knowledge of the impulse responses between each wire of the resonant metalens and the three antennas at hand, we then employ a modified version of time reversal, known as iterative time reversal [57] to realize a linear spatial inversion of the received signals. This allows us to reconstruct the exact positions and phases/amplitudes of the emitting monopoles: we have reconstructed the object from the far field with a resolution as small as the period of the medium, $\lambda/80$ (black points of Fig. 2.6d).

A brief letter of this study can be found in Ref. [39] and the rest of it has been published in the form of longer articles [40, 41]. Here, we did not discuss the effect of material losses. Obviously, losses diminish all of the quality factors, which impacts more the most subwavelength modes propagating in the structure. This reduces their resonant enhancement and finally their radiation efficiency. This, in turns, sets a limit to the resolution of the system, whether used for time reversal focusing under the diffraction limit or for purposes of subwavelength imaging from the far field. Also, similar far-field imaging was realized using a split ring–based resonant metalens for imaging of subwavelength-varying magnetic objects [59], which gave similar results. Finally, we point out that our imaging experiment is relatively academic here, since the object has to be active and emit waves. We will prove in the last section of this chapter that this is not a necessary condition, and that a more complex imaging procedure can be realized only from measuring far-field-to-far-field scattered signals, which still can give images of an object placed in the near field of the resonant metalens with a resolution much better than the diffraction limit.

2.3.2 Moving from Microwaves to Acoustics: A Soda Can–Based Resonant Metalens

The previous microwave experiments have clearly demonstrated that one can easily access the polariton like modes from the far field and then use them for beating the diffraction limit. But, at this stage it was not obvious that this aspect was as general as we introduced in the first section of this chapter and might be specific to the case of the wire medium only. Therefore, we decided to move on to another field that also presents an easy access to the spatiotemporal-dependent wave field, namely acoustics. In this case, the wave originates from different microscopic phenomena and, for example, the acoustic waves are scalar waves which can be described by the pressure field only. But, they are also governed by the Helmholtz equation with a typical speed of 343 m.s^{-1} in air and 1490 m.s^{-1} in water. For practical reasons, we decided to work with audible acoustic waves in air where the wavelength is in the meter range compared to underwater ultrasonic waves which are millimeter sized. Moreover, the spanned frequency range is of the order of a few hundreds of hertz, which permits us to use experimental material that has a relatively low cost. The results that we present in this part have been published in Ref. [42].

Subwavelength control of acoustic waves has not been studied as much as in electromagnetics, but there have been few proposals in order to realize super-resolution imaging based on canalization [88] or the hyperlens [50]. Concerning focusing under the diffraction limit there have been propositions based on the analogue of the optical bull's eye [13] or based on the use of the negative index of refraction [87]. Nevertheless, none of these proposals clearly demonstrated subwavelength control of the acoustic waves. The only experimental proofs that clearly showed superfocusing come from the use of an acoustic sink [15], which requires an active source at the focal point, or a proposal which uses a phononic crystal where both the source and the image are in its near field [76]. The idea was therefore to transpose the concept of the resonant metalens to the acoustic community.

To do so, we first need to find the subwavelength unit cell that will couple to the continuum of propagating waves in air. The first basic idea was to reproduce the equivalent of the half-wavelength-long resonant wire used in microwaves, and we tried to use a pipe as a resonator. Indeed, it is well known in acoustics that, due to impedance mismatch at the edges of the pipe, resonances occur each time the half of the wavelength is a multiple of the pipe's length (for pipes opened at both ends). Similarly to the metallic wire in electromagnetics, this resonator can present a deeply subwavelength dimension in the transverse direction, while it is half-wavelength-long in the other dimension. Unfortunately, this resonator has been revealed to be too lossy because of the sliding effect along the walls. In acoustics, another subwavelength resonator was introduced a century ago by von Helmholtz [82] and is now called a Helmholtz resonator. It consists of a closed cavity linked to the outside by a small neck. The resonance is due to a collective oscillation of the air contained in the neck and the cavity's volume acts as a spring. The resonance frequency can therefore be tuned by shaping the cavity's size and the neck's length and width. Typically small openings and big cavity volumes decrease the resonant frequency, making the resonator really small compared to the wavelength. To avoid the losses that are due to the sliding as we experimented in the pipe, we chose to limit the neck's length to its lower possible value. For these reasons a good candidate for this experiment happened to be an everyday life object: a simple 33 cL aluminum soda can. This Helmholtz resonator presents a resonant frequency around 420 Hz, while its transverse dimension is 6.5 cm, which makes it a $\lambda/12$-size resonator.

Then, we need to build a 2D finite-size medium based on this resonant unit cell. We decided to work on a 7×7 close-packed square lattice of soda cans, as shown in Fig. 2.7a. The medium is surrounded by a set of eight computer-controlled speakers, and a motorized microphone is placed on top of the array of Helmholtz resonators. We first emit a short pulse (actually we used chirped emission and applied matched filtering to recover a short pulse) from one speaker and record the temporal signal received by the microphone placed 1 cm on top of the aperture of one can. The temporal signal and its frequency spectrum shown in Fig. 2.7b,c clearly reveal the multiresonant nature of the medium. The signal extends over hundreds of milliseconds compared to the initial pulse duration of 20 ms. On the frequency domain this manifests by the presence of many resonant peaks ranging from 250 Hz up to the resonant frequency of a single resonator, that is, 420 Hz.

We then repeat the same experiment when each speaker emits a short pulse and we move the microphone on top of the array. Knowing this entire set of impulse responses we can therefore mimic different monochromatic experiments. As already mentioned in the previous sections a resonant metalens supports several eigenmodes, each of them having its own resonance frequency. These eigenmodes also possess different radiation patterns and has to be excited with respect to them. To demonstrate this, we show in Fig. 2.7d the monochromatic field maps at various frequencies and with different emission patterns (a monopolar, a dipolar along the x direction, a dipolar along the y direction, and a quadrupolar). All of these maps clearly show the subwavelength nature of the modes supported by the medium. For example the first mode shows two nodes of the field while the entire dimension of the medium is roughly $\lambda/2$.

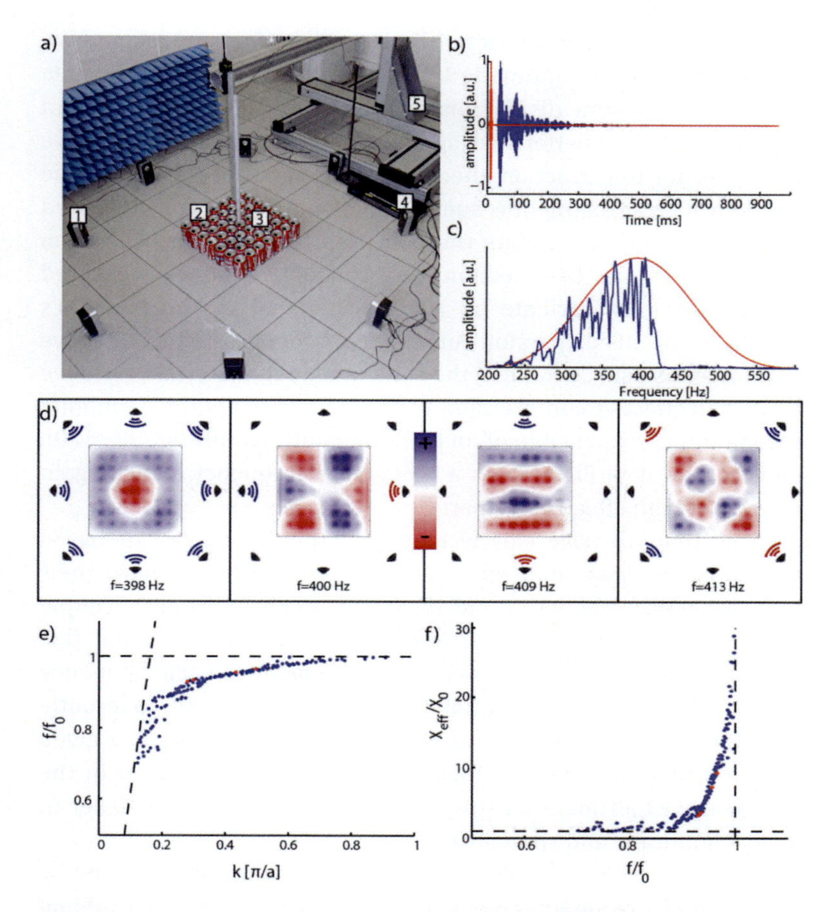

Figure 2.7 (a) A picture of the experimental setup. Eight commercial computer speakers (1) are controlled using a multichannel soundcard (4) and create sounds that excite the array of soda cans (2). Mounted on a 3D moving stage (5), a microphone records the pressure over the array. (b) Typical emitted pulse (red) and measured pressure (blue) on top of one can. (c) Corresponding spectra exhibiting many resonance peaks. (d) Measured monochromatic patterns at distinct frequencies and with different radiation patterns. (e) Experimental dispersion relation and (f) its equivalent in terms of effective compressibility.

The measured data does not limit to these four modes and we therefore made an automatic treatment of all the measurements in order to draw a dispersion relation. For each frequency and for each radiation pattern we spatially Fourier transform the wave field in order to extract an effective wavenumber of the mode. The result of such a treatment is summarized in Fig. 2.7e. The measured dispersion relation exhibits the polariton behavior that has been introduced in the first section. One can note that we measured eigenmodes that oscillate on a scale as small as the medium's period. From this dispersion curve, we can then extract an effective parameter as usually done in the context of metamaterials. Here we extract an effective compressibility as the air cavity of the Helmholtz resonators is responsible of an extra dynamic compressibility [20], and we plot it in Fig. 2.7f as a function of frequency, which again shows the high effective property.

Now that we have clearly identified that the soda can–based resonant metalens supports a set of eigenmodes that have their own resonance frequency and that these modes efficiently couple to the plane waves in the far field, we can expect to exploit this property. Again, the issue remains the same since each frequency encodes for a different spatial scale, and we need to coherently sum the different eigenmodes at a given position and at a given time in order to observe a focus. As we have already done in the microwave experiment we propose to use time reversal in order to focus temporally and spatially the waves.

We first start by a set of two control experiments while removing the Helmholtz resonators metalens. We record with the microphone placed at a given position the set of eight impulse responses when each of the speaker emits a short pulse. We then time-reversed each of those signals and simultaneously reemit them by their corresponding speaker. We then map the wave field around the initial microphone's position by moving it (and repeating the simultaneous emission each time the microphone has moved). As a result of such a procedure, we then plot the square of the maximum in time of each received signal. This is equivalent to show the maximum power received on each position. The result of such an operation is displayed in Fig. 2.8a. The obtained field maps show that eight-channel time reversal in a typical laboratory room permits

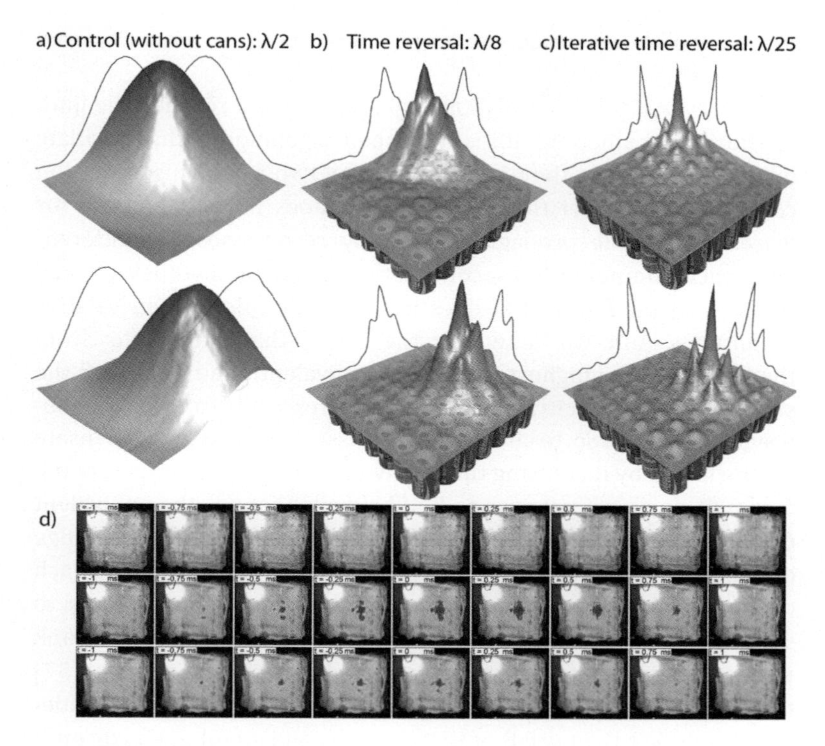

Figure 2.8 Subdiffraction focusing of sound. (a) Diffraction-limited focal spots obtained using time reversal without the array of cans. (b) The foci obtained using time reversal onto the same locations with the array of Helmholtz resonators. (c) Foci obtained with inverse filter signals demonstrating focal spots as thin as $\lambda/25$. (d) A few frames of a slow-motion movie of glass beads on top of a Mylar sheet when focusing acoustic waves: nothing happens on the control experiment (top), while beads move when creating subwavelength spots using time reversal (middle) and inverse filter signals (down).

one to focus wave on isotropic focal spots whose dimension is half a wavelength, limited by the diffraction. We then perform the same procedure in the presence of the cans. This time, the set of emitting signals span a longer time range as a signature of the modes resonances, and the time reversal permits one to focus the acoustic waves on focal dimensions as thin as $\lambda/8$ (Fig. 2.8b). This, again, is the result of the evanescent to propagating waves conversion offered by the resonant metalens.

But, even if this is a breakthrough, we have not reached the limit of the device since we initially probed eigenmodes that oscillate on scales as thin as the distance between two cans. This limitation comes from the fact that time reversal does not compensate for losses during the propagation. It only recombines the different frequency components by coherently adding them—they all add in phase at the focus—but does not play any role in their relative amplitudes. And, we know that the modes that most suffer from the losses are the highest Q ones, or equivalently the ones with the smallest group velocity, which exactly corresponds to the most sub-wavelength ones. So, to circumvent this issue we need to compensate for the losses by increasing the relative weight of these modes at the focus. To do so we propose to build signals that are the equivalent of an inverse filter [79]. This procedure first requires the knowledge of the set of all impulse responses between the 8 speakers and each desired focal position on top of the resonators, which we limited to 49 as the number of cans. Then we numerically computed a bank of 8×49 signals based on an iterative scheme of time reversal [57] that supposedly focus on each position with the lowest possible side-lobe levels. We then use 8 of those signals and simultaneously emit them with the speakers. Eventually, we map the wave field on top of the cans while emitting those signals, and we end up on the result shown in Fig. 2.8c with focal spots as thin as $\lambda/25$. Of course, because we cannot focus waves in between cans, the focusing resolution is actually limited by the period of the medium. Overall, we prove that we can beat the diffraction limit by a factor of 12.5 with a positioning accuracy of $\lambda/12$. We only present two maps but this focusing can indeed be performed on any position on top of a can.

Apart from its evident fundamental interest, this experiment opens up many avenues in terms of applications for sound and ultrasound. We believe that our approach is very promising for the design of arrays of actuators, micromechanical actuators in general, and, by reciprocity, of sensors. Indeed, using subwavelength coupled resonators offers three tremendous advantages. First, it introduces the possibility to engineer a matrix of actuators or sensors that are arranged on a subwavelength scale. Second, because our approach takes advantage of dispersion, it allows us to address independently many sensors using their temporal signature. Finally, as we will prove it next, it also enhances the acoustic displacement on a given location, because of the subwavelength dimensions of the focal spots.

As a principle proof-of-concept, we performed a visual experiment: we suspended a 20 μm thin sheet of metallized Mylar on top of the array of cans. This sheet is supposedly transparent to acoustic waves because it follows the displacement of the surrounding air. We deposited a few glass beads (diameter around 120 μm) on top of this sheet. When we emitted time reversal signals (or inverse filters ones), we saw the glass beads moving on the desired subwavelength area on top of the soda cans' array. Anyone who visited the laboratory at that time could enjoy this visual experiment, but translating it into publishable results was a bit challenging.

We used a high-frame-rate camera in order to catch the beads' displacements. A white light projector illuminated the sheet of Mylar, oriented a few degrees from its normal, while the camera was placed exactly at normal incidence. Because of the small angle between the projector and the camera, the Mylar sheet appeared dark, except for the direct image of the projector's bulb in the absence of beads. On the contrary, after depositing the beads the Mylar sheet appeared very shiny. This effect occurs thanks to the retroreflection of the light on the glass beads that are placed on top of a metallized surface. This means that when beads are stuck to the Mylar sheet the image on the camera appears white, while it appears dark when the beads jump away.

We utilized this experimental procedure in order to make movies of the field created when we focus onto various locations on top of the array of cans. Figure 2.8d shows a few frames from films obtained for the three different types of emissions we previously performed. While nothing happens on the control experiment without cans, one can notice that we have actually been able to darken the image on a small area by using time reversal (and inverse filters signals) during roughly 1 ms. This means that the beads located on a $\lambda/12$ squared area jumped on top of the Mylar sheet and this has been controlled from speakers placed in the far field of this medium. This entails that our approach can be utilized for subwavelength-size actuators and microelectromechanical systems (MEMS).

Despite these promising applications specific to acoustic and/or elastic waves, this acoustical experimental demonstration of a resonant metalens permitted us to validate the approach presented in the first section of this chapter. This experiment first permitted to tackle the effects of losses since our first try on pipes did not work at all. Second, it proved that the idea of polariton-like modes is not specific to the wire medium in electromagnetics but really generalizes to any subwavelength resonators organized on a subwavelength scale, as we initially presented in Section 2.1 of this chapter. We will now see that our approach can be used in optics as well.

2.4 Optical Resonant Metalens with Plasmonic Nanoparticles

This section is now devoted to an optical realization of a resonant metalens. Indeed, the subwavelength abilities of such a device find most of their applications in the nanoscale lithography, sensing or imaging. The microwave experiment as well as the acoustic one have paved the way toward this optical realization since we have learned a lot regarding the effects of dispersion, losses, and far-field conversion. We therefore here discuss the issues related to the manipulation of optical waves (which are necessarily harder to manipulate), design an optical resonant metalens with realistic parameters for current nanofabrication abilities, and provide simulation results that demonstrate subwavelength focusing/imaging in the visible range. This section mostly retakes the results published in Ref. [43].

2.4.1 Specificity of Light Manipulation

As stated before, imaging below the diffraction limit requires to measure the evanescent spectrum of an illuminated object, which contains its information on a subwavelength scale but vanishes exponentially away from it. In optics, this has been achieved with near-field scanning optical microscopy [2, 47, 65], but as stated before this procedure requires point-by-point scanning realized by moving a nanotip in the near field of the object. The need for real-time super-resolution imaging has led to many proposals of superlenses [19, 63], high-numerical-aperture lenses [70], structured illumination methods [6, 27], or sophisticated procedures that exploit nonlinearities [29]. All of these monochromatic approaches, however, still remain limited in resolution, mainly owing to the poor properties of the materials available in the optical range [64].

With the experience gained from our microwave and acoustics experiments, we show that it is also possible in the optical range to take advantage of polychromatic illumination: the dispersive properties of the optical metamaterials now helps to circumvent the limitations associated with monochromatic illumination. To that aim one needs to be able to temporally manipulate light, which becomes harder than that we have previously performed, since visible light oscillates at frequencies that avoid any possible sampling of the temporal wave field. Nevertheless, in the last few years, tremendous work has demonstrated that it is possible to spatially or/and temporally manipulate light wave fields [58].

On the one hand, a technology that permits one to control spatially the wave field is the so-called spatial light modulators (SLMs). These devices consist on a matrix of many controllable pixels that are used either on transmission or in reflection. The user is able to control the phase and/or the amplitude of the transmitted (reflected) light through each pixel. This way instead of having a fixed delay law as imposed by a geometrical lens, these SLMs permit one to shape the outgoing wavefront. By doing so, Vellekoop and Mosk demonstrated in 2007 that it is possible to focus light through a multiply scattering medium [80]. Since this work, many other groups have similarly demonstrated amazing applications [14, 58, 66] of such technics and the field is growing year after year.

On the other hand, to temporally control the wave field, the use of these SLMs in combination with pulsed lasers has also permitted several achievements in the also-growing field of femtosecond pulse shaping. The simplest technology consists of (i) separating the different spectral components of a pulse by the use of a diffraction grating, for example, (ii) phase- and amplitude-modulating each spectral component by the use of an SLM, and (iii) recombining all of the spectral components to build the desired pulse. This can find many applications in pulse compression in order to reach the attosecond short pulses [85], dispersion compensation in the context of fiber optic communications [12], light–matter interactions [71], and spectrally selective nonlinear microscopy [72].

Eventually, tremendous achievements have combined the pulse and wavefront–shaping techniques in order to spatiotemporally control light [4, 35, 56], as time reversal techniques have permitted us to do experiments in the low-frequency regime, where it is possible to easily measure the spatiotemporal wave field. For all of these reasons we believe that our proposal is experimentally realizable.

2.4.2 Designing the Plasmonic Resonant Metalens

Again, the aim of the design is to build a medium made out of subwavelength resonant unit cells in the proper frequency range. This leads to the polariton behavior of the medium that eventually gives rise to collective modes that oscillate on a scale much smaller that the free-space wavelength.

In optics, we can imagine two different types of subwavelength resonators, Mie particles and plasmonic ones. The Mie resonators [22] consist of small (compared to the free-space wavelength) particles made of a high-index material. The Mie particle exhibits a resonance when the particle size is around half the effective wavelength inside the particle. This type of resonators could present the great advantage of not suffering too much from losses. But given the available range of high-index materials in the visible spectrum, the Mie resonator is not really scalable down to the deeply subwavelength size. Another type of resonator is the plasmonic nanoparticle [37], which presents resonant peaks when the particle size is much smaller than the free-space wavelength. Such nanoparticles have excited a lot the optical community for absorbing light for solar cell applications [3, 23], spectroscopy [83], and guiding of light on a subwavelength scale [54]. This time the resonance is not due to the high index of the particle but due to a collective excitation of the electron cloud of the particle. The resonant wavelength can be tuned by changing the shape and/or the metal of the particle.

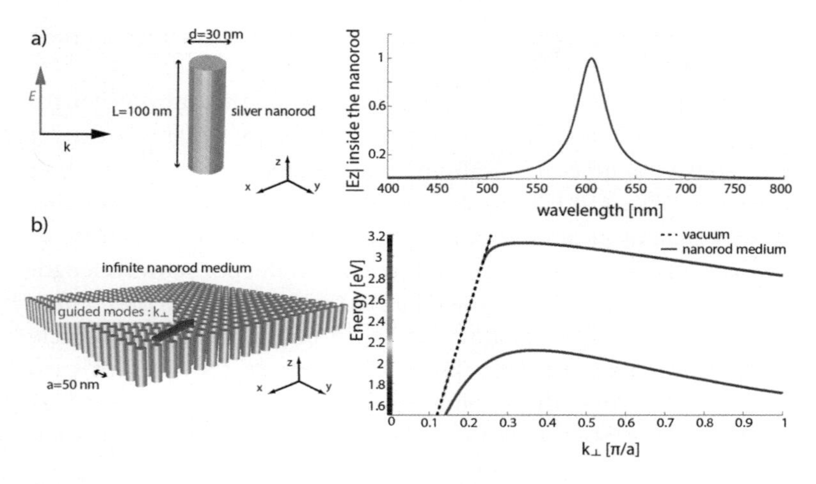

Figure 2.9 The plasmonic resonator–based medium. (a) Geometric parameters of the silver nanorod that exhibits a resonance peak around 600 nm. (b) When organized on a square lattice the nanorod medium supports subwavelength-propagating modes, with a dispersion relation given in the right panel.

For the purpose of the resonant metalens, to get the smallest resonator possible, we chose to work on a nanorod: Due to the anisotropy in its shape it presents a resonance along its long axis, while its transverse dimension can remain much thinner than the free-space wavelength. This way we will be able to stack many resonators within a wavelength. Silver has been chosen as the metal because it presents a minimum of losses in the near infrared [34] compared to gold and copper, for example. To stay within the limitations of current fabrication technologies, we set the diameter and height of the nanorod to, respectively, 30 nm and 100 nm, hence keeping an aspect ratio close to 3. To determine the resonant wavelength of such a plasmonic resonator we perform 3D simulation of a single nanorod. The permittivity that describes the silver is taken from Ref. [34], which takes into account the losses through an imaginary part of the permittivity. The result of such a simulation is summarized in Fig. 2.9a. When excited with a plane wave linearly polarized along its long axis, the nanorod exhibits a resonance peak at a wavelength near 600 nm, which manifests by a strong enhancement of the vertically polarized electric field in the near field of the rod.

Then, we build a periodic medium based on this subwavelength resonating unit cell. We chose to work on a square lattice of nanorods with a period of 50 nm, still keeping in mind fabrication feasibility (Fig. 2.9b). As explained in previous sections, organizing these resonant plasmonic nanoparticles on a subwavelength scale induces a collective coupling that results in deeply subwavelength modes, whose period can be as small as the medium's. In the context of optics this kind of modes have been observed in nanoparticle chain waveguides of Refs. [54, 84], albeit in the present case they are 2D. Their dispersion relation is given in Fig. 2.9b (see "Supplementary Information" of Ref. [43] for a detailed description on how to extract this dispersion relation). One can also note that we plotted a second propagation band that is associated to a second resonance of the nanorod that has not been probed in the single rod simulation.

This dispersion relation deserves few comments since it does not show the horizontal asymptote at the resonant frequency of the unit cell as opposed to the polariton behavior presented in the previous sections. Also, a mode with a zero group velocity is observed for a wavenumber which is not at the edge of the first Brillouin zone. There are two reasons that explain this behavior: (i) The optical properties of silver evolve in the spanned spectral range and the polariton dispersion is modified by this change in the permittivity of silver, and (ii) some near-field coupling affects the usual polariton behavior because, due to charge accumulation on the sides of the rods, one nanorod is affected by the field in its nearest neighbors.

Besides this specific effect observed with the silver nanorods, this medium provides the existence of the subwavelength modes (the dispersion relation is behind the light cone) that a resonant metalens requires. Because of their subwavelength nature, these wave fields cannot escape from the top or bottom interfaces and propagate in the transverse directions of the rods. As a consequence of the specific shape of this dispersion relation, the deep-subwavelength modes do not suffer from losses too drastically: they can escape the metalens faster because the group velocity does not vanish as much as in the classical polariton dispersion relation. Furthermore, the array of nanorods is designed such that the most subwavelength varying modes, which present the highest wavenumbers, appear at a wavelength close to 700 nm, the minimum of absorption of silver in the visible. These two properties will give the lens its deep-subwavelength resolution [40].

We now move from the infinite array of resonant nanorods to the resonant metalens by introducing finiteness along the transverse dimensions. As already described in the previous sections, this results in two consequences: First there is a quantization of the eigenmodes supported by the medium, and second, these quantized eigenmodes experience resonances and radiate toward the far-field zone. To highlight these two consequences, we perform a numerical analysis of a square medium made of 9×9 nanorods, which will be the resonant metalens under study (Fig. 2.10a). We first excite the medium with a small dipole located 25 nm above one of the nanorods and polarized along the z axis. We emit a 5 fs long pulse centered at 700 nm, which covers the whole spectrum of the lens eigenmodes. As shown in Fig. 2.10b, the radiated field extends over 200 fs, a consequence of the resonant nature of the medium. This is also confirmed by its spectrum which displays several peaks characteristic of resonances.

To verify that those peaks correspond to the eigenmodes of the nanorods medium, we represent in Fig. 2.10c, a map of the near field of the medium (both in amplitude and phase) for various frequencies. These maps correspond to the z component of the electric field in a plane located 25 nm away from the medium on the opposite side to that of the exciting dipole. Each of these near-field maps displays the presence of subwavelength-varying fields with a spatial scale given by the dispersion relation of the infinite medium. This observation proves that the emitting dipole excites the eigenmodes (which do not have a node at the source position) of the medium and that each frequency corresponds to different spatial scales, thanks to the highly dispersive nature of the resonant medium.

The second consequence of the finite dimensions of the medium is the conversion of these subwavelength-varying fields to propagating waves, as already observed when probing the far field. Because each mode supported by the medium has its own

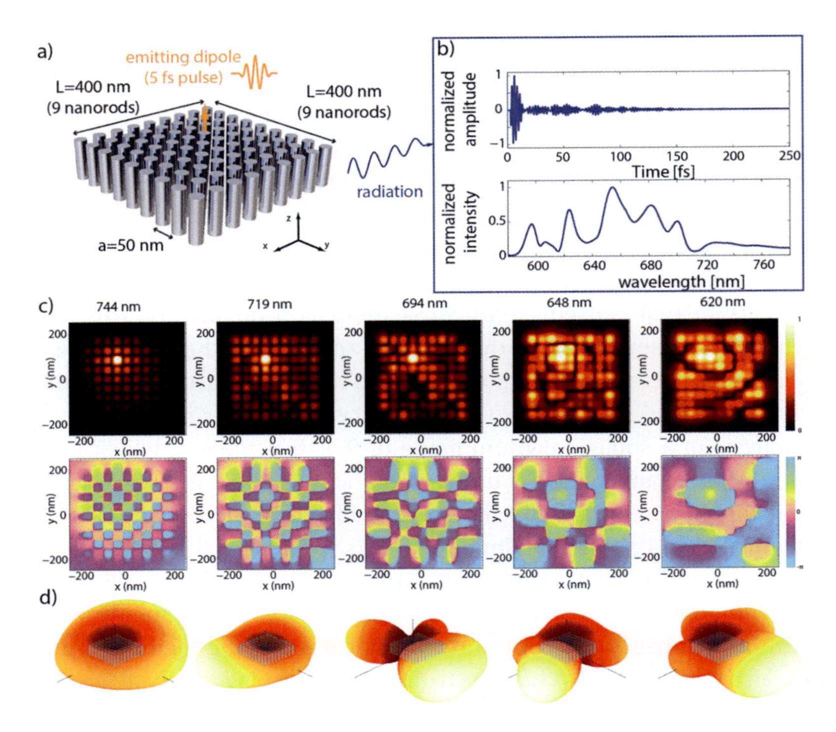

Figure 2.10 Resonant metalens illuminated with polychromatic light. (a) Geometry of the nanorod medium with finite dimensions. When excited with a small dipole emitting a short pulse (5 fs), we measure in the far field a radiated z polarized electric field that extends over 200 fs (b) and exhibits the presence of many resonance peaks. (c) The measured near fields, 25 nm away from the other interface of the medium, show subwavelength-varying modes. (d) Each mode radiates toward the far field with its own directivity pattern.

resonant frequency, their spatial information reaches the far field as a frequency signature giving rise to the so-called temporal degrees of freedom. Furthermore, the modes have distinct radiation patterns, as shown in Fig. 2.10d, which this time results in spatial degrees of freedom. It is the conjunction of these spatial and temporal degrees of freedom which gives the lens its properties, demarcating it from monochromatic approaches. Adopting the polychromatic approach in conjunction with a resonant metalens multiplies by the number of independent modes the amount of degrees of freedom available to the experimentalist for imaging or focusing purposes compared to the monochromatic case. A noteworthy fact is that these degrees of freedom can be scrambled by a multiply scattering medium and hence controlled from the temporal domain, the spatial one, or both, as in Refs. [4, 35, 38]. Another comment concerns the highly symmetric shape of the metalens under study here: a nonsymmetric shape should lift the degeneracy of the modes of the symmetric one and hence result in slightly more degrees of freedom.

2.4.3 Far-Field Subwavelength Focusing of Light Using Time Reversal

To test the possibilities offered by the lens, we first use it to generate deep-subwavelength focal spots from the far field. To that aim, we again use the concept of time reversal focusing. In optics, time reversal was used by Mark Stockman [49], who, inspired by Ref. [46], demonstrated theoretically the possibility to create subwavelength hot spots on random ultrathin silver films. Here, contrary to Ref. [49], designing our lens permits us to engineer the subwavelength modes it supports. Furthermore the illumination is realized azimuthally as opposed to their transmission scheme which forced the medium to be impractically thin.

Using time reversal techniques supposes, however, that the set of temporal impulse responses between the desired focal position and the set of sources is known. This step is easy to perform numerically but is more challenging experimentally. We propose three different ways to achieve this acquisition. First, one can use a time-resolved near-field scanning optical microscope to measure those signals, but this technique might suffer from perturbing effects from the tip. Second, one can imagine exploiting the nonlinearity of the silver nanorods in an analogous manner to the work of Ref. [5]. One can, for instance, use a SLM as a pulse shaper, as in Ref. [4], and maximize the second harmonic generation at a given position on the nanorod array with an adaptive algorithm, as in Ref. [5]. This could be realizable from the far field since the second harmonic waves are not trapped in the structure (the medium does not support modes) and a single focus detection is resolution unlimited (for high enough signal-to-noise ratios). Third, one can characterize the sample by

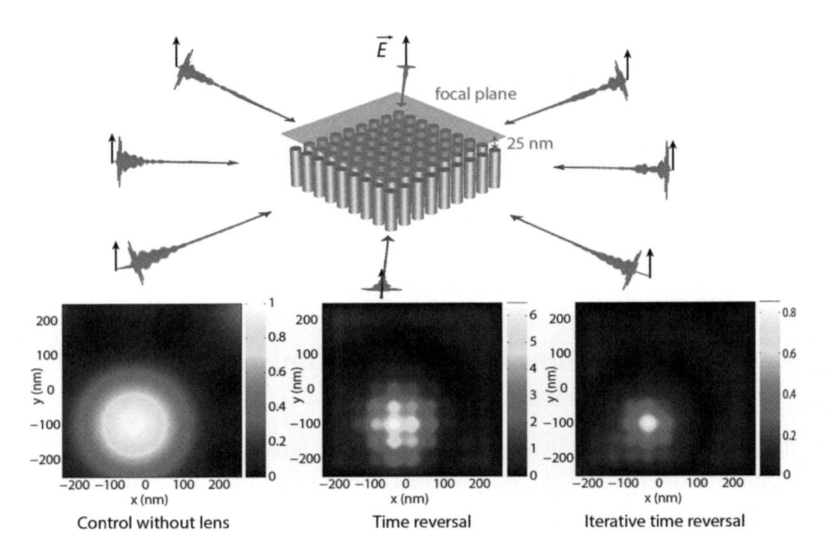

Figure 2.11 Far-field subwavelength focusing of shaped light pulses with the plasmonic far-field superlens. Emitting from eight directions distinct time-varying light signals we map the intensity in the near field of the medium. We obtain an isotropic diffraction-limited focal spot in the absence of the lens, while we obtain a $\lambda/5$ wide spot with the emission of time reversal signals and a $\lambda/23$ wide spot with the emission of iterative time reversal signals.

the use of a high-resolution microscope (e.g., a scanning electron microscope) and then use a simulation to numerically obtain the experimental signals that have to be emitted.

The knowledge of the impulse responses at hand and to take advantage of the distinct radiation patterns, we emit light from eight distinct directions around the resonant metalens (Fig. 2.11). Numerically, we thus emit the time-reversed version of the radiated fields in the eight directions when a small dipole like the previous one emitted a short pulse. With this method, we obtain a $\lambda/5$ (130 nm) wide spot shown in Fig. 2.11, while it is limited to $\lambda/2$ by the diffraction limit in the absence of the medium. Naturally, we can focus onto each rod by changing the emitted shaped pulses. Shaping the pulse in phase and amplitude provides hence even greater flexibility for focusing than polarization shaping methods [1]. Furthermore, because we focus light on a subwavelength scale, an enhancement of the intensity at the focus can be observed at the focal position. All the focal spots are normalized by the emitted energy. The control experiment gives an intensity of the electric field at the focal position ,which is therefore normalized to 1 since all the frequency components sum constructively (in time and space) at the focal position. When we focus on a thinner scale, thanks to the nanorod medium, we measured an electric field intensity at the focal position which is six times higher than the case without nanorods.

Time reversal focusing has demonstrated its ability to focus on a subwavelength scale from the far field, but it is not the best candidate in terms of resolution, as we have already observed in the previous section. Indeed, it does not compensate for the relative losses experienced by the modes at distinct frequencies. Much more adequate signals in terms of resolution, which behave as spatiotemporal pseudo-inverse filters are again obtained through an iterative scheme of time reversal [57]. This approach requires one to have the knowledge of the impulse responses from the 8 chosen directions to the 81 nanorods of the array. Applying this procedure we end on a set of temporal signals that supposedly focus on a thinner scale. We then take those signals and simulate the situation where they are emitted from the far field. The result of this procedure demonstrates a focusing of light from the far field on a $\lambda/23$ wide spot (30 nm). The increase in resolution has a cost in terms of deposited energy, and we measured an intensity at the focal position which is comparable to the one obtained without the nanorods. Indeed, the inverse filter signals optimize the resolution by increasing the frequency components that suffer more from losses.

2.4.4 Polychromatic Interferometric Far-Field Subwavelength Imaging

In the previous section, we have proven that the designed lens, used alongside polychromatic light, can focus light from the far field onto dimensions of the order of 30 nm. This means that the nanorod medium has efficiently projected toward the far field the deep-subwavelength information of the initial exciting dipole. We now use this property to demonstrate that it can as well realize the reciprocal

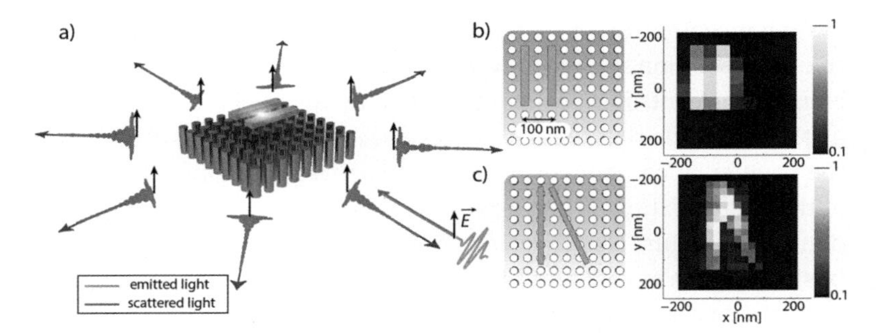

Figure 2.12 Far-field subwavelength imaging of a low-contrast object. (a) Acquisition from the far field of the signals that carry the subwavelength information of the object placed in the near field of the nanorod medium. (b) Geometry of an object (left) and the reconstructed super-resolved image which shows a resolution of 100 nm (right). (c) Geometry of an object that is not entirely aligned with the metalens (left) and subwavelength-resolved image of this object, displaying a 80 nm resolution.

operation: imaging objects from the far field with a subwavelength resolution. We assume the full knowledge of the 8×81 transient impulse responses $G_{ik}(t)$ (i corresponds to the emitting direction and k corresponds to a position on top of one nanorod, that is, a pixel of the image) between the 8 chosen far-field directions and the 81 nanorods. To verify the possibility to image objects with a resolution thinner than the diffraction limit, we place an object in the focal plane of the resonant metalens, that is, 25 nm above it (Fig. 2.12a). The object consists of two bars of a dielectric with an index of refraction of 1.5, with a square cross section of 30 nm. The distance between the centers of the two objects is 100 nm. We voluntarily opt for a low-refractive-index contrast object, which is typically much harder to image than a relatively high-contrast one. The imaging procedure, described in Fig. 2.12a, consists solely of far-field operations. Basically, we record the phase and amplitude of the light fields scattered in eight distinct directions when we emit a pulse from one direction. This would experimentally require interferometric measurements. This corresponds to the knowledge of the signal $S_{ij}(t)$ (i and j correspond, respectively, to the emitting direction and the receiving one). To take advantage of the spatial degrees of freedom offered by the lens, we repeat this operation for the eight emitting directions. Then, from this set of 8×8 temporal signatures $S_{ij}(t)$ that carry the object information, we start the reconstruction of the object.

The imaging procedure takes advantage of the subwavelength focusing property at emission and reception, as it is often done in ultrasonic echography imaging. This step is done numerically by convolution between the signal $S_{ij}(t)$ and the signal $G_{ik}(t)$ (for focusing at the emission) and then we convolve the result of this operation with $G_{jk}(t)$ in order to take advantage of focusing at the reception step. The signal corresponding to the kth pixel of the image is therefore the sum of the 8×8 such operations. The pixel's value in the image is eventually the maximum in time of this temporal signal. We repeat these operations for the 81 pixels of the image. The final image is eventually the intensity renormalized to have a color scale varying from 0 to 1.

In practice, to increase the resolution of the imaging procedure we do not use the impulse responses $G_{ik}(t)$ for the convolution products but the signals $H_{ik}(t)$ that correspond to the inverse filter signals. That way, we take advantage of the thinnest resolution possible obtained in the focusing scheme. This step can be highly improved by finding the ideal set of signals. We also replace the signals $S_{ij}(t)$ by the signals $S_{ij}(t) - S_{ij}^0(t)$, where $S_{ij}^0(t)$ corresponds to the scattered signals in each direction in the absence of the object. That way we suppress diffractive effects before starting the imaging procedure. Those subtraction signals present an energy which is 6 times lower (in average) than the signals before subtraction, but they still remain measurable with nowadays experimental techniques.

In Fig. 2.12b, we map the reconstructed image of the object where each pixel corresponds to a point on top of each nanorod. Meanwhile, using an interpolated set of Green's function (289×8), we try to image another object, similar to the first one, albeit not totally aligned with the nanorods (Fig. 2.12c).

The image obtained in Fig. 2.12b clearly shows that the two objects separated by a distance of 100 nm can be resolved from the far field. Moreover, using the object with a tilted dielectric rod, we verify that an object that is not aligned with the nanorods can be imaged, and we estimate our limiting resolution to about 80 nm.

Since we are still far from our focusing resolution of 30 nm, it is clear that this imaging resolution could be improved using more sophisticated inversion procedures [7], which could include the modification of the metalens response by the object to image in case of strong interactions. Also, even though we use here a set of temporal Green's function on top of the metalens to realize our reconstruction, there are obviously other ways to achieve it. For instance one could translate the tip of a near-field microscope on top of the metalens, and record the backscattering from the lens when pulses illuminate it. This would give an a priori knowledge of the metalens which could be used for imaging purposes in a manner analogous to ours.

Despite the simplicity of the objects imaged, these first results prove that the subwavelength information of an object can be registered in the far-field region, thanks to the polychromatic information given by our resonant metalens. Our approach amounts to a parallel and far-field type of near-field scanning optical microscopy. We underline here that it can be implemented in real time since it only requires few illuminations of the sample from the far field, which can be done using SLMs and/or movable mirrors. Eventually, we also mention that although we chose to use only 81 nanorods for computational limitations, it is possible to design much larger lenses, while keeping the physics unchanged. This, of course, would result in the presence of much more modes, which in turn will require to control and measure the far field from more directions.

2.5 Conclusion

In this chapter, we have given a simple and qualitative overview of the concept of resonant metalens, a superlens made out of finite-size, locally resonant metamaterials that can be used alongside a broadband or polychromatic approach to beat the diffraction limit for imaging or focusing purposes. We have introduced the concept in a first section, and linked it to that of locally resonant metamaterials. We have given the two key principles of the resonant metalens, namely that it permits (i) to code the spatial information of a source or object placed in its near field in the complex spectrum of the field propagating in the lens and (ii) to convert this near-field information into propagating waves that can be collected efficiently in the far field. We have given potential applications of the resonant metalens in terms of time reversal subwavelength focusing from the far field and also in terms of super-resolution imaging from the far field. We have then briefly discussed two experimental demonstrations of the approach in various domains of wave physics. First, a proof-of-concept was proposed and proved subwavelength focusing and imaging from the far field with microwaves and a resonant metalens made out of resonant metallic wires. Second, we shown that an array of soda cans can also, because they are Helmholtz resonators for audible acoustic waves, be a very good resonant metalens for sound. We have demonstrated with this lens deep-subwavelength focusing of sound from the far field and underlined its potential applications for sensors and actuators due to its ability to generate extremely high acoustic displacement on subwavelength spots. Finally, we have transposed the concept to the visible part of the electromagnetic spectrum by employing numerical simulations and a resonant metalens constituted of plasmonic resonant nanorods. We have shown how such a lens could be used, with a femtosecond laser or a white light supercontinuum, to focus light far below the diffraction limit with applications for instance in nanolithography or sensing and spectroscopy at the very deep-subwavelength scale. We have also proposed a pulse echo or backscattering far-field imaging scheme that could allow to realize super-resolved images of objects. Meanwhile we have given clues on how to bring this concept to an optical table, notably by using SLMs and random media coupled to this resonant metalens.

Of course there are still many criticisms that could be raised against the concept, for instance, it solely allows imaging on a 2D plane, or the optical application of the idea is not trivial, despite our proposals to simplify it. We agree with these criticisms and do not believe that the approach is perfect and suitable for all applications. Especially, we think that for imaging of the "living," fluorescence-based methods such as STED, PALM, STORM, and SIM, are much more appropriate [10, 27–29, 68], even though they require to inject fluorescent molecules into living organisms and use rather high optical intensities. We believe nonetheless that the resonant metalens has potential applications due to its relative ease of realization, its broad generality, and its simplicity. Furthermore, we think that these works permit us to bring new ideas and to question very well established concepts. For instance, metamaterials have always been studied mainly for their ability to produce negative-index media, being the holy grail to design a negative-index superlens that can image below the diffraction limit. Here we assert and prove that it is not necessary to have a negative index to beat the diffraction limit from the far field. On the contrary, using the high effective properties of locally resonant metamaterials actually permits to obtain much better results than negative refraction allows, at least in the three domains studied here: the microwave, the audible acoustic one, and the visible range. If the superlens idea was definitely worth the interest it has gathered, notably due to the enormous efforts it has generated toward the study and understanding of metamaterials, let us hope that the resonant metalens can also bring a stone to the building and lead scientists to work in different and complementary directions. Eventually, the concept also sheds light on the fact that the monochromatic approach, which has always been the way to go in optics, especially since the discovery of the laser, may not be the optimal one when one intends to beat the diffraction limit and more generally to control waves. As it has continuously been proven for the past 20 years with the concept of time reversal, using the polychromatic approach and the temporal approach brings a lot more frequency or temporal degrees of freedom for imaging or focusing applications, especially in complex and dispersive media such as metamaterials, multiple scattering media, and photonic and phononic crystals. Since spatial and temporal degrees of freedom in a given system do not add but multiply [38], we bet that there is much more to expect from broadband approaches than from monochromatic ones. With optics entering now the era of supercontinuum and ultrashort sources, we believe that the concept of the resonant metalens, if not the ultimate solution in this field, can at least inspire exciting research.

References

1. Aeschlimann, M., et al. (2007). Adaptive subwavelength control of nano-optical fields, *Nature*, **446**(7133), pp. 301–304.
2. Ash, E. A., and Nicholls, G. (1972). Super-resolution aperture scanning microscope, *Nature*, **237**, pp. 510–512.
3. Atwater, H. A., and Polman, A. (2010). Plasmonics for improved photovoltaic devices, *Nat. Mater.*, **9**(3), pp. 205–213.
4. Aulbach, J., Gjonaj, B., Johnson, P. M., Mosk, A. P., and Lagendijk, A. (2011). Control of light transmission through opaque scattering media in space and time, *Phys. Rev. Lett.*, **106**(10), p. 103901.
5. Aulbach, J., Bretagne, A., Fink, M., Tanter, M., and Tourin, A. (2012). Optimal spatiotemporal focusing through complex scattering media, *Phys. Rev. E*, **85**, p. 016605.
6. Bailey, B., Farkas, D. L., Taylor, D. L., and Lanni, F. (1993). Enhancement of axial resolution in fluorescence microscopy by standing-wave excitation, *Nature*, **366**, pp. 44–48.
7. Belkebir, K., and Sentenac, A. (2003). High-resolution optical diffraction microscopy, *J. Opt. Soc. Am. A*, **20**, pp. 1223–1229.
8. Belov, P. A., et al. (2003). Strong spatial dispersion in wire media in the very large wavelength limit, *Phys. Rev. B*, **67**(11), p. 113103.
9. Belov, P. A., Hao, Y., and Sudhakaran, S. (2006). Subwavelength microwave imaging using an array of parallel conducting wires as a lens, *Phys. Rev. B*, **73**(3), p. 033108.
10. Betzig, E., et al. (2006). Imaging intracellular fluorescent proteins at nanometer resolution, *Science*, **313**(5793), pp. 1642–1645.
11. Cassereau, D., and Fink, M. (1992). Time-reversal of ultrasonic fields. III. Theory of the closed time-reversal cavity, *IEEE Trans. Ultrason. Ferroelectr. Freq. Control*, **39**(5), pp. 579–592.
12. Chang, C., Sardesai, H., and Weiner, A. (1998). Dispersion-free fiber transmission for femtosecond pulses by use of a dispersion-compensating fiber and a programmable pulse shaper, *Opt. Lett.*, **23**, pp. 283–285.
13. Christensen, J., Fernandez-Dominguez, A. I., de Leon-Perez, F., Martin-Moreno, L., and Garcia-Vidal, F. J. (2007). Collimation of sound assisted by acoustic surface waves, *Nat. Phys.*, **3**(12), pp. 851–852.
14. Cui, M., and Yang, C. (2010). Implementation of a digital optical phase conjugation system and its application to study the robustness of

turbidity suppression by phase conjugation, *Opt. Express*, **18**, pp. 3444–3455.

15. De Rosny, J., and Fink, M. (2002). Overcoming the diffraction limit in wave physics using a time-reversal mirror and a novel acoustic sink, *Phys. Rev. Lett.*, **89**, p. 124301.
16. Deymier, P. (2013). *Acoustic Metamaterials and Phononic Crystals* (Springer, Berlin/Heidelberg).
17. Durant, S., Liu, Z., Steele, J., and Zhang, X. (2006). Theory of the transmission properties of an optical far-field superlens for imaging beyond the diffraction limit, *J. Opt. Soc. Am. B*, **23**, pp. 2383–2392.
18. Engheta, N., and Ziolkowski, R. W. (2006). *Electromagnetic Metamaterials: Physics and Engineering Explorations* (IEEE-Wiley, New York).
19. Fang, N., Lee, H., Sun, C., and Zhang, X. (2005). Sub-diffraction-limited optical imaging with a silver superlens, *Science*, **308**(5721), pp. 534–537.
20. Fang, N., et al. (2006). Ultrasonic metamaterials with negative modulus, *Nat. Mater.*, **5**(6), pp. 452–456.
21. Fano, U. (1961). Effects of configuration interaction on intensities and phase shifts, *Phys. Rev.*, **124**(6), p. 1866.
22. Fenollosa, R., Meseguer, F., and Tymczenko, M. (2008). Silicon colloids: from microcavities to photonic sponges, *Adv. Mater.*, **20**, pp. 95–98.
23. Ferry, V. E., Munday, J. N., and Atwater, H. A. (2010). Design considerations for plasmonic photovoltaics, *Adv. Mater.*, **22**, pp. 4794–4808.
24. Fink, M. (1997). Time reversed acoustics, *Phys. Today*, **50**(3), pp. 34–40.
25. Goodman, J. W. (2005). *Introduction to Fourier Optics* (Roberts & Company Publishers, Englewood).
26. Goos, F., and Hänchen, H. (1947). Ein neuer und fundamentaler Versuch zur Totalreflexion, *Ann. Phys.*, **436**(7–8), pp. 333–346.
27. Gustafsson, M. G. L. (2000). Surpassing the lateral resolution limit by a factor of two using structured illumination microscopy, *J. Microsc.*, **198**(2), pp. 82–87.
28. Gustafsson, M. G. (2005). Nonlinear structured-illumination microscopy: wide-field fluorescence imaging with theoretically unlimited resolution, *Proc. Natl. Acad. Sci. U.S.A.*, **102**(37), pp. 13081–13086.
29. Hell, S. W., and Wichmann, J. (1994). Breaking the diffraction resolution limit by stimulated emission: stimulated-emission-depletion fluorescence microscopy, *Opt. Lett.*, **19**, pp. 780–782.

30. Hopfield., J. J. (1958). Theory of the contribution of excitons to the complex dielectric constant of crystals, *Phys. Rev.*, **112**, pp. 1555–1567.
31. Jacob, Z., Alekseyev, L., and Narimanov, E. (2006). Optical hyperlens: far-field imaging beyond the diffraction limit, *Opt. Express*, **14**, pp. 8247–8256.
32. Joannopoulos, J. D., Johnson, S. G., Winn, J. N., and Meade, R. D. (2008). *Photonic Crystals: Molding the Flow of Light* (University Press, Princeton).
33. John, S. (1987). Strong localization of photons in certain disordered dielectric superlattices, *Phys. Rev. Lett.*, **58**(23), pp. 2486–2489.
34. Johnson, P. B., and Christy, R. W. (1972). Optical constants of the noble metals, *Phys. Rev. B*, **6**, pp. 4370–4379.
35. Katz, O., Small, E., Bromberg, Y., and Silberberg, Y. (2011). Focusing and compression of ultrashort pulses through scattering media, *Nat. Photonics*, **5**(6), pp. 372–377.
36. Lagendijk, A. (1993). Vibrational relaxation studied with light, *Ultrashort Processes in Condensed Matter*, **314**, pp. 197–236.
37. Lal, S., Link, S., and Halas, N. J. (2007). Nano-optics from sensing to waveguiding, *Nat. Photonics*, **1**(11), pp. 641–648.
38. Lemoult, F., Lerosey, G., de Rosny, J., and Fink, M. (2009). Manipulating spatiotemporal degrees of freedom of waves in random media, *Phys. Rev. Lett.*, **103**, p. 173902.
39. Lemoult, F., Lerosey, G., de Rosny, J., and Fink, M. (2010). Resonant metalenses for breaking the diffraction barrier, *Phys. Rev. Lett.*, **104**, p. 203901.
40. Lemoult, F., Fink, M., and Lerosey, G. (2011). Far-field sub-wavelength imaging and focusing using a wire medium based resonant metalens, *Waves in Random and Complex Media*, **21**(4), pp. 614–627.
41. Lemoult, F., Fink, M., and Lerosey, G. (2011). Revisiting the wire medium: an ideal resonant metalens, *Waves in Random and Complex Media*, **21**(4), pp. 591–613.
42. Lemoult, F., Fink, M., and Lerosey, G. (2011). Acoustic resonators for far-field control of sound on a subwavelength scale, *Phys. Rev. Lett.*, **107**(6), p. 064301.
43. Lemoult, F., Fink, M., and Lerosey, G. (2012). A polychromatic approach to far-field superlensing at visible wavelengths, *Nat. Commun.*, **3**, p. 889.
44. Lemoult, F., Kaina, N., Fink, M., and Lerosey, G. (2013). Wave propagation control at the deep subwavelength scale in metamaterials, *Nat. Phys.*, **9**, pp. 55–60.

45. Lerosey, G., et al. (2004). Time reversal of electromagnetic waves, *Phys. Rev. Lett.*, **92**(19), p. 193904.
46. Lerosey, G., de Rosny, J., Tourin, A., and Fink, M. (2007). Focusing beyond the diffraction limit with far-field time reversal, *Science*, **315**, p. 1120.
47. Lewis, A., Isaacson, M., Harootunian, A., and Muray, A. (1984). Development of a 500 Å resolution microscope, *Ultramicroscopy*, **13**, pp. 227–231.
48. Li, J., and Chan, C. T. (2004). Double-negative acoustic metamaterial, *Phys. Rev. E*, **70**(5), p. 055602.
49. Li, X., and Stockman, M. I. (2008). Highly efficient spatiotemporal coherent control in nanoplasmonics on a nanometer-femtosecond scale by time reversal, *Phys. Rev. B*, **77**, p. 195109.
50. Li, J., Fok, L., Yin, X., Bartal, G., and Zhang, X. (2009). Experimental demonstration of an acoustic magnifying hyperlens, *Nat. Mater.*, **8**(12), pp. 931–934.
51. Liu, Z., et al. (2000). Locally resonant sonic materials, *Science*, **289**, pp. 1734–1736.
52. Liu, Z., Lee, H., Xiong, Y., Sun, C., and Zhang, X. (2007). Far-field optical hyperlens magnifying sub-diffraction-limited objects, *Science*, **315**(5819), p. 1686.
53. Lord Rayleigh, F. R. S. (1879). Investigations in optics, with special reference to the spectroscope, *Philosophical Magazine Series 5*, **8**(49), pp. 261–274.
54. Maier, S. A., et al. (2003). Local detection of electromagnetic energy transport below the diffraction limit in metal nanoparticle plasmon waveguides, *Nat. Mater.*, **2**(4), pp. 229–232.
55. Martin, Y., Zenhausern, F., and Wickramasinghe, H. K. (1996). Scattering spectroscopy of molecules at nanometer resolution, *Appl. Phys. Lett.*, **68**(18), pp. 2478–2477.
56. McCabe, D. J., et al. (2011). Spatio-temporal focusing of an ultrafast pulse through a multiply scattering medium, *Nat. Commun.*, **2**, p. 447.
57. Montaldo, G., Tanter, M., and Fink, M. (2004). Real time inverse filter focusing through iterative time reversal, *J. Acoust. Soc. Am.*, **115**(2), pp. 768–775.
58. Mosk, A. P., Lagendijk, A., Lerosey, G., and Fink, M. (2012). Controlling waves in space and time for imaging and focusing in complex media, *Nat. Photonics*, **6**(5), pp. 283–292.
59. Ourir, A., Lerosey, G., Lemoult, F., Fink, M., and de Rosny, J. (2012). Far field subwavelength imaging of magnetic patterns, *Appl. Phys. Lett.*, **101**, p. 111102.

60. Pendry, J. B., Holden, A. J., Stewart, W. J., and Youngs, I. (1996). Extremely low frequency plasmons in metallic mesostructures, *Phys. Rev. Lett.*, **76**(25), pp. 4773–4776.
61. Pendry, J. B., Holden, A. J., Robbins, D. J., and Stewart, W. J. (1999). Magnetism from conductors and enhanced nonlinear phenomena, *IEEE Trans. Microw. Theory*, **47**(11), pp. 2075–2084.
62. Pendry, J. B., Martin-Moreno, L., and Garcia-Vidal, F. J. (2004). Mimicking surface plasmons with structured surfaces, *Science*, **305**(5685), pp. 847–848.
63. Pendry, J. B. (2000). Negative refraction makes a perfect lens, *Phys. Rev. Lett.*, **85**(18), pp. 3966–3969.
64. Piestun, R., and Miller, D. A. B. (2000). Electromagnetic degrees of freedom of an optical system, *J. Opt. Soc. Am. A*, **17**, pp. 892–902.
65. Pohl, D. W., Denk, W., and Lanz, M. (1984). Optical stethoscopy: image recording with resolution $\lambda/20$, *App. Phys. Lett.*, **44**, pp. 651–653.
66. Popoff, S. M., et al. (2010). Measuring the transmission matrix in optics: an approach to the study and control of light propagation in disordered media, *Phys. Rev. Lett.*, **104**(10), p. 100601.
67. Purcell, E. M. (1946). Spontaneous emission probabilities at radio frequencies, *Phys. Rev.*, **69**, p. 681.
68. Rust, M. J., Bates, M., and Zhuang, X. (2006). Stochastic optical reconstruction microscopy (STORM) provides sub-diffraction-limit image resolution, *Nat. Methods*, **3**(10), p. 793.
69. Salandrino, A., and Engheta, N. (2006). Far-field subdiffraction optical microscopy using metamaterial crystals: theory and simulations, *Phys. Rev. B*, **74**, p. 075103.
70. Sheppard, C. J., and Choudhury, A. (2004). Annular pupils, radial polarization, and superresolution, *Appl. Opt.*, **43**(22), pp. 4322–4327.
71. Shima, S. H., and Zanni, M. T. (2009). How to turn your pump–probe instrument into a multidimensional spectrometer: 2D IR and Vis spectroscopiesvia pulse shaping, *Phys. Chem. Chem. Phys.*, **11**, pp. 748–761.
72. Silberberg, Y. (2009). Quantum coherent control for nonlinear spectroscopy and microscopy, *Annu. Rev. Phys. Chem.*, **60**, pp. 277–292.
73. Silveirinha, M. G., Fernandes, C. A., and Costa, J. R. (2008). Additional boundary condition for a wire medium connected to a metallic surface, *New J. Phys.*, **10**, p. 053011.
74. Simovski, C. R., Belov, P. A., Atrashchenko, A. V., and Kivshar, Y. S. (2012). Wire metamaterials: physics and applications, *Adv. Mater.*, **24**(31), pp. 4229–4248.

75. Smith, D. R., Vier, D. C., Koschny, T., and Soukoulis, C. M. (2005). Electromagnetic parameter retrieval from inhomogeneous metamaterials, *Phys. Rev. E*, **71**, p. 036617.

76. Sukhovich, A., et al. (2009). Experimental and theoretical evidence for subwavelength imaging in phononic crystals, *Phys. Rev. Lett.*, **102**, p. 154301.

77. Sydoruk, O., Zhuromskyy, O., Radkovskaya, A., Shamonina, E., and Solymar, L. (2009). Magnetoinductive waves I: theory, in *Theory and Phenomena of Metamaterials* (Taylor and Francis, Boca Raton, FL).

78. Synge, E. H. (1928). A suggested method for extending microscopic resolution into the ultra-microscopic region, *Philosophical Magazine Series 7*, **6**(35), pp. 356–362.

79. Tanter, M., Aubry, J. F., Gerber, J., Thomas, J. L., and Fink, M. (2001). Optimal focusing by spatio-temporal inverse filter. I. Basic principles, *J. Acoust. Soc. Am.*, **110**, pp. 37–47.

80. Vellekoop, I. M., and Mosk, A. P. (2007). Focusing coherent light through opaque strongly scattering media, *Opt. Lett.*, **32**(16), pp. 2309–2311.

81. Veselago, V. G. (1968). The electrodynamics of substances with simultaneously negative values of ε and μ, *Sov. Phys. Usp.*, **10**(4), pp. 509–514.

82. von Helmholtz, H. (1885). *On the Sensations of Tone as a Physiological Basis for the Theory of Music* (Longmans, Green and Co., London).

83. Wang, J., Boriskina, S. V., Wang, H., and Reinhard, B. M. (2011). Illuminating epidermal growth factor receptor densities on filopodia through plasmon coupling, *ACS Nano*, **5**, pp. 6619–6628.

84. Weber, W. H., and Ford, G. W. (2004). Propagation of optical excitations by dipolar interactions in metal nanoparticle chains, *Phys. Rev. B*, **70**, p. 125429.

85. Weiner, A. M. (2000). Femtosecond pulse shaping using spatial light modulators, *Rev. Sci. Instrum.*, **71**(5), pp. 1929–1960.

86. Yablonovitch, E. (1987). Inhibited spontaneous emission in solid-state physics and electronics, *Phys. Rev. Lett.*, **58**(20), pp. 2059–2062.

87. Zhang, S., Yin, L., and Fang, N. (2009). Focusing ultrasound with an acoustic metamaterial network, *Phys. Rev. Lett.*, **102**, p. 194301.

88. Zhu, J., et al. (2011). A holey-structured metamaterial for acoustic deep-subwavelength imaging, *Nat. Phys.*, **7**(1), pp. 52–55.

Chapter 3

Ultrathin Metalens and Three-Dimensional Optical Holography Using Metasurfaces

Xianzhong Chen,[a,b] Lei Zhang,[c] Cheng-Wei Qiu,[c] and Shuang Zhang[a]

[a] *School of Physics and Astronomy, University of Birmingham, Birmingham B15 2TT, UK*
[b] *School of Engineering and Physical Sciences, Heriot-Watt University, Edinburgh EH14 4AS, UK*
[c] *Department of Electrical and Computer Engineering, National University of Singapore, Singapore 117583, Singapore*
x.chen@hw.ac.uk

3.1 Introduction

Optical devices—such as lenses—have restrictions in their potential for miniaturization and in many cases are ultimately limited in feature size by the wavelength of light. On the other hand, for optical systems to continue to be established as economically viable in a range of emerging application areas, it is necessary to continue the trend of miniaturization and integration. The general function of

Plasmonics and Super-Resolution Imaging
Edited by Zhaowei Liu

ISBN 978-981-4669-91-7 (Hardcover), 978-1-315-20653-0 (eBook)
www.panstanford.com

most optical devices can be described as the modification of the wavefront of light by altering the three fundamental properties of light (phase, amplitude, and polarization). One of the key techniques in many optical components is a spatially varying phase response, as illustrated in optical lens and phase modulation holograms. The functionality of a traditional optical device is usually realized by reshaping the wavefront of the light that relies on gradual phase changes along the optical paths, which are accomplished by either controlling the surface topography or varying the spatial profile of the refractive index. However, it is hard to accumulate sufficient phase change once the device size is further reduced to the micro-even the nanoscale due to the finite permittivity and permeability of natural materials.

Metamaterials can usually be engineered to exhibit extraordinary electromagnetic properties [1–7] that may not be found in nature or its constituent components, thus providing an unconventional alternative to optical design. Metasurfaces, the emerging field of metamaterials, which consist of a single layer of artificial "atoms," have recently captured the attention of the scientific community since they do not require complicated three-dimensional (3D) nanofabrication techniques [8, 9] but can steer light in equally dramatic ways. Unlike the phase change by the accumulated optical path in traditional optical elements, the abrupt phase change takes place within a two-dimensional (2D) plane, that is, metasurfaces [10, 11], meaning that a new freedom for controlling light propagation is introduced. By challenging established conventional optics with a transformative understanding of the phase discontinuities at the metasurfaces, there are several research groups, apart from our team, like Capasso's, Shalaev's, and Zhang's groups, who have done a lot of pioneering work in this emerging area. A plethora of applications have been proposed and demonstrated by using metasurfaces such as wave plates for generating vortex beams [10, 12], ultrathin metalenses [15–17], aberration-free quarter-wave plates (QWPs) [14], the spin Hall effect of light [20, 23], polarization-dependent unidirectional surface plasmon polariton excitation [21, 22], spin-controlled photonics [23], and optical holography [24, 25]. The thickness of these new optical devices is less than one-tenth of the wavelength, a highly desirable feature for the further miniaturization and integration of optical devices. In this chapter, we are going to focus on ultrathin metalens and 3D optical holography based on plasmonic metasurfaces.

3.2 Ultrathin Metalens

3.2.1 Background

The optical lens, as an indispensable tool, has been widely exploited in various scientific communities, and its operation is well understood on the basis of classical optics. Traditionally, reshaping the wavefront of the light relies on gradual phase changes, which are accomplished by either controlling the surface topography or varying the spatial profile of the refractive index. Although diffractive and gradient index lenses bring distinguished functionality in imaging and spectroscopy beyond what can be achieved with refractive optics, their applications are considered limited by the contrast of the refractive index attainable from conventional materials and methods of fabrication. Plasmonic lenses based on nanoaperture or nanoslit array with varying geometries in a metal film have been proposed and experimentally demonstrated [26–29]. For the design of most optical components including lenses, it is important that the phase change can vary smoothly in the range of $[0, 2\pi]$. However, in those previously reported works, the phase was acquired accumulatively for waves propagating through plasmonic or photonic waveguide modes supported by the nanoapertures, and the attainable phase range is well below 2π with realistic thickness of the metal film. Furthermore, it is still a technical challenge to create thin subwavelength slits with extremely high aspect ratios.

Recently, the concept of interfacial phase discontinuities has been proposed [10] and the devices based on this new concept have been demonstrated experimentally at the infrared (IR) wavelength [10, 11]. The interface consisted of an array of plasmonic antennas that partially converted the linearly polarized incident light into its cross-polarization with a discontinuity in phase for both transmission and reflection, as shown in Fig. 3.1. Importantly,

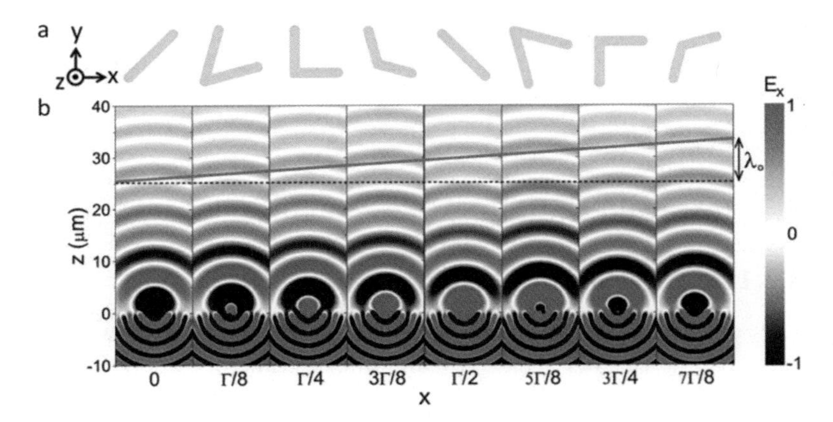

Figure 3.1 (a) Schematic unit cell of the plasmonic interface consisting of V-shaped antennas and (b) finite-difference time domain (FDTD) simulations of the scattered electric field for the individual antennas composing the array in (a). From Ref. [10]. Reprinted with permission from AAAS.

the phase variation across 2π can be readily achieved without sacrificing the uniformity of amplitude. Arbitrary phase profiles along the interface can be realized by varying the geometry of each individual plasmonic antenna. Based on this principle, a linear gradient of the phase discontinuity at the interface was realized, leading to anomalous reflection and refraction described by a generalized Snell's law [10, 11]. It was further shown that the phase discontinuities generated by a suitably designed plasmonic antenna interface could be utilized to create a vortex beam upon normal illumination by linearly polarized light [10]. Figure 3.1a shows the schematic unit cell of the plasmonic interface for demonstrating the generalized laws of reflection and refraction. The sample is created by periodically translating the unit cell in the *xy* plane. The antennas are designed to have equal scattering amplitudes and constant phase difference $\Delta\Phi = \pi/4$ between neighbors. Figure 3.1b shows finite-difference time domain (FDTD) simulations of the scattered electric field for the individual antennas composing the array in Fig. 3.1a. Plots show the scattered electric field polarized in the *x* direction for *y* polarized plane-wave excitation at normal incidence from the silicon substrate. The silicon substrate is located at $z \leq 0$. The antennas are equally spaced at a subwavelength separation $\Gamma/8$, where Γ is the unit cell length. The tilted red straight line in Fig. 3.1b is the envelope of the projections of the spherical waves scattered by the antennas onto the *xz* plane. On account of Huygens's principle, the anomalously refracted beam resulting from the superposition of these spherical waves is then a plane wave that satisfies the generalized Snell's law $n_t \sin\theta_t - n_i \sin\theta_i = \lambda_0 \mathrm{d}\Phi/2\pi \mathrm{d}x$ with a phase gradient $|\mathrm{d}\Phi/\mathrm{d}x| = 2\pi/\Gamma$ along the interface. Here θ_i and θ_t represent the angle of incidence and the angle of refraction, respectively; n_i and n_t are the refractive indices of the two media, and λ_0 is the vacuum wavelength.

The design of this novel class of focusing devices is free from monochromatic aberrations typically present in conventional refractive optics. The phase distribution created from a spherical lens focuses light to a single point only in the limit of paraxial approximation; a deviation from this condition introduces monochromatic aberrations such as spherical aberrations, coma, and astigmatism. To circumvent these problems, complex optimization techniques such as aspheric shapes or multilens designs are implemented. In this case, the hyperboloidal phase distribution imposed at the interface produces a wavefront that remains spherical even for nonparaxial conditions. This will lead to high numerical aperture (NA) focusing without aberrations.

3.2.2 Design Theory and Simulation

3.2.2.1 Required phase profile

To focus an incident plane wave, the flat lensing surface must undergo a spatially varying phase shift. In this way, secondary waves emerging from the metasurface constructively interfere at the focal plane similar to the waves that emerge from conventional lenses. For a given focal length f, the phase shift Φ_L imposed in every point $P_\mathrm{L}(r)$ on the flat lens must satisfy the following equation (Fig. 3.2a):

$$\Phi_\mathrm{L} = \mathbf{k}_0 \overline{P_\mathrm{L} S_\mathrm{L}} = \mathbf{k}_0 (\sqrt{r^2 + f^2} - f) \quad (3.1)$$

where $\mathbf{k}_0 = 2\pi/\lambda_0$ is the free-space wavevector and r is the distance to the origin.

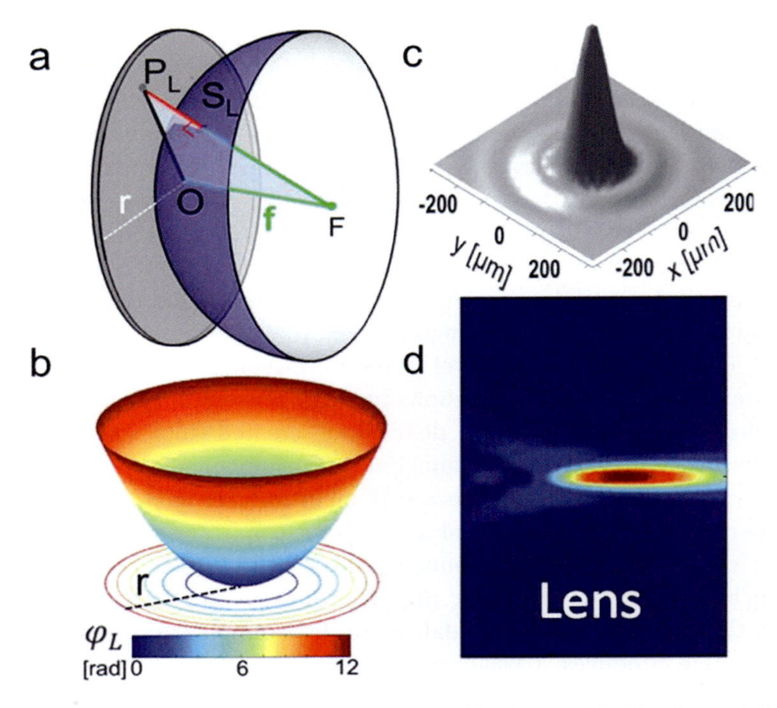

Figure 3.2 Schematic showing the design of flat lenses (a). The phase shift at a point P_L on the lens surface is proportional to the distance $\overline{P_L S_L}$, where S_L is the projection of P_L onto the spherical surface of radius equal to the focal length f. The resulting hyperboloidal radial phase distribution on the flat lens is shown in (b); (c) transverse cross section of the intensity profiles calculated using the analytical model for the focal lens with $f = 3$ cm; (d) theoretical calculations of the intensity distribution in the focal region for the flat lens. Reprinted with permission from Ref. [15]. Copyright (2012) American Chemical Society.

3.2.2.2 Simulation method

To facilitate the design of the metasurfaces, a simple analytical model based on dipolar emitters can be used [30]. The emission of the antennas can be well approximately by that of electric dipoles [31, 32]. The intensity of the field ($|E|^2$) scattered from a metasurface for a particular distribution of amplitudes and phases of the antennas can be calculated by superposing the contributions from many dipolar emitters. This approach offers a convenient alternative to time-consuming FDTD simulations. The metasurface is modeled as a continuum of dipoles with identical scattering amplitude and a phase distribution given by Eq. 3.1. Figure 3.2c is the calculated intensity profiles in the transverse direction. The focusing behavior is clearly shown in the theoretical calculations of the intensity distribution in the focal region (Fig. 3.2d).

3.2.2.3 Dual-polarity metalens

3.2.2.3.1 *Phase discontinuity in circular polarization basis*

The concept of interfacial phase discontinuity can also be used to design a novel type of ultrathin flat lens with polarization-switchable polarities. Instead of converting one linear polarization to the other as in previous works [10, 11], the abrupt phase change occurs when a circularly polarized (CP) light converted to its opposite helicity. The metamorphosing phase shift, ranging from 0 to 2π, is realized by a metasurface consisting of an array of plasmonic dipoles with subwavelength separations. By adjusting the orientation angle ϕ of the individual dipole antennas, the required phase shift for the plasmonic lens can be obtained, as the local abrupt phase change is simply given as $\Phi = \pm 2\varphi$, with the sign determined by the combination of the incidence/transmission polarizations, with the + sign for left circular polarization (LCP)/right circular polarization (RCP) and the – sign for RCP/LCP.

Consider a light field is normally incident on a dipole with its orientation direction forming an angle φ with the x axis. The electric dipole momentum of the single dipole induced by the incident electric field can be expressed as [16]

$$\begin{bmatrix} p_x \\ p_y \end{bmatrix} = \alpha_e \begin{bmatrix} \cos^2\varphi & \sin\varphi\cos\varphi \\ \sin\varphi\cos\varphi & \sin^2\varphi \end{bmatrix} \begin{bmatrix} E_x \\ E_y \end{bmatrix} \tag{3.2}$$

where p_x, p_y, E_x, and E_y are the components of the electric dipole momentum and the electric field along x and y directions, and α_e is the electric polarizability.

For normal incidence, the dipole momentum in Eq. 3.2 with an incident CP state can be decomposed into two different CP states with a phase shift of 0 and $\exp(\pm i2\varphi)$, respectively,

$$P_{\mathrm{L(R)}} = \frac{1}{2}\alpha_{\mathrm{e}}\left(\mathbf{e}_x \pm i\mathbf{e}_y\right) + \frac{1}{2}\alpha_{\mathrm{e}}e^{\pm i2\varphi}\left(\mathbf{e}_x \mp i\mathbf{e}_y\right)$$
$$= \frac{1}{\sqrt{2}}\alpha_{\mathrm{e}}\left(\mathbf{e}_{\mathrm{L(R)}} \pm e^{\pm i2\varphi}\mathbf{e}_{\mathrm{R(L)}}\right) \quad (3.3)$$

where the subscripts R and L indicate the right- and left-handedness of the CP light, $\mathbf{e}_x$ and $\mathbf{e}_y$ represent the unit vectors along the x and y directions, and $\mathbf{e}_{\mathrm{L(R)}} = (\mathbf{e}_x \pm i\mathbf{e}_y)/\sqrt{2}$ represents the unit vector for the left-handed (+) and right-handed circular polarization (–), respectively. Thus, for an incident beam with circular polarization, the radiation from the dipole into the opposite radiation in the forward direction experiences a phase discontinuity $\Phi = \pm 2\varphi$, with the sign depending on the incidence/transmission polarization combinations, + for LCP/RCP and – for RCP/LCP. The abrupt phase change can cover the phase shift from 0 to 2π, with φ being tunable from 0 to π. The sign of the phase shift can be switched between positive and negative when changing the handedness of the incident light. Hence, the polarity of the lens will be changed accordingly.

This phase change can be considered as the Pancharatnam–Berry (PB) phase that is acquired when the polarization state of light travels around a contour on the Poincaré sphere [33, 34]. Specifically, for an incident beam with circular polarization σ, the phase difference between the scattered waves of opposite circular polarization σ^- scattered by two dipole antennas of different orientations φ_1 and φ_2 can be viewed as half of the solid angle enclosed between two paths on the Poincaré sphere, $\sigma \rightarrow L(\varphi_1) \rightarrow \sigma^-$, and $\sigma \rightarrow L(\varphi_2) \rightarrow \sigma^-$, where σ and σ^- are represented by the north and south poles of the Poincaré sphere, respectively, and $L(\varphi_i)$ are the linear polarization states residing on the equator of the Poincaré sphere [33].

3.2.2.3.2 *Design and simulation*

To achieve the phase profile equivalent to a conventional lens, the following expression [18] governs the relation between the rotation angle φ and the location of the dipole antenna, r:

$$\phi(r) = \pm 0.5k_0\left(\sqrt{f^2 + r^2} - |f|\right) \quad (3.4)$$

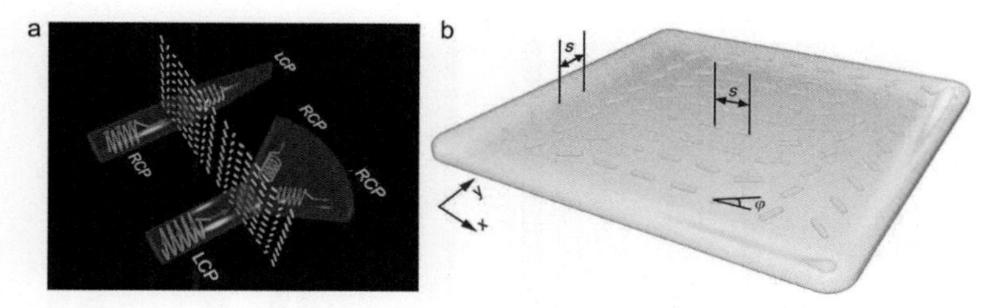

Figure 3.3 Schematic of metalens with interchangeable polarity and diagram of the bipolar plasmonic lens. (a) The focusing properties of the same metalens can be switched between a convex lens and a concave lens by controlling the helicity of the incident light. Reprinted with permission from Ref. [18], © 2013 Wiley. (b) The dual-polarity cylindrical metalens consists of an array of plasmonic dipole antennas on a glass substrate with orientations varied along the focusing direction (x). The distance between neighboring dipoles, $S = 400$ nm, is the same along the two in-plane directions. φ is the rotation angle of the dipole relative to the x axis. The abrupt phase shift is solely determined by the orientation of the dipoles [16].

Note that the + and – signs in Eq. 3.4 correspond to a positive (convex) and negative (concave) polarity, respectively, for an RCP incident wave, and the opposite holds for an LCP incident wave.

To implement the proposed dual-polarity lens, the rotation angle of the dipole antennas should vary according to Eq. 3.4. To understand clearly about the physical nature, we initially take acylindrical lens as an example.

Figure 3.3 shows the schematic of a metalens with interchangeable polarity and the diagram of the designed plasmonic bipolar lens which consists of dipole nanoantennas with the directional orientation corresponds to the + sign in Eq. 3.4. The dipoles are arranged in a 2D array with a subwavelength period of S in both x and y directions. The full wave numerical validation is performed using a commercial software package (CST Microwave Studio) to simulate the propagation of a CP wave through the plasmonic lens at normal incidence (Fig. 3.4). In the simulation, we calculated a miniaturized lens as the size of the actually fabricated one is beyond the capability of our numerical simulation. The simulated lens in Fig. 3.4 has a focal length of 10 μm and an aperture of 8 μm × 8 μm, consisting of 21 dipole antennas along the x direction with

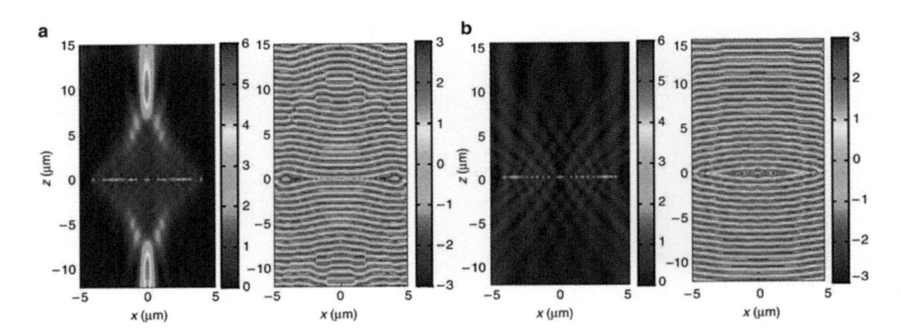

Figure 3.4 Simulation on a dual-polarity plasmonic lens. Full wave simulation is performed by CST Microwave Studio for the propagation of a circularly polarized wave at 740 nm through the lens at normal incidence. (a) Intensity and phase distribution indicate that the lens function as a positive lens for RCP incident light. (b) With LCP incident light, the same lens changes its polarity and turns into a negative lens. In both plots, only the fields with the circular polarization opposite to that of the incident wave are plotted [16].

$S = 400$ nm. For an incident beam with RCP, the plasmonic lens functions as a converging (positive) lens. Two focal planes are clearly visible at $z = \pm 10$ μm away from the plasmonic lens surface, as shown in Fig. 3.4a. There are two real focal lines, one for the transmitted beam and the other for the reflected beam, on each side of the plasmonic lens. Thus by manipulation of the phase discontinuity along an interface with a suitably designed dipole antenna array for the incident CP light, a light wave can be fully concentrated.

When the polarization of the input light is switched from RCP to LCP, the simulation shows that the polarity of the proposed dual-polarity lens is indeed transformed from positive (convex) to negative (concave). This is clearly indicated by the spatial distribution of the intensity and the phase (Fig. 3.4b). Unlike conventional cylindrical lenses, a single flat lens can be metamorphosed to converging and diverging lenses, which only depends on the helicity of the CP light.

3.2.2.3.3 *Sample fabrication*

The antenna structures are defined in the electron beam resist on an ITO-coated glass substrate by using standard electron beam

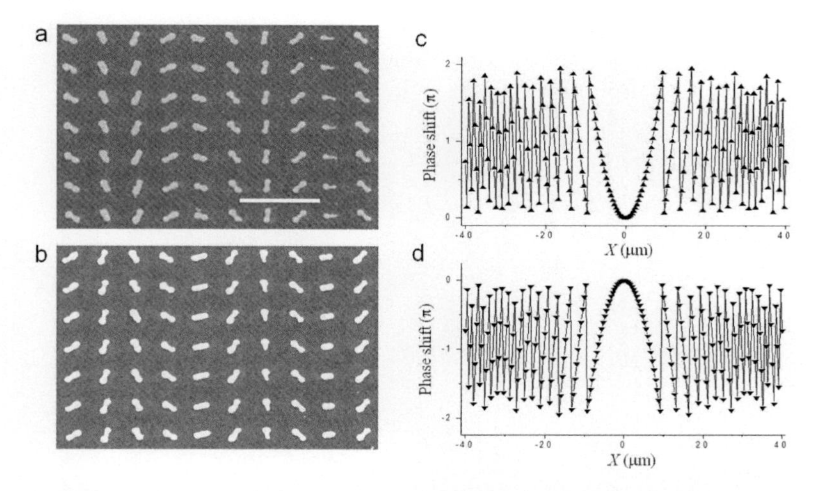

Figure 3.5 Scanning electron microscopy (SEM) images of selected areas of the plasmonic lenses and the expected profile of phase discontinuity. SEM images of plasmonic lens on an ITO-coated glass substrate with (a) negative polarity (lens A) and (b) positive polarity (lens B) for an incident light with RCP. The scale bar in the SEM image represents 1 μm. (c, d) The expected phase discontinuity for the positive lens and negative lens, respectively, for RCP incidence. Note that for LCP incidence, the phase discontinuity is reversed [16].

lithography (EBL), followed by a lift-off procedure. Based on the interfacial phase discontinuity, two plasmonic lenses, lens A and lens B, with a negative and a positive polarity for an incident beam with RCP polarization are fabricated. The dipole antennas are made from gold with a thickness of 40 nm. Scanning electron microscopy (SEM) images for part of the resulting patterns for lens A and lens B designed at 740 nm are shown in Figs. 3.5a,b. Each lens has an aperture of 80 μm × 80 μm and a focal length $f_A = -60$ μm and $f_B = +60$ μm for an incident wave with RCP, respectively. The dipole antennas are 200 nm long and 50 nm wide, exhibiting a longitudinal resonance around 970 nm and a transverse resonance around 730 nm. The rotation angles for the dipoles far from the lens center change more rapidly than those near the center. Figure 3.5c,d gives the expected abrupt phase changes for the two lenses for RCP

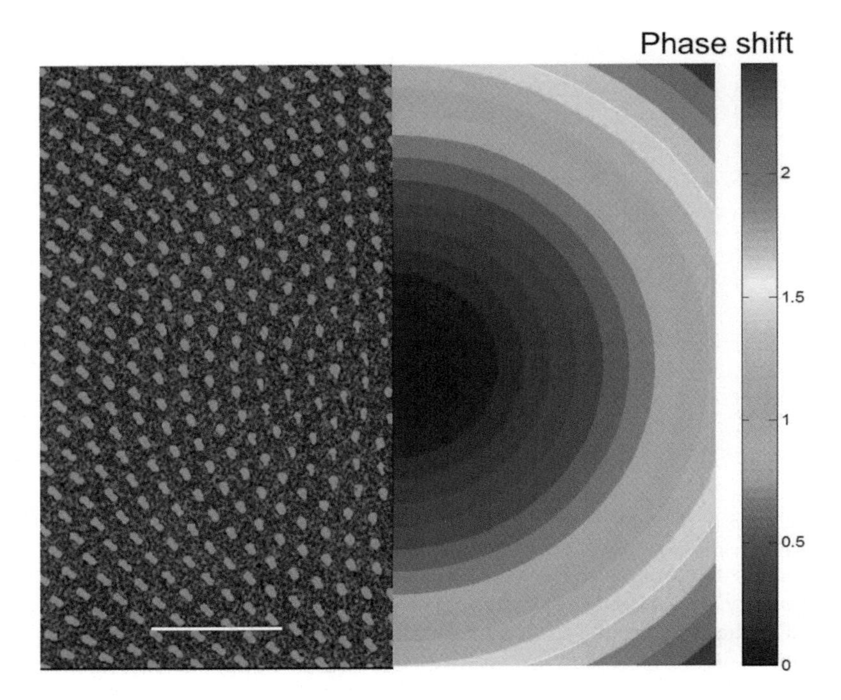

Figure 3.6 SEM image of the fabricated 2D dual-polarity plasmonic lens (top view) with a focal length of 80 μm (left). The corresponding phase shift profile is displayed on the right. The dipoles that are arranged on the same annulus have the same orientation. The distance between two neighboring annuluses is 400 nm along both the radial and azimuthal directions. The scale bar is 2 μm. Reprinted with permission from Ref. [18], © 2013 Wiley.

incidence, which show opposite phase profiles due to the opposite rotation directions of the dipole antennas in these two lenses.

The dual-polarity metalens fabricated above only manipulated the light along one direction owing to one-directional phase variation by the dipole array, resulting in distortion of the image of an arbitrary object due to the different magnifications along two directions. Figure 3.6 shows the fabricated ultrathin flat lens with polarization-dependent radial phase variation to display 3D focusing spots either in the real or the virtual focal plane. The metalens is composed of dipole antennas arranged in a number of evenly spaced concentric rings, with the radius of the rings

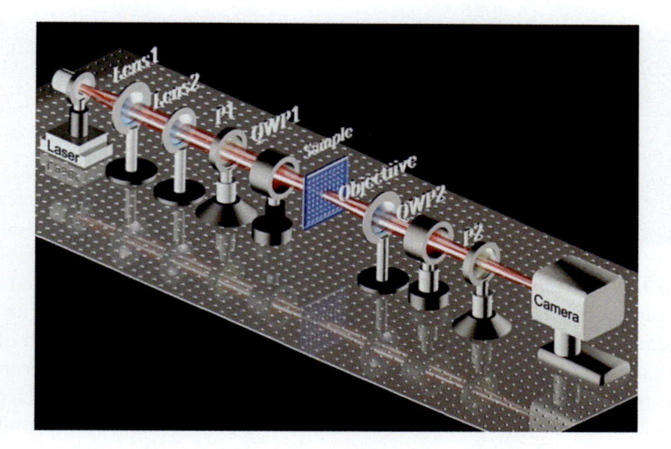

Figure 3.7 Schematic of the optical measurement setup. The polarization directions of the two polarizers are parallel to each other. The incident CP light is generated by the linear polarizer P1 and the quarter-wave plate QWP1. The opposite circular polarization in transmission is detected by a quarter-wave plate QWP2 and the linear polariser P2. The microscope objective is mounted on a 3D stage. To image an object, light from a white laser source is incident on the back side of the lens. The transmission through the sample (object and plasmonic lens) is collected with a 20× /0.40 objective and imaged on a CCD camera [16].

increasing by a step size of 400 nm. Within each ring, the dipole antennas have the same orientation and the separation between two neighboring dipoles along the ring is 400 nm. An SEM image of the resulting patterns for the lens (left) designed at a wavelength of 740 nm and the corresponding phase shift profile (right) are shown in Fig. 3.6. The lenses have a diameter of 180 μm with a focal length $f = 80$ μm.

3.2.2.3.4 *Measurement setup*

Figure 3.7 shows the schematic of the optical measurement setup, including several lenses, two QWPs, two linear polarizers, an objective, and a charge-coupled device (CCD) camera. The polarization directions of the two polarizers are parallel to each other. The incident CP light and the opposite circular polarization in transmission are generated by a QWP and a polarizer on each side of the plasmonic lens. For the imaging measurement, the microscope

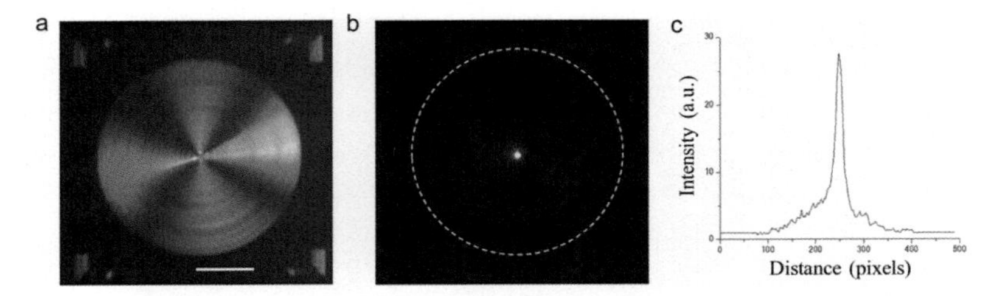

Figure 3.8 Optical microscope images of (a) the metalens illuminated with white light and (b) the focal point at real focal plane illuminated with an RCP incident laser beam at 740 nm. The scale bar is 50 μm. The position of the lens is marked by the white dashed circles. (c) Experimental measurement of the intensity distribution at the focal plane along one direction. Each pixel is equal to 0.37 μm. The spot diameter at full width of half maximum is 7.2 μm. Reprinted with permission from Ref. [18], © 2013 Wiley.

objective is mounted on a 3D stage. To image an object, light from a laser source at a wavelength of 740 nm or 810 nm is incident on the back side of the fabricated metalens. The transmission through the sample (object and plasmonic lens) is collected with a 20× /0.40 objective and imaged on the CCD camera.

3.2.2.3.5 *Characterization of the lens*

3.2.2.3.5.1 *Focal plane*

The performance of the focusing of the plasmonic lens by using a CP laser beam at the visible range is experimentally demonstrated. A positive lens causes the incident laser beam to converge at a focal plane on the transmission side of the lens forming a real focal point, while a negative lens causes the incident laser beam to emerge from the lens as though it is emanated from a virtual focal plane on the incident side of the lens. The focusing performance of the metalens is characterized with a laser beam at $\lambda = 740$ nm. By gradually tuning the distance between an objective lens and the plasmonic metalens, the optical intensity distribution is examined along the propagation direction to determine the focal point. Figure 3.8 shows the microscope image of the metalens illuminated by white light (Fig. 3.8a) and the focal point at the focal plane (Fig. 3.8b) by using the laser beam. From the intensity distribution of the laser beam at

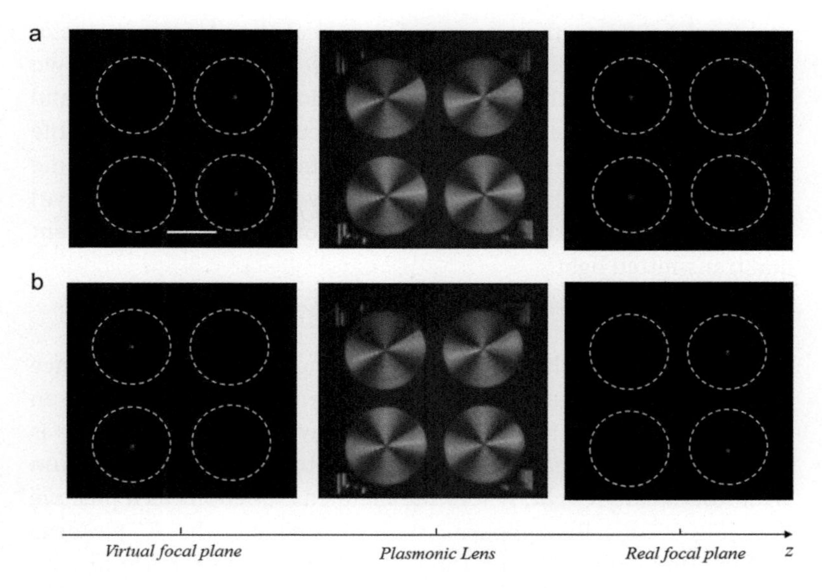

Figure 3.9 Lens polarity is switched by changing the helicity of the incident light. Optical microscope images at virtual focal plane (left), lens surface (middle), and real focal plane (right) for the incident light with (a) RCP and (b) LCP. The scale bar is 50 μm. CP laser beam is incident on the lenses from the left along the z direction, with the lens located at $z = 0$. Positions of lenses are marked by the white dashed circles. The polarity of the two lenses on the left is different from that of the lenses on the right for the same CP light. Reprinted with permission from Ref. [18], © 2013 Wiley.

the focal plane (Fig. 3.8c), a focused spot with a diameter of 7.2 μm at full-width at half-maximum (FWHM) is experimentally obtained.

To validate the dual polarity of such circular phase discontinuity lens, four metalenses are fabricated side by side, including two positive lenses on the left and two negative lenses on the right for RCP incident light. Figure 3.9 shows the optical microscopy images for two different incident/transmission polarization combinations: RCP/LCP (Fig. 3.9a) and LCP/RCP (Fig. 3.9b). As shown in Fig. 3.9a (left) for the RCP incident beam, we observe two bright focal points for the lenses on the right at $z = -80$ μm, which are virtual focal points as they lie in the incident side of the plasmonic lens. This confirms that lenses on the right are negative (concave) for the incident light with RCP polarization. On the other hand, two real focal points on the transmission side of the plasmonic lens at $z = 80$ μm are observed for the two lenses on the left. As shown in Fig. 3.9b, when the circular polarizations of the incident and transmitted beam are interchanged, the focusing behavior for all the lenses is reversed, unambiguously verifying that switching in the focusing properties from positive (negative) to negative (positive) is solely attributed to the helicity change of the CP for the incident and transmitted light.

3.2.2.3.5.2 *Imaging*

The major functionality of a lens is imaging. For an object distance greater than the focal length of the metalens, it is expected that an inverted real image is formed for a positive (convex) lens. This is confirmed by the measurement with an object distance of 150 μm with RCP incident light, where the metalens functions as a positive lens (Fig. 3.10b). By comparing with the image of the "T" patterns without the plasmonic lens, the magnification of the image formed by the metalens can be obtained. Experimentally, we measured the magnification of the image to be 1.3, which shows reasonable agreement with a simple ray calculation of 1.14. The slight difference in magnification is due to the substrate flatness and air gap accuracy. Due to the resolution limitation of the measurement system, the upright virtual image for LCP incident light with a predicted magnification of 0.34 cannot be resolved. To observe the upright virtual image, we decrease the air gap to 55 μm. As the air gap is within the focal length of the plasmonic metalens, upright virtual images are observed for both circular polarizations of incident light, as shown in Fig. 3.10c,d. The magnifications in Figs. 3.10c and 3.10d are 3.1 and 0.58, respectively, which coincide with a magnification of 3.2 for the positive lens (convex) with RCP incident light and a magnification of 0.59 for the negative lens (concave) with LCP incident light. Thus, the reconfigurable imaging functionality of the same plasmonic metalens as a convex lens or a concave lens is clearly demonstrated by controlling the helicity of the incident CP light.

3.2.2.3.6 *Discussion*

The conversion efficiency between two circular polarization states is an important parameter in the performance of the lens. In this

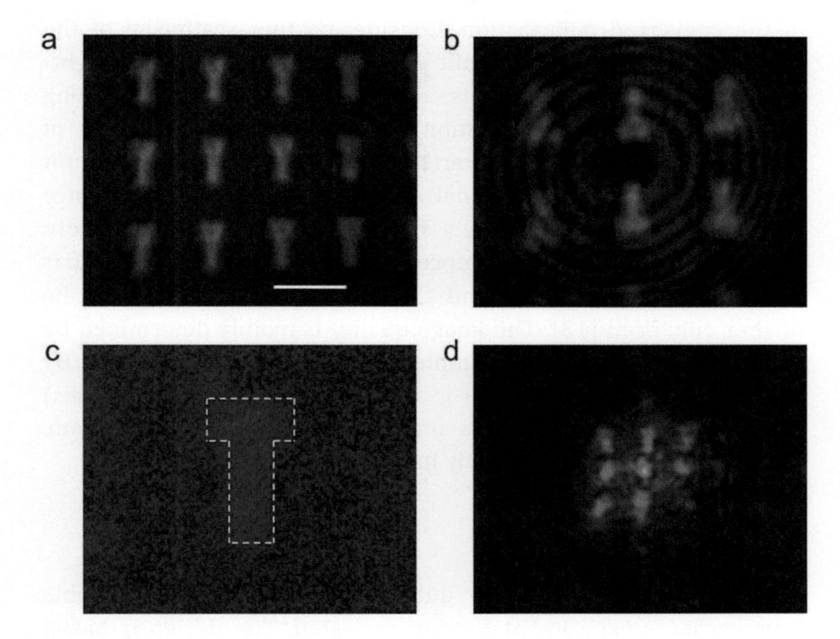

Figure 3.10 Experimental demonstration of imaging. (a) A "T" pattern array with a pitch of 20 μm along both in-plane directions is used as an object. The scale bar is 20 μm. The object and the lens are separated by an air gap whose thickness is controlled by spacers at the edge of the samples. (b) A real and inverted image with a magnification of 1.3 is achieved for RCP incident light, in which the polarity of lens is positive (convex lens). The air gap is 150 μm. (c) A virtual and upright image with a magnification of 3.1 on a positive lens with an air gap of 55 μm. The edge of the image is marked by the white dashed lines. (d) For the same air gap as in (c), a virtual and upright image with a magnification of 0.58 is obtained when switching the incident light from RCP to LCP, with the polarity of the same lens being negative (concave lens). Reprinted with permission from Ref. [18], © 2013 Wiley.

presented work the conversion efficiency between the focused power and the incident power is measured around 5%. With further optimization of the design parameters, for example, by increasing the density of the dipoles, and better alignment of the resonance wavelength of the antennas to the operating wavelength of the lens, the dipole antennas can achieve a significantly higher transmission in the converted polarization. Considering that scattering of the dipole antenna is evenly split between left-handed and right-handed circular polarizations, and further split equally along the transmission and reflection directions, the upper limit of efficiency is 25% if the material loss is neglected. If each unit cell consists of two orthogonal dipoles with detuned resonance frequencies and consequently a significant phase difference, there will be a constructive interference for conversion from one circular polarization to the other, and the conversion efficiency can be further enhanced [13]. The image quality is mainly determined by the quality of the fabricated sample and the measurement system. By minimizing the fabrication error and defects (e.g., missing dipoles) during the fabrication process and optimizing the imaging system, the image quality will be greatly improved.

3.2.2.3.7 *Conclusion*

A plasmonic flat lens with dual polarity operating at visible frequencies is experimentally demonstrated. The design is based on the interfacial phase discontinuity that occurs when CP light is converted into the opposite circular polarization. By controlling the polarizations of the incident and transmitted beams, the focusing properties of the same plasmonic lens can be altered between a convex lens and a concave lens, as in stark contrast to conventional lenses with a fixed polarity. Although a plasmonic lens with dual polarity based on a nanoslit array was previously claimed [28], there was no direct experimental observation of the concave lensing effect, that is, presence of virtual focal plane or demagnified virtual image of an object formed by the plasmonic lens. In contrast, we have unambiguously shown both convex and concave functionalities of the same plasmonic lens by observing focusing at real and virtual focal planes, and the magnified and demagnified image when an object is placed close to the lens. Since the plasmonic lens is made of simple plasmonic dipoles with variable orientation, it does not involve complicated design of plasmonic nanostructures. The dual-polarity plasmonic flat lens opens an avenue for new applications of phase discontinuity devices and could also have an impact on integrated nanophotonic devices.

3.3 3D Optical Holography Using Metasurfaces

3.3.1 Background

Shaping the phase distribution of a wave is very important for reconstructing 3D images, a technique which has been known as *holography* for several decades. A hologram contains the complete information of the object beam, in contrast to photography or multiview parallax techniques which store only information for at most a limited number of viewing directions [35]. The holographic recording itself is not an image; rather it consists of seemingly random patterns of varying intensity or phase. The hologram can be generated either by interference of a reference beam with the scattered beam from a real object or by numerical computation to calculate the phase information of the wave at the hologram interface and encoding of the phase information into surface structures by lithography or a spatial light modulator (SLM). The latter method is usually referred to as computer-generated holography (CGH) [36, 37].

Recently, holography-based techniques for controlling the amplitude and phase of free-space beams have been used to achieve surface plasmon holographic displays [38, 39], beam shaping [40], data storage [41], digital holographic microscopy [42, 43], optical trapping, and micromanipulation in atom traps or diffractive laser tweezers [44, 45]. Two-dimensional holography, or projection, has also been experimentally demonstrated using metamaterials [46–52]. However, none of these techniques have achieved 3D CGH image reconstruction in the visible range, even though the essence of holography lies in its capability to display 3D images. Here, we demonstrate 3D CGH image reconstruction by using an ultrathin plasmonic metasurface consisting of an array of subwavelength plasmonic antennas with carefully defined orientations. The realization of the hologram by such metasurfaces is very simple and elegant due to the extremely straightforward relationship for encoding the phase information into the configuration of the structures, that is, the orientation angle of the nanorods. This particular feature of our metasurface is highly desired in holography as it is robust against fabrication tolerances and variation of metal properties because of the much simpler structure geometry and the geometric nature of the phase. More importantly, due to subwavelength control of the spatial phase profile, metasurfaces provide a solution for increasing the angular range of perspective for digital holograms and enhancing the space–bandwidth product of holographic systems.

3.3.2 Design and Simulation

3.3.2.1 Computer-generated hologram design

Point source algorithms and Fresnel diffraction theory are used for the generation of the CGH. In our phase-only hologram design, proper choice of random phase is added to the point sources constituting the 3D object to mimic the diffuse scattering body, which is in contrast to the specular holography. Thus, standard normal (Gaussian)-distributed pseudorandom phase $X \approx 2\pi \cdot N(\mu, \sigma^2)$, with $\mu = 0$ and $\sigma = 1$ (in the unit of radians) is added to each pixel of the target object to achieve a more uniform amplitude distribution.

Through the angular spectrum method (ASM), we can numerically reconstruct the 3D objects to verify the design. Also, to mimic the possible phase shifts due to fabrication errors, a certain range of random phase can be added to the hologram, and by ASM we have numerically verified that such a hologram is very robust against phase noise.

3.3.2.2 Design of a metasurface hologram

For our demonstration we utilize the abrupt phase change that occurs for CP light when converted to its opposite helicity [16, 18, 21, 24]. The phase shift at the interface, ranging from 0 to 2π, is realized by a metasurface consisting of an array of plasmonic dipole antennas with subwavelength separation. The local phase of light transmitted through the metasurface is geometrical and solely controlled by the orientation angle ϕ of the individual dipole antennas as $\Phi = \pm 2\varphi$, with the sign determined by the particular combination of incident/transmitted polarization, that is, + sign for RCP/LCP and – sign for LCP/RCP. Importantly, the scattering amplitudes of the antennas for converting the incident light to its opposite helicity are uniform. This greatly eases the encoding procedure of phase-only holograms; therefore, circular polarization–based metasurfaces that consist of simple plasmonic nanorods can be used to record the hologram without an extra look-up table.

By using a CGH algorithm [53], the 3D object is approximated as a collection of point sources and both the recording and image reconstruction procedures are achieved without the need for a reference beam. Furthermore, the objects to be reconstructed are designed to have rough surfaces that give rise to diffuse reflection. As such, the amplitude information of the hologram can be entirely eliminated without degrading the image quality. Each pixel of the hologram only contains a single subwavelength plasmonic nanorod whose orientation encodes the desired continuous local phase profile for CP light illumination. Due to the subwavelength pixel pitch, the zero-order on-axis 3D reconstruction can potentially achieve very high resolution and wide field of view (FOV). Moreover, the dispersion-less nature of the circular polarization–based metasurface allows the hologram to be reconstructed at a broad range of wavelengths.

Figure 3.11 illustrates the hologram structure and reconstruction procedure of the 3D image. When the pixelated nanorod pattern is illuminated with CP light, it generates the desired continuous local phase profile for the transmitted light with opposite handedness. The number of pixels on the hologram is determined by the conservation of the space bandwidth product and the sampling law so as to provide sufficient CGH object information and avoid aliasing effects. The holographic 3D image appears in the Fresnel range of the hologram. Note that if the polarizations of the incident and transmitted beams are both reversed, the sign of the phase acquired is flipped, which would result in a mirrored holographic image on the other side of the metasurface.

The procedure of designing and displaying 3D holograms consists of the following steps: digital synthesis and numerical calculation of the 3D CGH, encoding of the phase information into the pixelated nanorods by their orientation, and reconstruction of the image by a conventional optical transmission scheme. In general, the 3D objects are set up by 3D computer graphics software or by 3D mathematical functions in MATLAB.

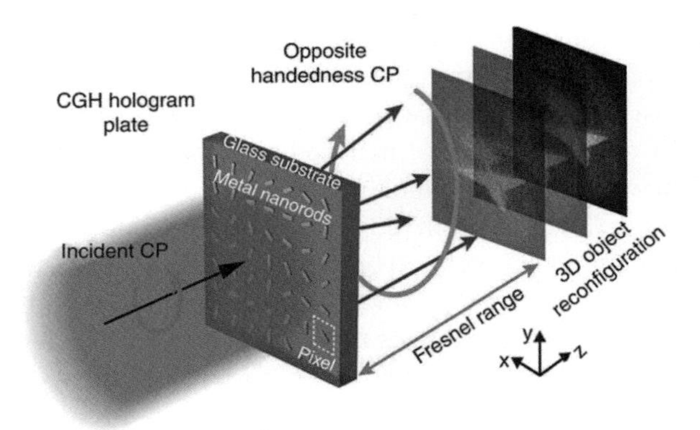

Figure 3.11 Hologram structure and reconstruction procedure. Each nanorod plays the role of a pixel of diffractive element, which can generate the required continuous local phase profile with normal incidence of CP light, and only opposite-handed CP light is collected. The reconfigured 3D models are designed to appear within the Fresnel range [24].

By using a point source algorithm, the calculated 3D objects are approximated as a collection of discrete point sources. The complex amplitude $H(x, y)$ on the hologram plane can be then calculated by superimposing the optical wavefronts of all the point sources. With the proper incorporation of a Gaussian distributed random phase for each point source, which emulates diffuse reflection of the object surface, a uniform amplitude distribution can be obtained in the hologram plane, and thus good-quality holographic reconstruction with only the phase information can be achieved.

3.3.3 Characterization of a Metasurface Hologram

For the experimental proof of the concept we fabricate various hologram samples, consisting of metallic nanorods, by EBL followed by deposition of 40 nm gold and a lift-off process. We first carry out the 3D holography of a solid jet model (Fig. 3.12a). The target 3D solid jet model is created by computer graphics software. In the next step the metasurface hologram for the jet pattern is designed for a wavelength of 810 nm. The target object is chosen to be submillimeter size, which is comparable to that of the hologram

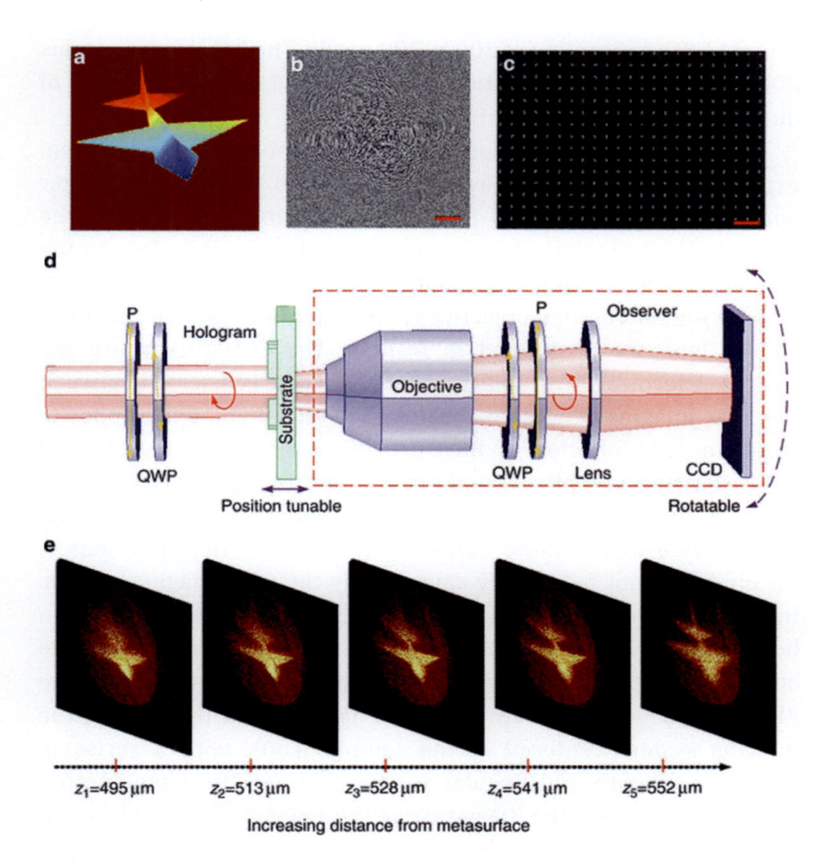

Figure 3.12 Experimental investigation of 3D holography by the metasurface. (a) The 3D models for a jet generated in a graphics program. The size of the jet from left wing to right wing is 330 μm and from head to tail is 232 μm. The dimension along the z direction is 48.2 μm. (b) Calculated hologram of the jet. The scale bar is 20 μm. (c) SEM image of a part of the corresponding metasurface. The scale bar is 1 μm. The gold nanorods in the hologram plate are ~150 nm long and ~75 nm wide, and the pixel size s between neighbouring two rods along the x and the y direction is 500 nm. The entire hologram is made of 800 × 800 pixels. (d) Optical setup for observation of holographic image with tunable focus positions and rotational angular perspective. (e) Evolution of the appearance of jets for different focusing positions along the z direction with RCP illumination [24].

sample, with a sampling of 200 × 200 pixels. The hologram contains 800 × 800 pixels, with a lattice constant of 500 nm and nanorods of dimensions 150 nm × 75 nm (Fig. 3.12b,c).

To determine the performance of the hologram we use an experimental setup as shown in Fig. 3.12d. A linear polarizer and a QWP are positioned in front of the sample to prepare the desired circular polarization state for the illumination. Due to the submillimeter size of the reconstructed image, a 10× (NA = 0.3) magnifying microscope objective lens, in combination with another convex lens are positioned in front of the CCD to capture the images. A second linear polarizer and QWP pair is inserted between the two imaging lenses to ensure that only light with opposite handedness is collected. To measure different viewing angles, all of the observational optical components, that is, components placed after the sample, are arranged on a rotation-stage centered on the sample. Due to the finite depth of focus of the objective lens, a 3D image cannot be directly captured by the CCD imaging system. Instead, the depth information of the 3D constructed image can be analysed by gradually tuning the distance between the sample and the objective. In addition, the perspective information of the 3D holographic image can be further investigated by rotating the imaging system centered at the sample. Thus, with a series of 2D images, depth and angular perspective can be demonstrated separately, allowing the 3D nature of the holographic images to be verified.

Results for the holographic images of the solid jet model at different object planes are shown in Fig. 3.12e. At first the imaging system is deliberately designed to have a small depth of focus. This helps to verify the 3D nature of the holographic image, as the sharp focus allows certain parts of the full 3D holographic image to be viewed selectively. In the first polarization configuration RCP light of 810 nm wavelength is normally incident onto the sample and LCP light is detected. A real holographic image of the jet appears on the transmission side of the metasurface, as shown by the evolution of the 2D images at different distances relative to the metasurface. For $z_1 = 495$ μm away from the metasurface, only the head of the jet forms a sharp image on the CCD camera, whereas the tail is slightly blurred. At a different distance of $z_2 = 552$ μm, the tail now forms a sharp image on the CCD camera and the head appears blurred. The difference between z_1 and z_2 agrees reasonably well with the depth of the target 3D holographic object along the z direction, which is 48.2 μm. The phase profile of the circular polarization–based metasurface relies on the selection of different circular input/detected polarization [24]. When both of them are reversed, the phase profile is expected to be reversed as well. This indicates that there exists a virtual holographic image at the opposite side of the sample when simultaneously switching the polarizations of the illuminated and transmitted light, as schematically shown in Fig. 3.13a. This effect is experimentally verified by observing an image of the jet on the opposite side of the sample space (Fig. 3.13b).

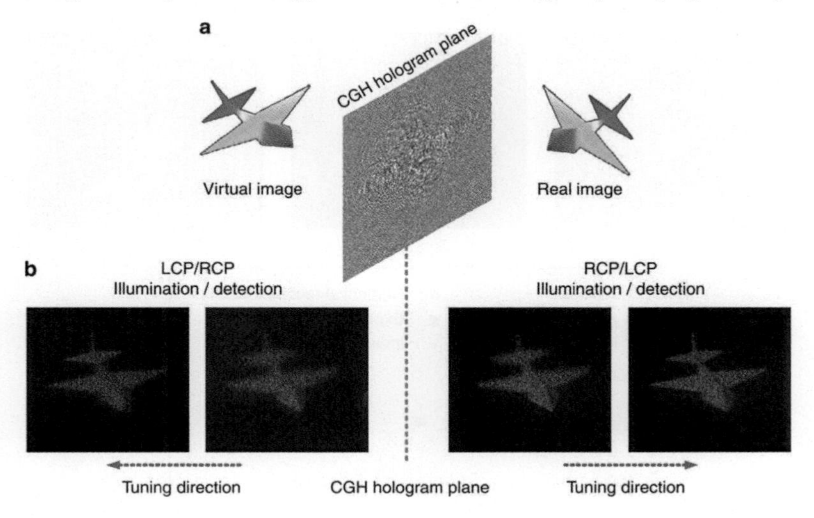

Figure 3.13 Demonstration of real and virtual holographic images. (a) Schematic illustration for the appearance of real and virtual holographic images of a jet on both sides of a metasurface plane. (b) Experimentally obtained images. The real image appears on the transmission side when illuminated and detected by the RCP/LCP combination, while the virtual image is on the opposite side when both illumination and detection polarizations are reversed. For both real and virtual images, the location where the head of the jet gives a sharp image and is therefore in the imaging focus plane is closer to the metasurface than where the tail looks clear. This verifies that the virtual image and real image are symmetric about the metasurface. The wavelength at which the images were [24]. Reprinted with permission from Ref. [24], © 2013 NGP.

The measurement further shows that the virtual holographic image is exactly symmetric about the position of the metasurface relative to the real holographic image detected previously. That is to say, by tuning the object plane away from the hologram (metasurface), we first obtain a clear image of the jet's head and then gradually the sharp focus shifted to the tail for both real and virtual holographic images.

The metasurface design in our experiment exhibits a dispersionless phase profile which results from the geometric Berry phase [12]. Consequently, it is expected that holograms based on such metasurfaces should be broadband, even though the hologram was designed for a specific wavelength (810 nm). Figure 3.14a–c

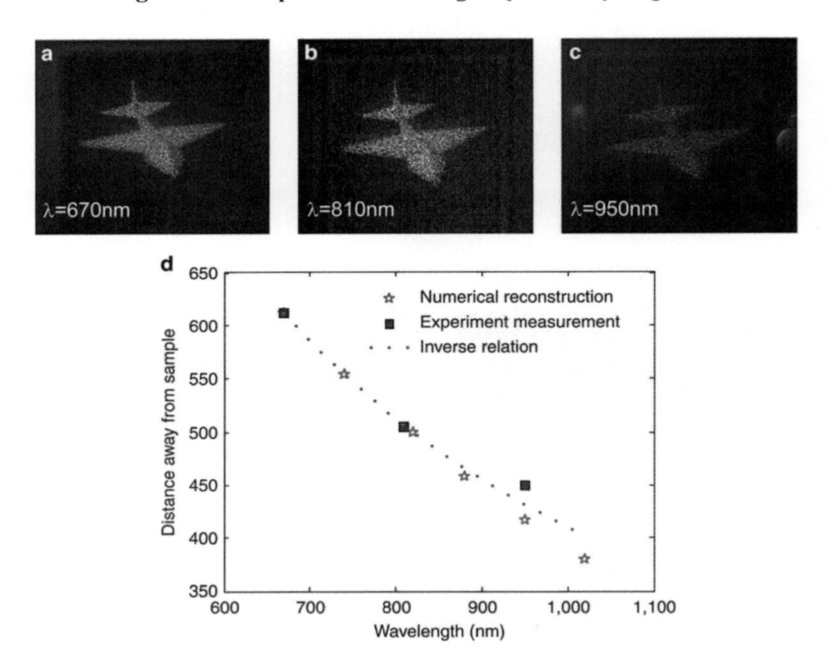

Figure 3.14 Holographic image reconstruction for different wavelengths. Recorded images for wavelengths of (a) $\lambda = 670$ nm, (b) $\lambda = 810$ nm, and (c) $\lambda = 950$ nm, respectively. (d) Distance of the reconstructed holographic image from the sample surface. The red stars denote the value calculated from angular spectrum method, the square dots represent the experimental measurement, and the blue dotted curve is calculated from the simple inverse relationship between the wavelength and z position [24].

shows the holograms at three different wavelengths: 670 nm, 810 nm, and 950 nm. Note that the imaging system is now slightly modified such that the depth of focus is increased. Consequently, all parts of the jet, from head to tail, now appear sharp on the CCD camera simultaneously. The reconstructed images at these three wavelengths are almost identical; only the distance to the metasurface is varied from 620 μm at $\lambda = 670$ nm wavelength to 450 μm at $\lambda = 950$ nm. The change of location can be explained by the change in phase accumulated during the wave propagation due to the change in the wavevector. Mathematically, the locations of the holographic images corresponding to two different wavelengths are related by

$$\frac{2\pi}{\lambda_1}\left(\sqrt{z_1^2+(x-x_1)^2+(y-y_1)^2}-z_1\right) = \frac{2\pi}{\lambda_2}\left(\sqrt{z_2^2+(x-x_2)^2+(y-y_2)^2}-z_2\right) \quad (3.5)$$

where (x, y) are the coordinates on the metasurface and (x_i, y_i, z_i) represent the location of a particular point in the 3D holographic image at wavelength λ_i. It is easy to verify that, under the paraxial approximation, the relation can be simply expressed as $x_1 = x_2$, $y_1 = y_2$, and $\lambda_1 z_1 = \lambda_2 z_2$. Thus, the distance between the holographic image and the metasurface is approximately inverse-proportional to the wavelength. This simple relationship explains very well the experimental observation as well as a more rigorous calculation of the image positions by using the angular spectrum method (Fig. 3.14d).

We further investigate the performance of 3D metasurface holography for a five-turn hollow helix pattern (400 μm pitch, 150 μm diameter), with the helix axis along the z direction (perpendicular to the metasurface). The target 3D object, the CGH image of the overall hologram pattern, and the SEM image of its constituent dipole antennas are shown in Figs. 3.15a, Fig. 3.15b, and Fig. 3.15c, respectively. The hollow helix is specifically designed for the demonstration of different perspective views of the 3D image. First, by tuning the object plane along the z direction, the on-axis evolution of the five-turn helical image is measured (Fig. 3.15d). For each 2D image slice at least one complete helix pitch

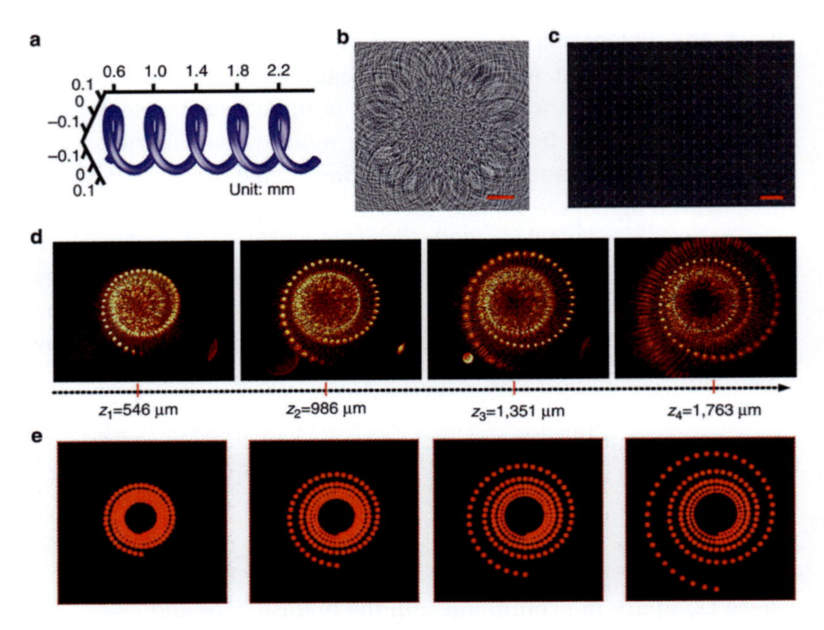

Figure 3.15 Holography of a 3D helix. (a) Geometry of the helix. (b) The calculated hologram and (c) SEM view of the constituent nanorods' pattern. The scale bars in (b) and (c) are 20 μm and 1 μm, respectively. (d) On-axis evolution of the total five turns of the helix by tuning the focusing position along the z direction. (e) Numerical calculation of the 2D perspective view by taking into consideration the position-dependent magnification [24].

(400 μm along the z direction) can be seen clearly. The recording of 3D images on a 2D plane in general shows some signature of perspective, that is, the size of the image changes with distance to the observer. As the perspective of our imaging system is not linear, the magnification for the sections of helix that falls out of focus show a nonlinear dependence on its z location. The distance-dependent magnification in our imaging system is obtained by a ray-tracing calculation (Fig. 3.16), which is subsequently used to reconstruct the perspective image on a 2D plane at the location of the CCD camera. The calculated 2D perspective images of the 3D helix, shown in Fig. 3.15e, show reasonable agreement with the observed images. Note that the calculated perspective images do not show the blurry

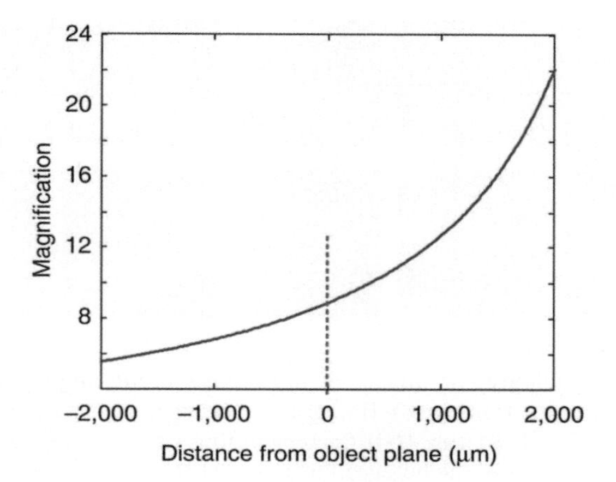

Figure 3.16 Position-dependent magnification of the optical system measurement system. The magnification of our imaging system is obtained from a ray-tracing calculation. It shows a nonlinear dependence on the distance of the holographic image to the object plane [24].

out-of-focus effect, as the calculation does not take into account the finite depth of focus in the real imaging system.

When rotating the imaging system while keeping the illuminating system and metasurface fixed, an oblique view of the 3D holographic image can be obtained. The holographic images for viewing angles ranging from -20° to 20° at steps of 10° are shown in Fig. 3.17a–e, which clearly shows the tilting effect and further verify the 3D nature of the holographic image. Note that at larger viewing angles of $\pm 20^\circ$, only part of the holographic image can be captured by the objective due to the finite beam divergence of the light forming the holographic image. Again, the experimental observations can be qualitatively explained by the 2D perspective images calculated by ray tracing shown in Fig. 3.17f–j. From our experiment we estimate the FOV to be in the range of $[-40^\circ, 40^\circ]$.

3.3.4 Discussion

As the 3D objects demonstrated here consist of diffuse reflecting surfaces or point sources, the amplitudes of the holograms are

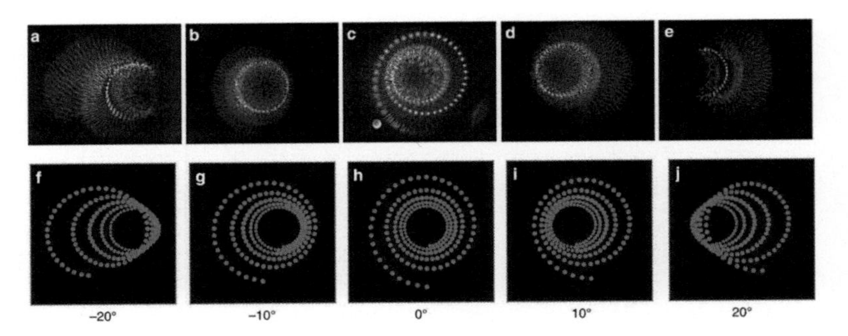

Figure 3.17 Holographic images at tilted observation angles but normal incidence illumination. (a–e) Holographic images observed at different angles with $z = 1351$ μm. (f–j) Corresponding ray-tracing calculations of the 2D perspectives of the holographic images [24].

highly uniform, thus eliminating the need for amplitude control without sacrificing the image quality. On the other hand, if multiple reflection properties of the object surfaces are desired, such as specular reflection and Phong shading, a complex amplitude modulation of the incident light can show better performance. Nevertheless, complex amplitude modulation may be obtained in the metasurface by additionally encoding the amplitude information into the length of the nanorods. This will lead to a modification of the resonance frequency and therefore different scattering amplitude.

Current holography technologies are mainly limited by the large pixel size of the holograms. Specifically, the FOV is determined by the maximum diffraction angle of the hologram, which is given by $\sin\theta = \lambda/2s$, where s is the pixel pitch of the hologram. For example, SLMs, despite being dynamic, have pixel sizes of at least 6.4 μm × 6.4 μm, which are more than 1 order of magnitude larger than the wavelength of light. Therefore for an SLM-based hologram the FOV is far from being desired for 3D holography. Holograms based on the diffractive optical element (DOE), superficial microrelief with depths on the order of the optical wavelength, usually have pixel sizes of at least 10 wavelengths, such that scalar diffraction theory can be applied [54]. Below this length scale, rigorous vector diffraction theory would be required, which is computationally time consuming. In addition, the multiphase level in DOE necessary to produce high-quality holography entails multiple steps of lithography with precise alignment. Recently, researchers have reported amplitude-based holograms using carbon nanotubes [50] with pixel sizes down to 400 nm $\times$ 400 nm. However, the binary nature of the holograms results in significantly less information per pixel than the continuous phase profiles in our metasurface holograms, thus compromising the quality of image reconstruction. In addition, the problems such as twin image and zero-order diffraction are present in such amplitude holograms. Bulk metamaterial–based holograms [46] are so far only designed to work at IR wavelengths due to the fact that they require complex fabrication processes involving multilayer alignment. Thus compared to other methods for generating holograms the metasurface approach presented here provides easy access to subwavelength pixel sizes with continuously controllable phase profiles. This advantage arises from the fact that the geometrical size of the plasmonic nanoantennas (here around 150 nm) is much smaller than the free-space wavelength used for generating the image (670–950 nm), and the capability of generating well-controlled continuous phase profile benefits from the simplicity in the encoding of phase information into the orientation of the plasmonic nanorods.

3.4 Conclusion

In summary, we have demonstrated on-axis plasmonic holography for distortion-free 3D images in the visible and near-IR range by using subwavelength pixelate plasmonic metasurfaces. A complete phase control is achieved without an extra look-up table or phase accumulation along an optical path. The subwavelength pixel size of the metasurface hologram represents a great advantage over other conventional methods such as CGH with SLMs or DOEs, and more importantly, the common issue of multiple diffraction orders accompanying 3D holographic images is avoided. The dispersion-less nature of our metasurface can result in broadband operation without sacrificing image quality. Such a scheme could potentially be used in high-resolution holographic data storage, optical information processing and other holography-based techniques.

Acknowledgments

X.C. acknowledges the start package from the School of Engineering and Physical Sciences, Heriot-Watt University. This work is partly supported by the Engineering and Physical Sciences Council of the United Kingdom under the scheme of NSF/EPSRC Materials Network. C.W.Q. acknowledges support from National University of Singapore (Grant No. R-263-000-688-112). S.Z. acknowledges support from the University of Birmingham and the European Commission under the Marie Curie Career Integration Program. We thank the contribution from T. Zentgraf, L. Huang, H. Mühlenbernd, G. Li, J. Li, H. Zhang, S. Chen, B. Bai, Q. Tan, G. Jin, and K.-W. Cheah.

References

1. Shelby, R. A., Smith, D. R., and Schultz, S. (2001). Experimental verification of a negative index of refraction, *Science*, **292**, pp. 77–79.
2. Valentine, J., Zhang, S. Zentgraf, T. Avila, E. U., Genov, D. A., Bartal, G., and Zhang, X. (2008). Three-dimensional optical metamaterial with a negative refractive index, *Nature*, **455**, pp. 376–379.
3. Pendry, J. B. (2000). Negative refraction makes a perfect lens, *Phys. Rev. Lett.*, **85**, pp. 3966–3969.
4. Fang, N. Lee, H., Sun, C., and Zhang, X. (2005). Sub-diffraction-limited optical imaging with a silver superlens, *Science*, **308**, pp. 534–537.
5. Taubner, T., Korobkin, D. Urzhumoy, Y., Shvets, G., and Hillenbrand, R. (2006). Near-field microscopy through a SiC superlens, *Science*, **313**, p. 1595.
6. Liu, Z., Lee, H., Xiong, Y. Sun, C., and Zhang, X. (2007). Far-field optical hyperlens magnifying sub-diffraction-limited objects, *Science*, **315**, p. 1686.
7. Pendry, J. B., Schurig, D., and Smith, D. R. (2006). Controlling electromagnetic fields, *Science*, **312**, pp. 1780–1782.
8. Jeon, S., Menard, E., Park, J.-U., Maria, J., Meitl, M., Zaumseil, J., and Rogers, J. A. (2004). Three-dimensional nanofabrication with rubber stamps and conformable photomasks, *Adv. Mater.*, **16**, pp. 1369–1373.
9. Rill, M. S., Plet, C., Thiel, M., Staude, I., Freymann, G. V., Linden, S., and Wegener, M. (2008). Photonic metamaterials by direct laser writing and silber chemical vapour deposition, *Nat. Mater.*, **7**, pp. 543–546.

10. Yu, N., Genevet, P., Kats, M. A., Aieta, F., Tetienns, J.-P., Capasso, F., and Gaburro, Z. (2011). Light propagation with phase discontinuities: generalized laws of reflection and refraction, *Science*, **334**, pp. 333–337.
11. Ni, X., Emani, N. K., Kildishev, A. V., Boltassev, A., and Shalaev, V. M. (2012). Broadband light bending with plasmonicnanoantennas, *Science*, **335**, p. 427.
12. Huang, L., Chen, X., Mühlenbernd, H., Li, G., Bai, B., Tan, Q., Jin, G., Zentgraf, T., and Zhang, S. (2012). Dispersionless phase discontinuities for controlling light propagation, *Nano Lett.*, **12**, pp. 5750–5755.
13. Kang, M., Feng, T., Wang, H. T., and Li, J. (2012). Wave front engineering from an array of thin aperture antennas, *Opt. Express*, **20**, pp. 15882–15890.
14. Yu, N., Aieta, F., Genevet, P., Kats, M. A., Gaburro, Z., and Capasso, F.(2012). A broadband, background-free quarter-wave plate based on plasmonicmetasurfaces, *Nano Lett.*, **12**, pp. 6328–6333.
15. Aieta, F., Genevet, P., Kats, M. A., Yu, N., Blanchard, R., Gaburro, Z., and Capasso, F. (2012). Aberration-free ultrathin flat lenses and axicons at telecom wavelengths based on plasmonicmetasurfaces, *Nano Lett.*, **12**, pp. 4932–4936.
16. Chen, X., Huang, L., Mühlenbernd, H., Li, G. X., Bai, B., Tan, Q., Jin, G., Qiu, C.-W., Zhang, S., and Zentgraf, T. (2012). Dual-polarity plasmonicmetalens for visible light, *Nat. Commun.*, **3**, p. 1198.
17. Ni, X., Ishii, S., Kildishev, A. V., and Shalaev, V. M. (2013). Ultra-thin, planar, Babinet-inverted plasmonicmetalenses, *Light Sci. Appl.*, **2**, p. e72.
18. Chen, X., Huang, L., Mühlenbernd, H., Li, G., Bai, B., Tan, Q., Jin, G., Qiu, C.-W., Zentgraf, T., and Zhang, S. (2013). Reversible three-dimensional focusing of visible light with ultrathin plasmonic flat lens, *Adv. Opt. Mater.*, **1**, pp. 517–521.
19. Sun, S., He, Q., Xiao, S., Xu, Q., Li, X., and Zhou, L. (2012). Gradient-index meta-surfaces as a bridge linking propagating waves and surface waves, *Nat. Mater.*, **11**, pp. 426–431.
20. Yin, X., Ye, Z., Rho, J., Wang, Y., and Zhang, X. (2013). Photonic Spin Hall effect at metasurfaces, *Science*, **339**, pp. 1405–1407.
21. Huang, L., Chen, X., Bai, B., Tan, Q., Jin, G., Zentgraf, T., and Zhang, S. (2013). Helicity dependent directional surface plasmonpolariton excitation using a metasurface with interfacial phase discontinuity, *Light Sci. Appl.*, **2**, p. e70.
22. Lin, J., Mueller, J. P. B., Wang, Q., Yuan, G., Antoniou, N., Yuan, X.-C., and Capasso, F. (2013). Polarization-controlled tunable directional coupling of surface plasmonpolaritons, *Science*, **340**, pp. 331–334.

23. Shitrit, N., Yulevich, I., Maguid, E., Ozeri, D., Veksler, D., Kleiner, V., and Hasman, E. (2013). Spin-optical metamaterial route to spin-controlled photonics, *Science*, **340**, pp. 724–726.
24. Huang, L., Chen, X., Mühlenbernd, H., Zhang, H., Chen, S., Bai, B., Tan, Q., Jin, G., Cheah, K.-W., Qiu, C.-W., Li, J., Zentgraf, T., and Zhang, S. (2013). Three-dimensional optical holography using a plasmonicmetasurface, *Nat. Commun.*, **4**, p. 2808.
25. Ni, X., Kildishev, A. V., and Shalaev, V. M. (2013). Metasurface holograms for visible light, *Nat. Commun.*, **4**, p. 2807.
26. Lin, L., Goh, X. M., McGuinness, L. P., and Roberts, A. (2010). Plasmonic lenses formed by two-dimensional nanometric cross-shaped aperture arrays for Fresnel-region focusing, *Nano Lett.*, **10**, pp. 1936–1940.
27. Gao, H., Hyun, J. K., Lee, M. H., Yang, J. C., Lauhon, L. J., and Odom, T. W. (2010). Broadband plasmonicmicrolenses based on patches of nanoholes, *Nano Lett.*, **10**, pp. 4111–4116.
28. Ishii, S., Kildishev, A. V., Shalaev, V. M., Chen, K. P., and Drachev, V. P. (2011). Metal nanoslit lenses with polarization-selective design, *Opt. Lett.*, **36**, pp. 451–453.
29. Verslegers, L., Catrysse, P. B., Yu, Z., White, J. S., Barnard, E. S., Brongersma, M. L., and Fan, S. (2009). Planar lenses based on nanoscale slit arrays in a metallic film, *Nano Lett.*, **9**, pp. 235–238.
30. Tetienne, J. P., Blanchard, R., Yu, N., Genevet, P., Kats, M. A., Fan, J. A., Edamura, T., Furuta, F., Yamanishi, M., and Capasso, F. (2011). Dipole modelling and experimental demonstration of multi-beam plasmonic collimators, *New J. Phys.*, **13**, p. 053057.
31. Kats, M. A., Genevet, P., Aoust, G., Yu, N., Blanchard, R., Aieta, F., Gaburro, Z., and Capasso, F. (2012). Giant birefringence in optical antenna arrays with widely tailorable optical anisotropy, *Proc. Natl. Acad. Sci. U.S.A.*, **109**, pp. 12364—12368.
32. Blanchard, R., Aoust, G., Genevet, P., Yu, N., Kats, M. A., Gaburro, Z., and Capasso, F. (2012). Modelling nanoscale V-shaped antennas for the design of optical phased arrays, *Phys. Rev. B*, **85**, p. 155457.
33. Berry, M. V. (1987). The adiabatic phase and Pancharatnam's phase for polarized light, *J. Mod. Opt.*, **34**, pp. 1401–1407.
34. Pancharatnam, S. (1956). Generalized theory of interference, and its applications. Part I. Coherent pencils, *Proc. Indian Acad. Sci. A*, **44**, pp. 247–262.
35. Dennis, G. (1948). A new microscopic principle, *Nature*, **161**, pp. 777–778.

36. Kelly, D. P., Monaghan, D. S., Pandey, N., Kozacki, T., Michalkiewicz, A., Finke, G., Hennelly, B. M., and Kujawinska, M. (2010). Digital holographic capture and optoelectronic reconstruction for 3D displays, *Int. J. Digital Multimedia Broadcast.*, **2010**, p. 759323.
37. Slinger, C., Cameron, C., and Stanley, M. (2005). Computer-generated holography as a generic display technology, *Computer*, **38**, pp. 46–53.
38. Ozaki, M., Kato, J.-I., and Kawata, S. (2011). Surface-plasmon holography with white-light illumination, *Science*, **332**, pp. 218–220.
39. Chen, Y. H., Huang L., Gan, L., and Li, Z. Y. (2012). Wavefront shaping of infrared light through a subwavelength hole, *Light Sci. Appl.*, **1**, p. e26.
40. Dolev, I., Epstein, I., and Arie, A. (2012). Surface-plasmon holographic beam shaping, *Phys. Rev. Lett.*, **109**, p. 203903.
41. Pegard, N. C., and Fleischer, J. W. (2011). Optimizing holographic data storage using a fractional Fourier transform, *Opt. Lett.*, **36**, pp. 2551–2553.
42. Barsi, C., Wan, W., and Fleischer, J. W. (2009). Imaging through nonlinear media using digital holography, *Nat. Photonics*, **3**, pp. 211–215.
43. Buse, K., Adibi, A., and Psaltis, D. (1998). Non-volatile holographic storage in doubly doped lithium niobate crystals, *Nature*, **393**, pp. 665–668.
44. Grier, D. G. (2003). A revolution in optical manipulation, *Nature*, **424**, pp. 21–27.
45. Midgley, P. A., and Dunin-Borkowski, R. E. (2009). Electron tomography and holography in materials science, *Nat. Mater.*, **8**, pp. 271–280.
46. Larouche, S., Tsai, Y.-J., Tyler, T., Jokerst, N. M., and Smith, D. R. (2012). Infrared metamaterial phase holograms, *Nat. Mater.*, **11**, pp. 450–454.
47. Levy, U., Kim, H.-C, Tsai, C.-H., and Fainman, Y. (2005). Near-infrared demonstration of computer-generated holograms implemented by using subwavelength gratings with space-variant orientation, *Opt. Lett.*, **30**, pp. 2089–2091.
48. Hu, D., Wang, X., Feng, S., Ye, J., Sun, W., Kan, Q., Klar, P. J., and Zhang, Y. (2013) Ultrathin terahertz planar elements, *Adv. Opt. Mater.*, **1**, pp. 186–191.
49. Walther, B., Helgert, C., Rockstuhl, C., Setzpfandt, F., Eilenberger, F., Kley, E.-B., Lederer, F., Tünnermann, A., and Pertsch, T. (2012). Photonics: spatial and spectral light shaping with metamaterials, *Adv. Mater.*, **24**, pp. 6300–6304.

50. Butt, H., Montelongo, Y., Butler, T., Rajesekharan, R., Dai, Q., Shiva-Reddy, S. G., Wilkinson, T. D., and Amaratunga, G. A. J. (2012). Carbon nanotube based high resolution holograms, *Adv. Mater.*, **24**, pp. OP331–OP336.
51. Chang, C. M., Tseng, M. L., Cheng, B. H., Chu, C. H., Ho, Y. Z., Huang, H. W., Lan, Y.-C, Huang, D.-W, Liu, A. Q., and Tsai D. P. (2013). Three-dimensional plasmonic micro projector for light manipulation, *Adv. Mater.*, **25**, pp. 1118–1123.
52. Zhou, F., Liu, Y., and Cai, W. (2013). Plasmonic holographic imaging with V-shaped nanoantenna array, *Opt. Express*, **21**, pp. 4348–4354.
53. Zhang, H., Tan, Q., and Jin, G. (2012). Holographic display system of a three-dimensional image with distortion-free magnification and zero-order elimination, *Opt. Eng.*, **51**, p. 075801.
54. Pommet, D. A., Moharam, M. G., and Grann, E. B. (1995). Limits of scalar diffraction theory for diffractive phase elements, *Opt. Lett.*, **11**, pp. 1827–1834.

Chapter 4

Plasmonic Structured Illumination Microscopy

Feifei Wei,[a] Joseph Louis Ponsetto,[a] and Zhaowei Liu[a,b,c]

[a]*Department of Electrical and Computer Engineering, University of California, San Diego, 9500 Gilman Drive, La Jolla, CA 92093, USA*
[b]*Center for Magnetic Recording Research, University of California, San Diego, 9500 Gilman Drive, La Jolla, CA 92093, USA*
[c]*Materials Science and Engineering Program, University of California, San Diego, 9500 Gilman Drive, La Jolla, CA 92093, USA*
zhaowei@ucsd.edu

Optical microscopes are widely used in biological research because light is a noninvasive probe for examining biological specimens. However, due to diffraction, the resolution of conventional optical microscopy is limited to the scale of a few hundreds of nanometers. Recently, various advanced imaging techniques with subdiffraction-limited resolution have been proposed and demonstrated. In this chapter, we will focus on one of the emerging super-resolution techniques, plasmonic structured illumination microscopy (PSIM), which combines tunable surface plasmon interference (SPI) with the structured illumination microscopy (SIM) technique to achieve more than two times' resolution improvement compared to conventional fluorescence microscopy. First, the chapter will start with a brief introduction of optical microscopy and various resolution-

Plasmonics and Super-Resolution Imaging
Edited by Zhaowei Liu

ISBN 978-981-4669-91-7 (Hardcover), 978-1-315-20653-0 (eBook)
www.panstanford.com

improving techniques, followed by the working principle of SIM and the basics of SPs. Then, the principle of PSIM, the tuning mechanism of SPI, as well as the image reconstruction algorithm will be discussed. After that, the numerical and experimental demonstration of the PSIM technique will be explained in detail. A second-generation PSIM technique using localized plasmons will also be introduced. Finally, the chapter closes with a discussion of the advantages and potential of this super-resolution PSIM technique.

4.1 Introduction

4.1.1 Optical Microscopy and Resolution Limit

The optical microscope is an instrument that produces a magnified optical image of an object using lenses. Very early optical microscopes only consisted of a single magnifying lens and were commonly used to study small insects. In the sixteenth to seventeenth centuries, compound optical microscopes were invented by assembling multiple lenses together in a cylindrical tube. As a result, the magnifying power of the optical microscope was greatly improved and very small specimens such as cells were discovered. Because of its impressive magnifying power, optical microscopy was widely used in biology studies ever since. Within the following 400 years, optical microscopy has been greatly improved due to advancements in optical system design and lens manufacturing processes. Various special optical microscopy techniques, such as dark-field microscopy, phase contrast microscopy, and fluorescence microscopy have been developed to increase the image contrast and facilitate the examination of certain types of samples [1]. Nowadays, despite the invention of other high-magnification equipment, such as electron microscopes, the optical microscope is still an indispensable tool for biological research because of the noninvasive, biosafe probing capability of visible light.

For an optical microscope, resolution is an important parameter characterizing the ability to resolve closely located point sources. Due to the band-pass nature of optical imaging systems, the finest resolvable feature of a visible optical microscope is limited to several hundreds of nanometers. The resolution is determined by the Abbe diffraction limit, $\lambda/(2\text{NA})$, in which λ is the wavelength of the light and NA is the numerical aperture of the objective [2]. Even today, hundreds of years after the invention of the optical microscopes, the diffraction limit is still the most formidable resolution barrier.

4.1.2 Traditional Methods of Improving the Resolving Power

There are various methods to improve the resolving power of an imaging system. First, since the diffraction limit is inversely proportional to the NA of the objective, the resolving power of optical microscopy can be improved by increasing the NA of the objective using high-refractive-index immersion liquids, such as oil. In this case, the maximum-achievable NA is constrained by the refractive index of the immersion media. Although this can successfully increase the NA of the objective from around 0.8 to around 1.7, this method can only provide limited improvement due to the limited choices of high-refractive-index immersion liquids. Another straightforward way of improving the resolving power of optical microscopy is to decrease the operational wavelength to the ultraviolet or even X-ray spectrum because the resolution limit is proportional to the wavelength of the light. However, these high-energy probes can damage the biosamples and therefore are not suitable for life specimen imaging.

Besides the aforementioned methods, utilizing the evanescent waves emerging from objects is another useful strategy to improve the resolving power of an imaging system. Near-field scanning optical microscopy (NSOM), invented in the 1980s, is a widely used technique within this category. It achieves resolution of less than 1/10 of the illuminating wavelength by detecting the evanescent waves with a subwavelength tip or aperture placed within the near field of the object and forms an image through stage scanning [3]. Due to the exponential decay of the evanescent wave intensity, NSOM uses a feedback mechanism to keep the probe within the near field of the object for effective detection. However, this requirement also limits NSOM's applications due to various challenges involved in placing and scanning a probe in the near field of objects.

4.2 Super-Resolution Fluorescence Microscopy and Surface Plasmons

4.2.1 Super-Resolution Fluorescence Microscopy Techniques

As more and more biological research requires sub-100 nm resolution, subdiffraction-limited super-resolution microscopy has become a tremendously important research field. In modern biology, fluorescence microscopy has been widely used to study specific biostructures through proper labeling. Through utilizing the response properties of fluorescence dyes, various super-resolution fluorescence microscopy techniques, such as stimulated emission depletion (STED) microscopy [4–7], single-molecule localization techniques [8–13], etc., have been experimentally demonstrated. For instance, STED microscopy is capable of achieving lateral resolution of tens of nanometers by illuminating the object with a co-axially aligned excitation beam and a STED beam. Because of the STED, only fluorescent molecules located in a subdiffraction-limited overlap region of the excitation beam and the dark center of the STED beam could fluoresce, with the overlap region area determined by the STED beam intensity. After scanning the synchronized excitation beam and STED beam across the sample, a 2D super-resolution image could be formed [4–7]. For stochastic optical reconstruction microscopy (STORM) and photoactivated localization microscopy (PALM), the diffraction limit is surpassed through accurately determining the photoswitchable fluorophore's location with postprocessing [8–13]. Since only several molecules fluoresce during each PALM or STORM raw image acquisition, the images of these molecules have little chance to overlap. As a result, the collected images can be used to extract the molecule locations to an accuracy of tens of nanometers through Gaussian fitting, with the localization accuracy eventually limited by the number of photons captured from the corresponding fluorophore each time [8–13]. The final super-resolution image can be reconstructed by combining a large number of processed images together. Although STED, PALM, and STORM are capable of achieving sub-50 nm lateral resolution, the time they take to acquire a 2D super-resolution image is relatively long, due to either slow raster-scanning processes or the larger number of raw images required, and scales linearly to the imaging area.

4.2.2 Structured Illumination Microscopy

Among super-resolution imaging techniques, structured illumination microscopy (SIM) is a method of special interest due to its wide-field, high-speed subdiffraction-limited imaging capability. Since the first demonstration in 2000 [15], SIM with twice the resolving power compared to conventional fluorescence microscopy has been applied to various biological studies [14–16]. Recently, as a result of improvements in illumination generation and manipulation, the speed of SIM was improved to 11 frames per second (fps) for $8 \times 8\ \mu m^2$ imaging area [17]. Different from NSOM, STED, STORM, and PALM, the speed of SIM is mainly influenced by the maximum frame rate of the imaging cameras.

Figure 4.1 schematically illustrates the working principle of SIM. Assuming an object (Fig. 4.1a) is illuminated by a periodic light pattern (Fig. 4.1b), the resulting image expresses as Moiré fringes (Fig. 4.1c). The Moiré fringes are composed of much larger features, compared to the object, so they are easier to be detected. If the illumination patter is given, the object can be deduced

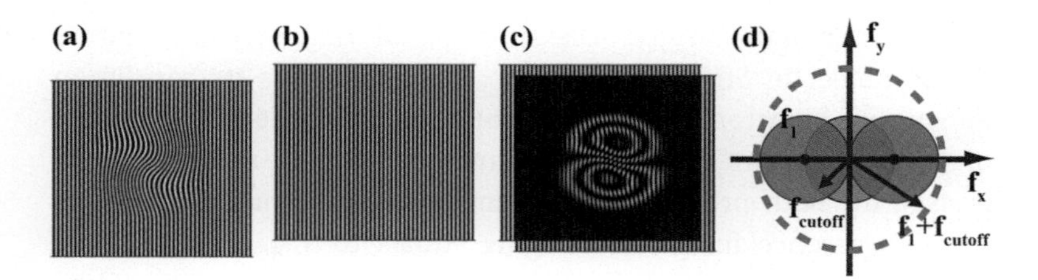

Figure 4.1 SIM principle. (a) A test object, (b) a periodic structured illumination pattern, (c) the Moiré fringes formed by superimposing (a) and (b), and (d) Fourier space representation of SIM resolution improvement. Through structured illumination, the high-spatial-frequency information (two red solid circles) can be detected and the cutoff spatial frequency of SIM can be extended to $f_{\text{cutoff}} + f_1$ (f_{cutoff} and f_1 represent the cutoff spatial frequency of the imaging system and the spatial frequency of the illumination pattern, respectively).

by the Moiré fringes through a numerical algorithm. It can also be explained by a straightforward Fourier optics analysis. For conventional fluorescence microscopy, the optical transfer function (OTF) is a low-pass filter in the spatial frequency domain and has a cutoff (f_{cutoff}) at $2\pi \cdot \text{NA}/\lambda_{\text{emission}}$. As a result, only the low-spatial-frequency information of the object (represented by the gray circle in Fig. 4.1d, with $f < f_{\text{cutoff}}$) can be detected in the far field, yielding a diffraction-limited resolution. SIM improves the resolution by utilizing the Moiré effect to shift some high-spatial-frequency information of the object (represented by the red circles in Fig. 4.1d) from outside the OTF to inside and recover it through postprocessing. The amount of the shift in Fourier space is determined by f_1, the spatial frequency of the illumination pattern. Because of the spatial frequency mixing introduced by structured illumination, the detected far-field image contains both the unshifted low-spatial-frequency information and the shifted high-spatial-frequency information of the object. From a set of diffraction-limited images taken under a series of laterally translated structured illuminations, all these spatial frequencies of the object can be recovered. After moving them back to their original positions in Fourier space and applying the inverse Fourier transform to the assembled extended-spatial-frequency distribution, a super-resolution image can be reconstructed [14–16]. The image reconstruction process will be explained in detail later. As shown in Fig. 4.1d, $f_1 + f_{\text{cutoff}}$ corresponds to the largest-accessible spatial frequency information of the object in SIM and $(f_1 + f_{\text{cutoff}})/f_{\text{cutoff}}$ gives its resolution enhancement factor. Since the excitation and emission wavelengths of fluorophores are generally close and normally the same objective is used for both projecting the illumination patterns and forming the fluorescence image, f_1 roughly equals to f_{cutoff}. As a result, SIM possesses roughly twice the resolving power compared to conventional fluorescence microscopy [14–16]. Although saturated structured illumination microscopy (SSIM) can further improve the resolution by exploiting the nonlinear fluorescence response of the dyes on the illumination intensity [18, 19], it is always accompanied by additional limitations, such as fast fluorescence bleaching and slow frame rate [19].

4.2.3 Background of Surface Plasmons

Surface plasmons (SPs) are electromagnetic excitations formed by the collective oscillation of the free electrons at a metal–dielectric interface [20]. The associated electromagnetic waves possess in-plane wavevectors larger than that of the photon in the same dielectric media and therefore exponentially decay away from the interface.

Based on Maxwell's equations and the boundary conditions, the dispersion of the SPs at a semi-infinite metal–dielectric interface can be derived as Eq. 4.1 and Eq. 4.2. In these equations, $\mathbf{k}_0$, ε_1, and ε_2 represent the wavevector of the free-space light, the dielectric constant of the metal, and the dielectric constant of the dielectric, respectively. A comparison of the typical dispersion curves of the SPs and photons, plotted in Fig. 4.2a, clearly shows that the SP wavevector ($\mathbf{k}_{\text{sp}}$) is always larger than that of the photon at the same frequency. The $\mathbf{k}_{\text{sp}}/\mathbf{k}_0$ ratio can be extremely large when the frequency of the excitation light approaches the SP resonant frequency. Besides the frequency of the excitation light, the $\mathbf{k}_{\text{sp}}/\mathbf{k}_0$ ratio can also be adjusted by the refractive index of the surrounding dielectric, as well as the metal film thickness [20].

$$\mathbf{k}_{\text{sp}} = \mathbf{k}_0\sqrt{\frac{\varepsilon_1\varepsilon_2}{\varepsilon_1+\varepsilon_2}} \tag{4.1}$$

$$\mathbf{k}_{z,i} = \sqrt{\varepsilon_i\mathbf{k}_0^2 - \mathbf{k}_{\text{sp}}^2} \quad i = 1,\, 2 \tag{4.2}$$

Due to the wavevector mismatch between the photons and the SPs, a coupling structure that is able to provide additional momentum is always needed in order to excite the SPs with light. For the prism coupling method (shown in Fig. 4.2b), the in-plane momentum of the photons inside a prism provides the needed momentum for exciting the SPs at the metal–air interface. Therefore, the prism should be made from materials with high refractive indices. In a grating coupling structure (see Fig. 4.2c), the diffraction from the grating bridges the momentum mismatch between the incident photon and the SPs. Hence, the SPs can only be effectively excited when the summation of the momentum provided by the grating structure and the illumination photon equals to that of the SPs.

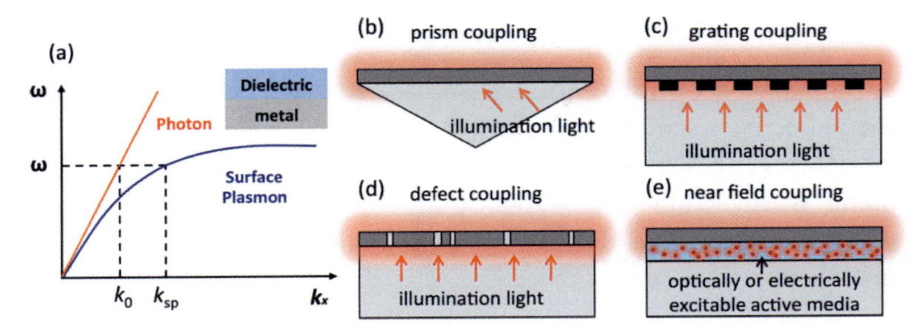

Figure 4.2 (a) The photon (red line) and the SP (blue line) dispersion curves. (b–e) Schematic representation of various coupling mechanisms. (b) A prism coupling, (c) a grating coupling, (d) a defect coupling, and (e) a near-field coupling structure, with the dark-gray areas representing the plasmonic material.

Due to the broadband wavevector provided by a subwavelength-scale defect, SPs can also be excited through the scattering of the illumination light at various boundaries or volumetric defects (see Fig. 4.2d), although with less efficiency compared to the case for grating coupling structures. Besides the above-mentioned methods, SPs can also be excited through near-field coupling between the plasmonic and active media layers (see Fig. 4.2e). In this case, the needed momentum is provided by the evanescent field emerging from the active media, such as photoluminescent or electrically luminescent materials.

Nowadays, SPs are widely used in nanophotonics and subwavelength-scale research due to their unique dispersion properties and relatively large wavevector. With the help of advancements in nanofabrication technologies, various exciting applications of SPs, such as subdiffraction-limited focusing, imaging, and lithography [21–33], subwavelength-scale waveguiding [34, 35], negative refraction [36], etc., have been proposed and demonstrated. Specifically, in the imaging field, many SP-related techniques, such as the far-field superlens (FSL) [27–29], hyperlens [30–32], metalens [37–40], and SP-assisted super-resolution microscopy [41–46] have been theoretically or experimentally demonstrated to break the diffraction limit. For instance, the FSL, consisting of a layer of metal film and a designed periodic grating structure enhances the evanescent waves emerging from the object and converts them to propagating light for far-field detection. After numerically processing the detected far-field signal, a super-resolution image could then be reconstructed [27–29]. Different from the FSL, the hyperlens is made from a concentrically layered hyperbolic metamaterial. Because of the designed material dispersion and shape of a hyperlens, the tangential wavevectors of the evanescent waves emerging from the object get compressed as they propagate radially outward due to angular momentum conservation. Therefore, both the diffraction-limited and the subdiffraction-limited spatial frequency information of the object can be delivered to the outer boundary of the hyperlens and form a magnified, super-resolution image of the object, which can be detected in the far field [30–32].

4.3 Principles of Plasmonic Structured Illumination Microscopy

As discussed in Section 4.2.2, the resolution enhancement factor of SIM compared to conventional fluorescence microscopy is limited by the spatial frequency of the illumination patterns. For illumination patterns formed through objectives, the spatial frequency could not exceed f_{cutoff} as well due to diffraction [14, 15]. To further increase the resolution enhancement factor, illumination patterns with higher spatial frequency are needed. Several SIM techniques utilizing the evanescent field interference formed on high-refractive-index substrates, such as glass and silicon, have been proposed and demonstrated either numerically or experimentally [47–49].

Besides the aforementioned methods, plasmonic material is another excellent candidate to form excitation features below the diffraction limit. Because the SP wavevector is larger than that of the illumination photon, its corresponding interference pattern possesses a higher spatial frequency compared to laser interference at the same time frequency. Therefore, by replacing conventional laser interference illumination with surface plasmon interference (SPI) illumination, PSIM enables the detection of the high-spatial-frequency information of the object beyond the SIM detection band and possesses more than two times' resolution improvement

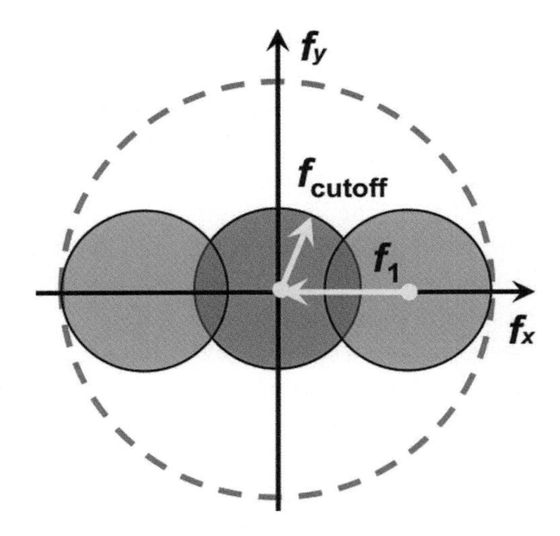

Figure 4.3 Fourier space representation of PSIM resolution improvement, in which $f_1 = 2\pi/\lambda_{sp} > f_{cutoff}$ and the dashed red circle represents the PSIM-detectable region. The low-spatial-frequency information detectable by conventional fluorescence microscopy and the additional high-spatial-frequency information accessible by PSIM are represented by the gray circle and the red solid circles, respectively. This example only shows 1D illumination along the x direction.

compared to conventional fluorescence microscopy, as shown by the radius ratio of the dashed circle and the gray circle in Fig. 4.3.

4.3.1 Surface Plasmon Interference Formation and Manipulation

Due to its smaller period compared to the corresponding light interference at a given frequency, SPI has been extensively studied during the past decade and applied in various fields, such as lithography [21–26]. To form an SPI pattern, counterpropagating coherent SP waves have to be excited at the same time and overlap in space. The SP waves can be excited through various configurations (see Fig. 4.2) as long as the momentum mismatch between the illumination light and the SP is compensated through the structure.

In 2004, Luo et al. demonstrated line array fabrications using SPI excited by a silver (Ag) grating mask with 300 nm period [21] (see

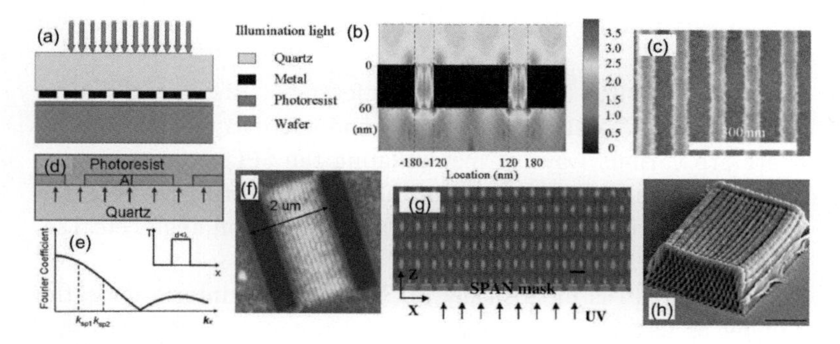

Figure 4.4 (a–c) SP resonant interference nanolithography demonstrated by Luo et al. (a) Schematic representation of a SP resonant interference nanolithography structure, (b) numerical simulation of the intensity distribution at the near field of the mask, and (c) SEM image of the fabricated pattern. (d–f) SPI nanolithography demonstrated by Liu et al. (d) Schematic configuration of the Al mask, (e) Fourier spectrum of a subwavelength-scale rectangular function, and (f) AFM image of the corresponding exposed patterns under 365 nm illumination. (g, h). 3D surface-plasmon-assisted nanolithography (3D-SPAN) demonstrated by Shao et al. (g). Intensity distribution in the photoresist after an Al mask under 365 nm illumination and (h) SEM image of the corresponding 3D structure fabricated by 3D-SPAN, with the scale bar representing 2 μm.

Fig. 4.4a–c). As shown in Fig. 4.4c, 50 nm wide lines were generated inside the photoresist beneath the mask after being exposed to the SPI formed under mercury g-line illumination (436 nm). Liu et al. also numerically studied the SPI formed by the aluminum (Al) grating mask around the same time [22], and later experimentally demonstrated SPI lithography using mercury i-line illumination (365 nm) [24] (see Fig. 4.4d–f). The atomic force microscopy (AFM) measurement in Fig. 4.4f shows that a line array with a period that roughly equals the SPI period (~120 nm) was generated in the SU8 resist due to the exposure of the SPI field [24]. Since the intensity distribution at the near field of a plasmonic mask is a 3D distribution, it could also be used to fabricate 3D structures as well. With the help of the SP field excited on top of the Al mask, Shao et al. showed that the contrast of the excited 3D interference pattern was significantly increased compared to the case using a phase-shifting mask [25]. As a result, a 3D periodic polymeric nanostructure can be fabricated in a typical photolithography setup, as is shown in Fig. 4.4h [25].

Besides forming SPI, tuning the inference pattern is also very important for SP-assisted lithography and microscopy applications. There are various ways of manipulating the SPI pattern, such as by changing the arrangements of the coupling structure [24], the illumination polarization [24], and the illumination angle, etc. [24, 26, 43].

Since the SPI is intrinsically 2D surface wave interference, the number as well as the propagation direction of the SP waves that interfere with each other can be controlled by the geometry of the coupling structure. Through utilizing different geometries of the SP coupling slits or edges, complicated SPI patterns can be formed (Fig. 4.5a–f) [24]. For a sample with fixed coupling structure geometry, the SPI pattern can be adjusted by the illumination polarization because the SP coupling efficiency is significantly higher for incident light with polarization perpendicular to the coupling structure than that along the coupling structure [24]. The influences of the illumination polarization are shown in Fig. 4.5g–l by comparing the fluorescence images of the dye-coated plasmonic structures under illumination with various polarizations. When illuminated by un-polarized light, the SP waves were excited with about the same efficiency for all the slits and form three-, four-, and five-wave interference in the middle region (Fig. 4.5g–i). However, when horizontally polarized light is used as the illumination, the excitation of the SP waves at the vertically oriented slits dominates, leading to two-, three-, and four-wave interference in the middle region instead, as confirmed by Fig. 4.5j–l [24].

Besides the geometry of the coupling structure and the illumination polarization, SPI can be controlled by the illumination angle as well [24, 26, 43]. Depending on whether the illumination light provides an in-plane wavevector that is parallel or perpendicular to the coupling structure, the illumination angle can either influence the propagation directions or the relative phases of the excited SP waves. For an excitation beam normal to the plasmonic film, the propagation directions of the excited SP waves are perpendicular to the fabricated coupling structures. However, if an in-plane wavevector along the coupling structure is introduced by changing

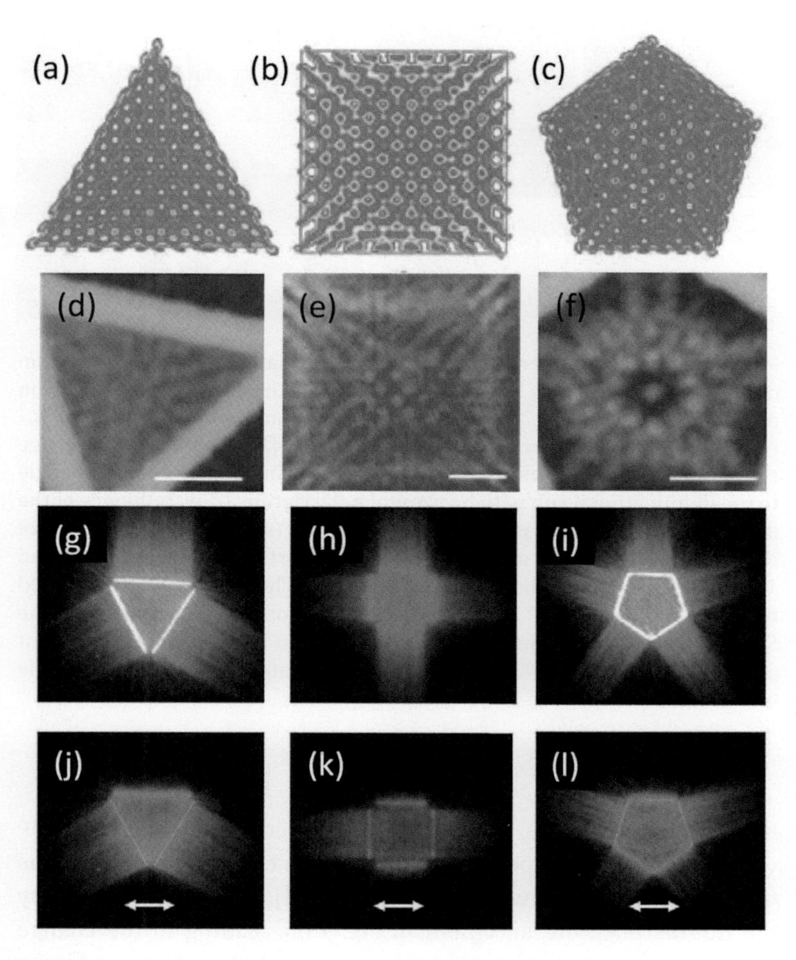

Figure 4.5 (a–f) Influences of the coupling structure geometry. (a–c) Simulation and (d–f) AFM measurements of the SPI lithography patterns formed by a 100 nm thick Al plate with different coupling structure geometries. Circularly polarized light was used in the simulations and the scale bars in (d–f) represent 500 nm. (g–l) Influences of the illumination polarization. The fluorescence images show the SPI patterns formed under (g–i) normal incident nonpolarized illumination and (j–l) normal incident horizontally polarized illumination. Due to diffraction, the interference fringes couldn't be resolved in the far field.

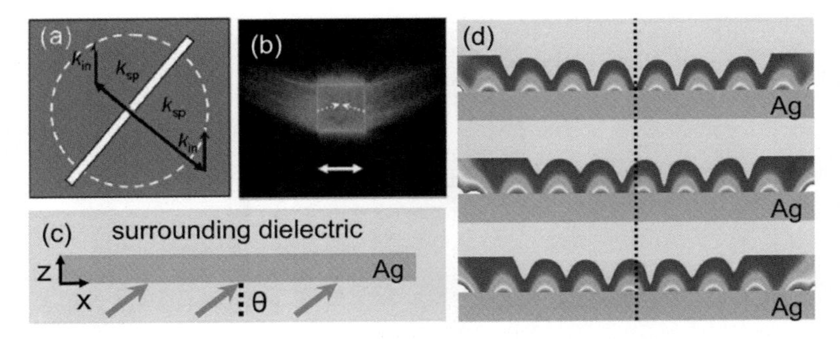

Figure 4.6 (a, b) Influence of the illumination angle to the propagation direction of excited SPs. (a) Diagram showing the propagation direction of SP waves excited by a nonperpendicular illumination beam, with $\mathbf{k}_{in}$ representing the wavevector of the incident light. (b) The fluorescent image revealing the SP propagation direction change by introducing incident angle to the illumination beam with horizontal polarization. (c, d) Influence of the illumination angle to the phase of the excited SPs. (c) Cross-sectional view of the plasmonic structure used in (d), which consists of a 1700 nm wide and 100 nm thick Ag film in the dielectric ($n = 1.33$). The structure is illuminated by 563 nm light. The SP waves will be excited on the two edges. (d) The SPI pattern formed under illumination with a 0°, 4.6°, and 8.0° incident angle (θ), respectively. The SPI laterally translates as the incident angle increases.

the illumination angle, the propagation direction of the excited SP wave will have a component parallel to the coupling structure due to momentum conservation (Fig. 4.6a) [24]. This is confirmed by the fluorescence measurement in Fig. 4.6b, leading to the change of the corresponding interference pattern [24]. For the oblique illumination that provides an in-plane wavevector perpendicular to the coupling edges or slits, it does not change the propagation direction of the excited SP waves, but introduces a relative phase φ between the SP waves excited by the adjacent edges or slits instead. This illumination angle dependent on relative phase eventually leads to a lateral translation of the SPI pattern. The simulations in Fig. 4.6d clearly show that the excited SPI pattern shifts laterally by a third of the period each time as the incident angle θ changes [43].

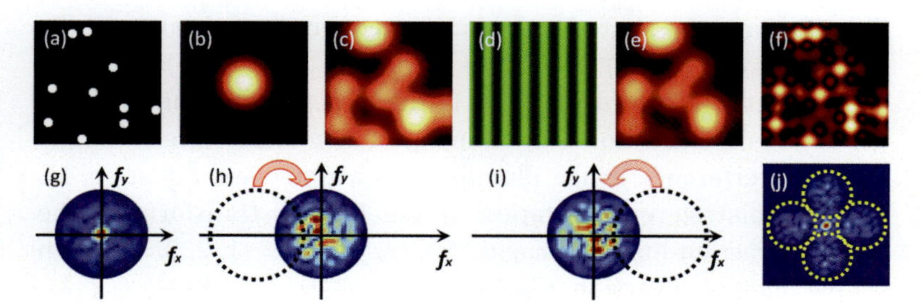

Figure 4.7 Fourier-transform-based SIM reconstruction. (a) Randomly distributed fluorescent beads serving as the object; (b) the PSF of the system; (c) the diffraction-limited fluorescence image; (d) 1D structured illumination pattern and (e) the corresponding fluorescence image under its illumination; (f) 2D super-resolution image reconstructed from six diffraction-limited subimages under structured illumination; (g) Fourier transform of (c), which represents the diffraction limited spatial frequency information of the object, $F_{\text{obj}}(f_x f_y) \cdot \text{OTF}$; (h, i) the shifted high-spatial-frequency information, $F_{\text{obj}}(f_x - f_0 f_y) \cdot \text{OTF}$ and $F_{\text{obj}}(f_x + f_0 f_y) \cdot \text{OTF}$; and (j) extended Fourier space information with regions inside the yellow dashed circles representing high spatial information accessible by structured illumination. Panel (f) was reconstructed from (j) through inverse Fourier transformation.

4.3.2 PSIM Image Reconstruction Method

To reconstruct a super-resolution image, the subdiffraction-limited spatial frequency information of the object needs to be extracted from multiple raw images through a numerical image reconstruction algorithm [14–16]. To derive the equations for reconstructing the final super-resolution image, we will start from the image formed by conventional fluorescence microscopy. The fluorescence image of an object ($obj(x, y)$, Fig. 4.7a) illuminated by 1D interference pattern ($I_i(x, y) = 1 + \cos(2\pi f_0 x + \varphi_i)$, as shown in Fig. 4.7d) can be represented by equation (3). The $psf(x, y)$, f_0, φ_i, and $im_i(x, y)$ represent the point spread function (PSF) of the microscope (Fig. 4.7b), the spatial frequency of the illumination pattern, the phase of the ith illumination pattern and the corresponding diffraction-limited fluorescence image (Fig. 4.7e), respectively. The Fourier transform of Eq. 4.3, after plugging in $I_i(x, y) = 1 + \cos(2\pi f_0 x + \varphi)_i$, becomes Eq. 4.4. It clearly shows that both the un-shifted diffraction-limited spatial frequency information of the object ($F_{\text{obj}}(f_x f_y) \cdot \text{OTF}$,

Fig. 4.7g), and the shifted high-spatial-frequency information of the object ($F_{\text{obj}}(f_x - f_0 f_y) \cdot \text{OTF}$ Fig. 4.7h, and $F_{\text{obj}}(f_x + f_0 f_y) \cdot \text{OTF}$, Fig. 4.7i), are encoded in $im_i(xy)$ due to the structured illumination. To extract them, diffraction-limited images taken under three laterally shifted interference fringe illuminations are needed. Equation 4.5 shows a matrix representation of the Fourier transform of the three diffraction-limited images, $im_i\,(x, y)$, $i = 1, 2, 3$. Through simple matrix inversion, $F_{\text{obj}}\left(f_x, f_y\right) \cdot \text{OTF}$, $F_{\text{obj}}(f_x - f_0 f_y) \cdot \text{OTF}$, and $F_{\text{obj}}(f_x + f_0 f_y) \cdot \text{OTF}$ could all be solved, as shown by Eq. 4.6. To achieve a 2D resolution improvement, this process should be conducted for at least horizontal and vertical interference orientations, and the extracted high-spatial-frequency information should be shifted to its original location and assembled together with the diffraction-limited information of the object (summation shown in Fig. 4.7j). Finally, by performing inverse Fourier transform to the assembled extended spatial frequency of the object, a super-resolution image could be reconstructed, as shown in Eq. 4.7 and Fig. 4.7f.

$$im_i(x, y) = [obj(x, y) \cdot I_i(x, y)] * psf(x, y) \tag{4.3}$$

$$\begin{aligned} F_{im,i}(f_x, f_y) = [&F_{\text{obj}}(f_x, f_y) \cdot \text{OTF} + 0.5F_{\text{obj}}(f_x + f_0, f_y)e^{-j\phi} \cdot \text{OTF} \\ &+0.5F_{\text{obj}}(f_x - f_0, f_y)e^{j\phi} \cdot \text{OTF}] \end{aligned} \tag{4.4}$$

$$\begin{pmatrix} F_{im,1}(f_x, f_y) \\ F_{im,2}(f_x, f_y) \\ F_{im,3}(f_x, f_y) \end{pmatrix} = \begin{pmatrix} 1 & 0.5e^{-\phi_1} & 0.5e^{\phi_1} \\ 1 & 0.5e^{-\phi_2} & 0.5e^{\phi_2} \\ 1 & 0.5e^{-\phi_3} & 0.5e^{\phi_3} \end{pmatrix}$$

$$\begin{pmatrix} F_{\text{obj}}(f_x, f_y) \cdot \text{OTF} \\ F_{\text{obj}}(f_x + f_0, f_y) \cdot \text{OTF} \\ F_{\text{obj}}(f_x - f_0, f_y) \cdot \text{OTF} \end{pmatrix} = A \begin{pmatrix} F_{\text{obj}}(f_x, f_y) \cdot \text{OTF} \\ F_{\text{obj}}(f_x + f_0, f_y) \cdot \text{OTF} \\ F_{\text{obj}}(f_x - f_0, f_y) \cdot \text{OTF} \end{pmatrix} \tag{4.5}$$

$$\begin{pmatrix} F_{\text{obj}}(f_x, f_y) \cdot \text{OTF} \\ F_{\text{obj}}(f_x + f_0, f_y) \cdot \text{OTF} \\ F_{\text{obj}}(f_x - f_0, f_y) \cdot \text{OTF} \end{pmatrix} = A^{-1} \begin{pmatrix} F_{im,1}(f_x, F_y) \\ F_{im,2}(f_x, F_y) \\ F_{im,3}(f_x, F_y) \end{pmatrix} \tag{4.6}$$

$$\begin{aligned} I_{re} = F^{-1}[&F_{\text{obj}}(f_x, f_y) \cdot \text{OTF} + F_{\text{obj}}(f_x, f_y)\text{OTF}(f_x - f_0, f_y) \\ &+F_{\text{obj}}(f_x, f_y) \cdot \text{OTF}(f_x + f_0, f_y) + F_{\text{obj}}(f_x, f_y) \cdot \text{OTF}(f_x - f_y, f_0) \\ &+F_{\text{obj}}(f_x, f_y) \cdot \text{OTF}(f_x, f_y + f_0)] \end{aligned} \tag{4.7}$$

Besides the aforementioned algorithm, an iterative algorithm, termed "blind-SIM" [50], can also be used for image reconstruction. The blind-SIM method is a minimization algorithm that utilizes multiple images taken under different illuminations to estimate the fluorescent object without prior knowledge of the precise illumination patterns. For the SIM case, by minimizing the cost function (Eq. 4.8) under the positivity constraint of obj, $I_{i=1...5}$ and the $\sum_{i=1}^{6} I_i = 6I_0$ constraint (I_0 is a constant matrix) [63], the algorithm jointly recovers the object (obj) and the illumination patterns ($I_{i=1...5}$). In Eq. 4.8, the $im_{i=1...6}$ are the six raw images taken under laterally translated 1D SPI illuminations oriented along both x and y directions and $\|\,\|$ represents the Euclidean norm. Starting from initial guesses (such as the diffraction-limited image of the object and uniform illumination), the obj and $I_{i=1...5}$ can be solved iteratively through the conjugate gradient method and the inexact line search, with the workflow shown in Fig. 4.8.

$$F(obj, I_{i=1,...,5}) = \sum_{i=1}^{5} \|im_i - (obj \cdot I_i) * \text{psf}\|^2 + \left\| im_6 - \left(obj \cdot \left(6I_0 - \sum_{i=1}^{5} I_i\right)\right) * psf \right\|^2 \quad (4.8)$$

Since the Fourier space algorithm directly solves the high-spatial-frequency information of the object and reconstructs the super-resolution images through inverse Fourier transform, it takes less computational effort to calculate the final image compared to the iterative blind-SIM algorithm. However, the illumination patterns have to be sinusoidal and perfectly periodic. Any distortions in the illumination patterns could generate artifacts in the reconstructed images. On the contrary, although the iterative blind-SIM algorithm takes more computational time, it can recover a super-resolution image of the object even when there are distortions in the illuminations [50]. For a general form of blind-SIM, the illuminations could be any patterns as long as the summation of all the illumination patterns roughly equals a uniform intensity distribution, such as a series of random speckle patterns. Therefore, the blind-SIM algorithm significantly increases the illumination quality tolerance, which is clearly beneficial in practice.

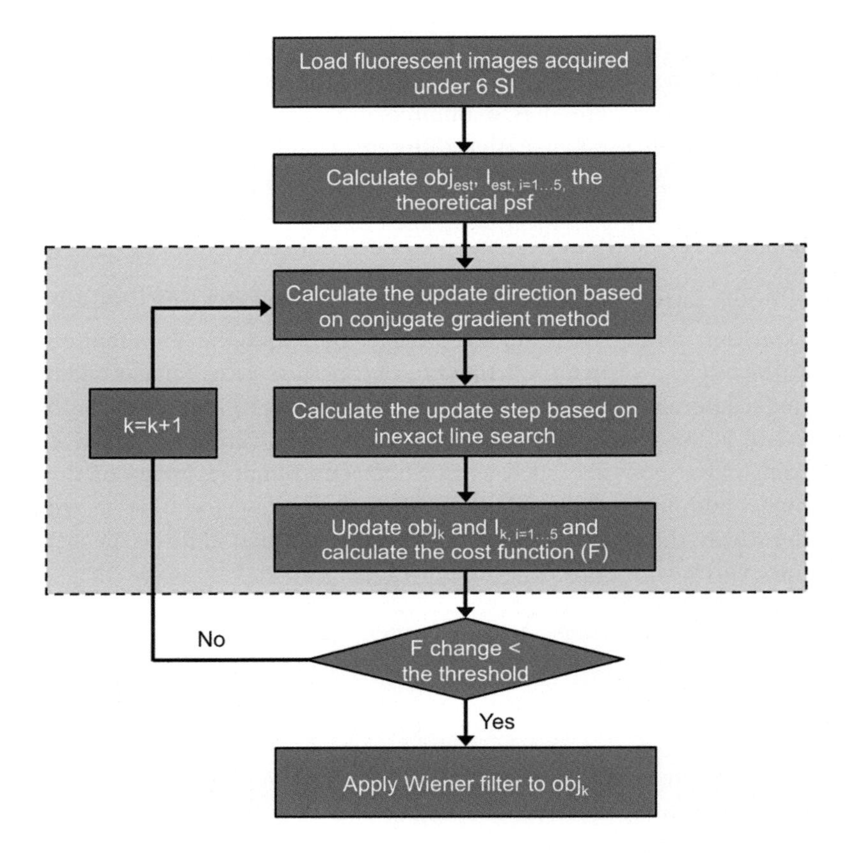

Figure 4.8 The workflow chart of the iterative blind-SIM algorithm.

4.4 PSIM Demonstration

4.4.1 Numerical Demonstration of PSIM

To demonstrate the imaging capability, Wei et al. provided two specific designs for PSIM purpose, utilizing thick-metal-film/dielectric and thin-metal-film/dielectric structures [43]. In both designs, a Ag film was chosen as the plasmonic material because it supports SPs across the entire visible spectrum and has relatively low loss [43].

For the thick-metal-film/dielectric structure proposed, edges were used as the coupling structure to bridge the momentum

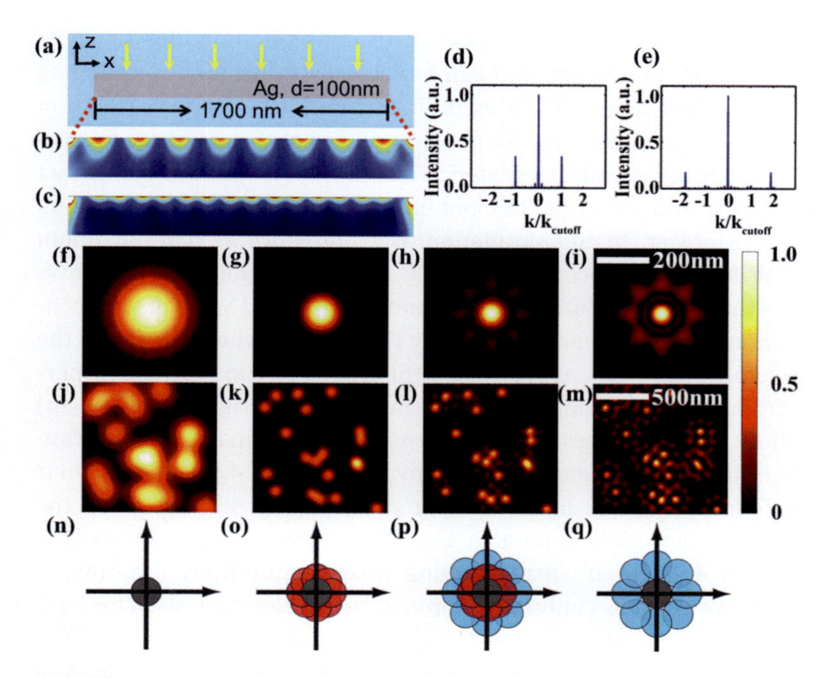

Figure 4.9 (a–e) The SPI simulation and characterization. (a) Cross-sectional view of the simulation structure, (b, c) SPI patterns excited by 563 and 390 nm illumination and (d, e) their corresponding Fourier spectrum. (f–q) PSIM demonstration using quantum dots with broadband absorption and 580 nm emission. The NA of the objective is 1.4. (f–i) PSF comparison. PSF of (f) epifluorescence microscopy, (g) PSIM with 563 nm, (h) PSIM with both 563 and 390 nm, and (i) PSIM with 390 nm excitation light. (j–m) The corresponding images of randomly distributed quantum dots and (n–q) Fourier space representations. In (n–q), the gray circles correspond to the diffraction-limited low-spatial-frequency information of the object; the red and blue circles correspond to the high-spatial-frequency information acquired due to the SPI illumination excited by 563 and 390 nm light, respectively.

mismatch between the illumination photons and the supported SPs [43]. Light with wavelengths of 390 nm and 563 nm was used to illuminate the slab of Ag film and excite the SPI patterns (Fig. 4.9a–c). Their corresponding Fourier transforms (Fig. 4.9d,e) have peaks around $1\mathbf{k}_{\text{cutoff}}$ and $2\mathbf{k}_{\text{cutoff}}$, where $\mathbf{k}_{\text{cutoff}}$ is the cutoff $\mathbf{k}$-vector of the objective at the fluorescence emission wavelength (580 nm).

These SPI patterns can be used as structured illumination to excite the fluorescent object, and can be laterally translated by changing the incident angle of the light. To reconstruct a super-resolution image, for each SPI orientation, a sequence of three images formed under three laterally translated SPI illuminations was used to extract both the diffraction-limited and high-spatial-frequency information of the object. In the simulation, the full-width at half-maximum (FWHM) of the reconstructed PSF of the PSIM microscopy shows a $\sim$threefold resolution improvement compared with that of conventional fluorescence microscopy (Fig. 4.9f–i), which surpasses the twofold improvement provided by SIM. As shown in the imaging performance simulation, the reconstructed PSIM images (Fig. 4.9k–m) show significant resolution improvements compared to the conventional fluorescence microscopy image (Fig. 4.9j) due to their extended coverage in the Fourier domain (Fig. 4.9o–q). Since the radii of the covered area in Fig. 4.9q is about the same as that in Fig. 4.9p, their corresponding image resolutions are similar. Moreover, because only one illumination wavelength is used for Figs. 4.9i and 4.9m, their image speed is twice as fast as that for Figs. 4.9h and 4.9l [43]. The reconstruction artifacts shown in Figs. 4.9i and 4.9m are caused by the missing spatial frequency information in Fourier space and could be reduced with additional postprocessing algorithms [43].

Because the SPs supported by thin-metal-film/dielectric structures possess even larger **k**-vector, Wei et al. also proposed to use thin plasmonic structures for PSIM to further improve the resolution [43]. Figure 4.10a shows the structure of a unit cell of the designed plasmonic structure, in which a patterned Cr layer is used to excite the two SP modes supported by the thin Ag film at the same time. As a result, an interference pattern containing multiple spatial frequencies, shown in Fig. 4.10a, is formed [51]. Its corresponding Fourier transform (Fig. 4.10b) shows peaks around $1\mathbf{k}_{\text{cutoff}}$, $2\mathbf{k}_{\text{cutoff}}$, and $3\mathbf{k}_{\text{cutoff}}$ (emission wavelength 508 nm), which match the first three theoretical peaks of the interference of the supported two SP modes. For this specific design, the intensity of the fourth peak is weak and therefore its contribution can be ignored. Figure 4.10c–j show the imaging performance simulations using the SPI formed by this plasmonic structure as well as the corresponding Fourier space

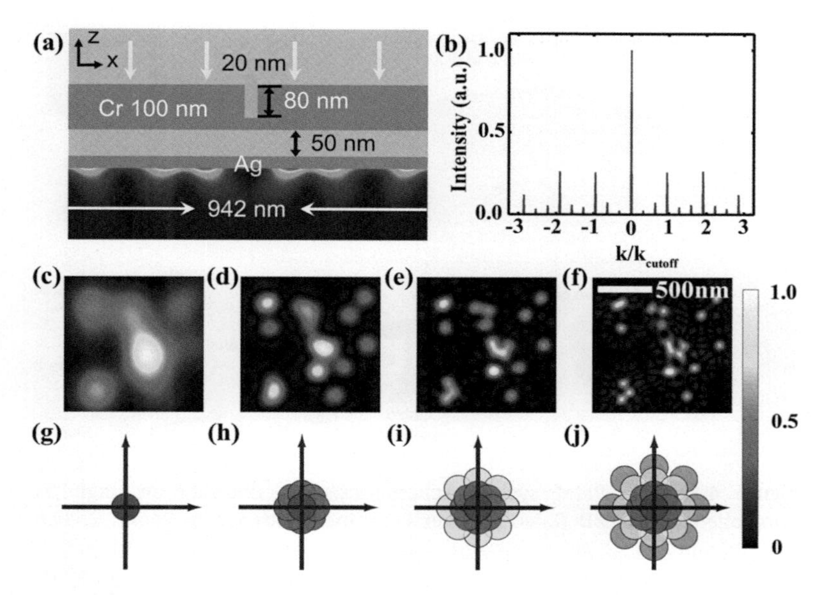

Figure 4.10 (a) Cross section of the simulated structure and the corresponding SPI pattern formed under 442 nm light illumination and (b) its corresponding Fourier transform, showing the first 3 orders of spatial frequency of the SPI pattern. (c–j) PSIM demonstration using fluorescent particles with 508 nm emission and objective with 0.85 NA. (c) Traditional epifluorescence microscopy image, (d–f) reconstructed PSIM images, and (g–j) corresponding Fourier space representations. Circles with red, yellow, and blue colors represent the information obtained due to different spatial frequency components of the illumination patterns.

representations. A comparison of the FWHM of the PSF of PSIM and conventional fluorescence microscopy shows a ~fourfold resolution improvement.

4.4.2 Experimental Demonstration of PSIM

Later, Wei et al. also provided a proof-of-concept experimental demonstration of the PSIM technique [52]. They built a PSIM imaging system (Fig. 4.11) to generate SPI patterns on top of a plasmonic substrate for fluorescence dye excitation and projected the emitted fluorescence signal onto an electron-multiplying charge-coupled device (EMCCD) using an objective combined with a tube

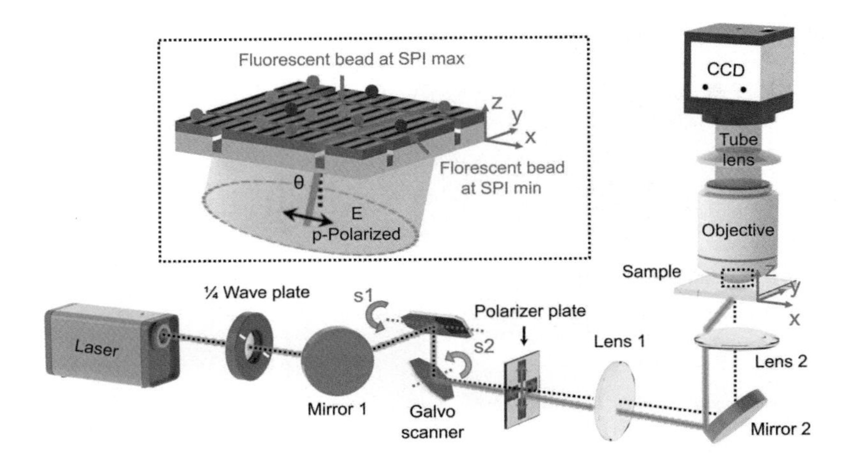

Figure 4.11 (a) PSIM system schematics. The illumination angle is controlled by the 2D Galvo scanner combined with lenses 1 and 2. The sample and the s2 mirror placed at the focal plane of lens 2 and lens 1, respectively. Inset: magnified view of the PSIM substrate with fluorescence objects placed atop.

lens. In their experiments, the designed plasmonic substrate was patterned with 2D periodic subwavelength slits, serving as the SP wave coupling structure. When illuminated by a laser beam (Fig. 4.11, inset), two counterpropagating SP waves are excited by the adjacent slits and form an interference pattern in the middle. The lateral shift of the SPI is realized by tuning the illumination angle with the illumination module, shown in the bottom part of Fig. 4.11.

The plasmonic structure shown in Fig. 4.12a–d was fabricated using standard nanofabrication techniques. A relatively thick (250 nm) Ag film patterned with a narrow-slit array (period: 7.6 μm; width: 100 nm) was used for exciting SPI and blocking the direct transmission of the illumination laser. To confirm the generation and lateral translation of the SPI, 100 nm diameter fluorescent beads were placed on top of the plasmonic structure, serving as the SPI probes, as shown by the schematics in Fig. 4.12e. A single bead's fluorescence intensity variation with respect to the relative illumination angle in *xz* and *yz* planes was recorded (Fig. 4.12f,g, red dots) and fitted by sinusoidal functions (Fig. 4.12f,g, solid blue lines),

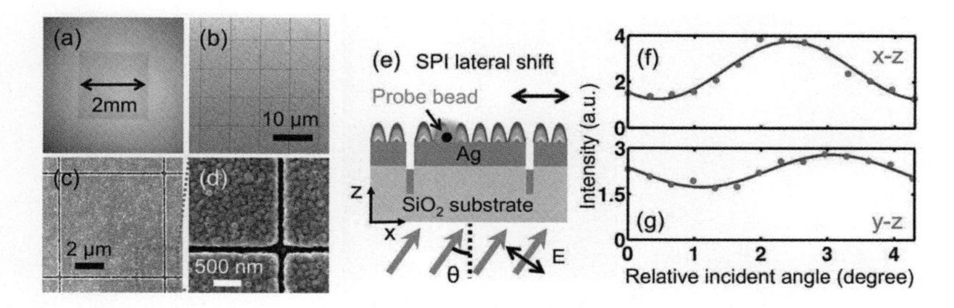

Figure 4.12 (a) Bright-field optical image of a large-area patterned PSIM substrate; (b–d) SEM images of (b) a patterned slit array, (c) one unit cell, and (d) the slit region; (e) schematic illustration of controlling the lateral translation of the SPI with the incident angle; and (f, g) fluorescence intensity of a single fluorescence bead (excitation/emission 540 nm/560 nm) versus the relative incident angle in (f) *xz* and (g) *yz* planes, with the experimental data and the sinusoidal fit represented by red dots and solid blue lines, respectively.

which proves that the lateral translation of the SPI can be controlled by the illumination angle.

Based on six diffraction-limited images, taken under laterally translated SPI illumination along the x and the y direction, respectively, the super-resolved image was reconstructed. The super-resolution image of three beads (Fig. 4.13b) has a much higher resolution compared to the conventional fluorescence image in Fig. 4.13a. Moreover, their corresponding Fourier spectra clearly show that the high-spatial-frequency information inaccessible by conventional fluorescence microscopy was recovered in PSIM. The ratio of the FWHM (Fig. 4.13f) of a single bead shows a factor of $\sim$2.6, which matches the theoretical enhancement factor estimated from the wavevector of the SP mode and the NA of the objective (NA = 1.0).

To show the wide-field imaging capability, they also provided a comparison of the conventional fluorescence image (Fig. 4.14a), its deconvolution (Fig. 4.14b), and the PSIM image (Fig. 4.14c) over a large area [52]. With the help of SPI illumination, closely located beads that couldn't be resolved in the conventional fluorescence images (Fig. 4.14d,h) and their deconvolution images (Fig. 4.14e,i) are resolved in the reconstructed PSIM images (Fig. 4.14f,j). The resolution improvement is also confirmed by the corresponding

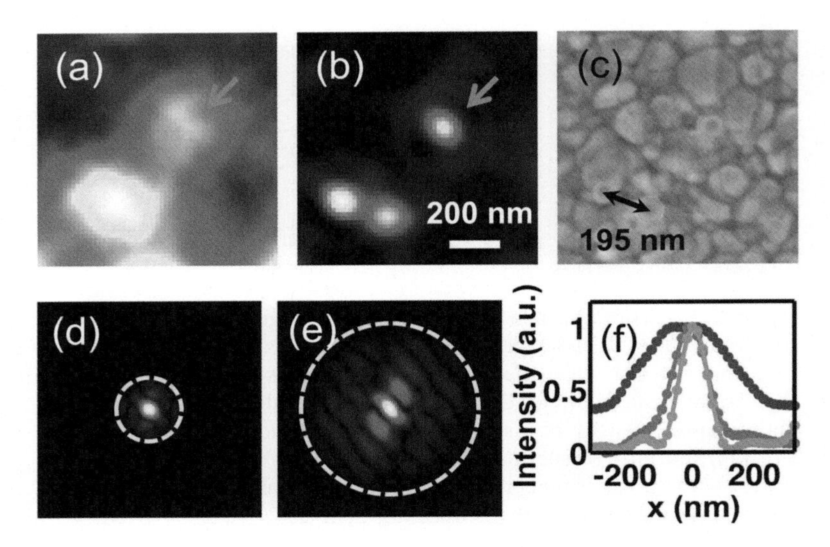

Figure 4.13 PSIM resolution demonstration using 100 nm (diameter) beads (excitation/emission 540 nm/560 nm). (a) Conventional fluorescence image, (b) PSIM image, and (c) SEM image with beads highlighted in red. (d, e) Corresponding Fourier spectra, with the yellow dashed circles representing the maximum-accessible spatial frequency for the fluorescence microscope and the PSIM system, respectively. (f) Comparison of the fluorescence intensity cross section of a single bead (green arrows), with the red, green, and blue lines representing those of the PSIM image along two perpendicular directions and the conventional fluorescence image, respectively.

intensity cross-sectional comparison shown in Fig. 4.14l,m, in which only the intensity cross section of the PSIM images (red curves) shows two peaks with separation, for instance, 122 nm in Fig. 4.14l, smaller than the Abbe limit of fluorescence microscopy ($\sim$290 nm) and traditional SIM microscopy ($\sim$140 nm), respectively.

4.4.3 Discussion

Because of the smaller period of SPI compared to laser interference, the demonstrated PSIM is capable of achieving resolution higher than that of SIM. For PSIM, its resolution can be calculated by $\lambda_{\text{emission}}/(2\text{NA} + 2\text{NA}_{\text{eff}})$, with NA_{eff} representing $\mathbf{k}_{\text{sp}}/\mathbf{k}_{\text{emission}}$ ($\mathbf{k}_{\text{sp}}$: wavevector of the SP; $\mathbf{k}_{\text{emission}}$: wavevector of the emission light).

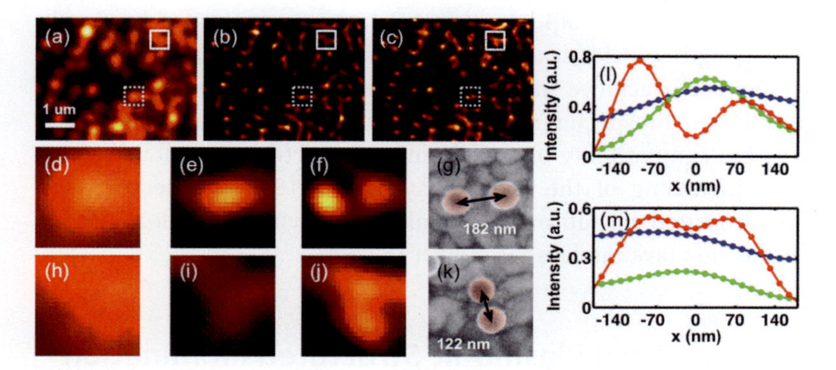

Figure 4.14 PSIM wide-field super-resolution demonstration by using closely located beads. (a) Conventional fluorescence, (b) the corresponding deconvolution, and (c) the reconstructed PSIM images. (d–f, h–j) Magnified images within the white dashed and white solid boxes in (a–c). (g, k) Corresponding SEM characterization (beads highlighted in red) and (l, m) fluorescence intensity cross-sectional comparison, with the blue, green, and red lines representing those of the conventional fluorescence image, the deconvolution, and the PSIM image, respectively.

With proper material design [23, 29], a very large $\mathbf{k}_{sp}$ could be engineered to further increase the NA_{eff} and eventually the PSIM resolving power. In conventional SIM, both the field of view and the optical resolution are determined by the collection objective. However, in PSIM the field of view and optical resolution are mainly determined by objective and NA_{eff} of the plasmonic substrate, respectively. This provides a very unique opportunity for PSIM to cope with a large field of view and high spatial resolution at the same time.

Moreover, PSIM is also capable of providing a high signal-to-noise ratio (SNR) by confining the fluorescence dye excitation to the interface region. Because of the exponential decay of the SP field, the emission of the fluorescence dye located far from the interface is negligible. Therefore, PSIM is especially suitable for examining subdiffraction-limited objects that locate close to the interface due to its super-resolution and background elimination capability.

In Wei et al.'s proof-of-concept demonstration, a slit array was used as the coupling structure to extend the imaging area to the whole patterned region. Since the PSIM is a wide-field imaging technique, and the SPI lateral translation was controlled by simply changing the incident angle of the illumination light, PSIM has the potential to achieve high imaging speed and be used for super-resolution biospecimen dynamics studies. Moreover, to increase the bio-compatibility of the plasmonic structure as well as reduce the quenching of the dyes, a layer of 5–10 nm dielectrics can be deposited on top of the metal film, serving as both a surface protection layer and a spacer layer [53, 54].

4.5 Localized Plasmonic Structured Illumination Microscopy

A recently developed alternative approach to plasmonics-based super-resolution imaging is localized plasmonic structured illumination microscopy (LPSIM) [55]. By designing a plasmonic substrate which uses localized plasmons to create the fluorescence excitation pattern, some limitations of propagating SPs can be avoided. Figure 4.15 shows the expanded resolution potentially attainable using this LPSIM method. In this section, a discussion of the physical principles of LPSIM will be given, followed by a numerical demonstration of the imaging performance of the technique.

The key advantage of LPSIM lies in the geometry-bound nature of localized plasmons. Localized plasmon resonances can

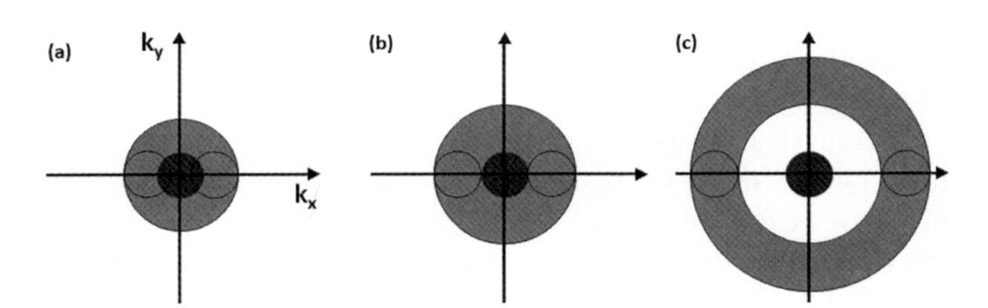

Figure 4.15 A k-space illustration of resolving power. The dark central circle represents the spatial frequencies detectable with a diffraction-limited imaging setup. (a) OTF of the PSIM technique and (b, c) two potential LPSIM schemes using excitation patterns with large **k**-vectors, dramatically enhancing resolving power.

be physically understood as collective electron charge oscillations within conductive materials, excited by incident electromagnetic radiation. These phenomena have been studied at length and have found practical use in a wide range of fields. Thanks to recent advances in nanoscale fabrication, generating nanoscale plasmonic patterns has become a practical tool. With LPSIM, a 2D array of localized plasmon antennas, metal particle array, generate a periodic field pattern which serves as the structured illumination for the imaged object, as shown in Fig. 4.16. These field patterns, unlike those of SIM or PSIM, can have almost arbitrary spatial frequency,

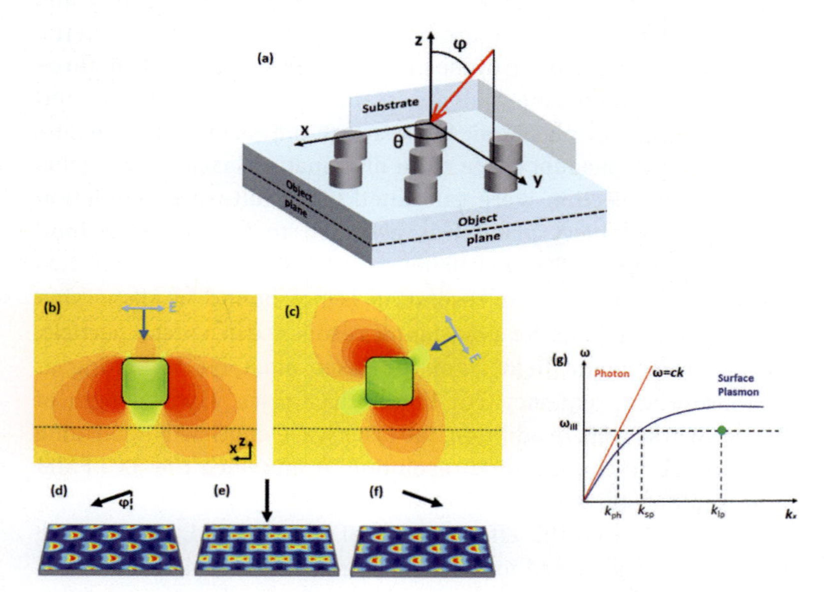

Figure 4.16 Proposed imaging scheme. (a) Schematic for LPSIM pattern translation. The red arrow represents p-polarized laser light, directed toward the plasmonic substrate at varying angles to create near-field excitation patterns a short distance behind the antennas, in the object plane. (b, c) Field strength around a nanodisc excited by laser light at incident angles of $\theta = 0^\circ$, $\varphi = 0^\circ$ and 60°, respectively. (d–f) Object plane patterns for incident angles $\theta = 0^\circ$ and $\varphi = 60^\circ$, 0°, and -60°, respectively, as used in reconstruction. (g) Dispersion advantage of LPSIM over existing methods. SIM and PSIM are limited by the red and blue dispersion curves, respectively. For LPSIM, the position of the green dot shown is determined only by the antenna geometry.

dependent primarily on fabrication capability. So, the spatial imaging resolution obtainable with LPSIM is not fundamentally limited by free-space or propagating plasmon dispersion relations. This means k-space information can be collected from arbitrary regions based on the localized plasmon antenna array design. As an initial demonstration, we design our antenna array geometry for a 3X resolution improvement over diffraction-limited standard microscopy.

The nanoantenna design shown in this section calls for silver discs 60 nm in diameter and 60 nm in height. They are spaced 150 nm apart (center to center) and embedded in a glass substrate. The hexagonal lattice structure has threefold rotational symmetry, which allows for scanning of the plasmonic excitation fields in three orientations so as to collect all necessary Fourier space information for a 2D image. The plasmonic field pattern can easily be changed by controlling the incident angle of an illuminating laser beam. Fields shown in this section were calculated using full-wave simulation software. The object plane was assumed to be 40 nm behind the antenna array. The wavelength of this laser matters for two reasons, even though our resolution is not bound by dispersion relations. Firstly, to excite a plasmonic resonance in a metal particle, the electromagnetic field from the laser must be oscillating at an appropriate frequency. For these silver discs, a wavelength of 405 nm works well. In addition, the laser wavelength will determine the initial diffraction-limited resolution, from which the 3X LPSIM improvement starts.

To demonstrate the effectiveness of this method, the imaging performance was tested for a single point particle, a distribution of quantum dots, and a nonsparse object with varying feature sizes. These fluorescent objects were placed against the LPSIM substrate, where they were subjected to the near-field patterns generated by the localized plasmon excitations. For these simulations, the fluorescent emission wavelength was chosen to be 430 nm. The field intensity was assumed to be in the linear regime of the fluorophore molecules. It was further assumed that the objects themselves would not dramatically distort the plasmonic field patterns. The fields are roughly translated along three directions in the object plane by changing the laser's incident angle. The recorded image is captured in the far field, taking into account a standard diffraction-limited Airy disc PSF. The NA of our objective optics was taken to be 1.4, as is common in oil immersion objectives. Thus, the high-k super-resolution information must be decoded and incorporated into the final images during the reconstruction step, in the same way as it is done with PSIM.

As shown in Fig. 4.17, the FWHM and two-point separation ability was tested using the LPSIM method. For this simulation, a small 5 nm point particle was used as an object. Shown in Fig. 4.17a, the standard, diffraction-limited image of a single quantum dot is effectively the PSF of a standard microscopy system. Figure 4.17b, in comparison, shows the PSF of the LPSIM technique, which reveals a major improvement in resolution. Some asymmetry is introduced, in the form of weak artifacts surrounding the central spot. This is a result of some small gaps in the overall collected k-space information, as shown in Fig. 4.17c. These artifacts can be dealt with by some simple additional postprocessing steps, if desired. The profile of the LPSIM PSF is reduced threefold relative to the diffraction limit, for an FWHM of just 52 nm. This is a dramatic improvement that surpasses the performance of both SIM and PSIM. No deconvolution is used to enhance this result, although in many relevant specific cases, deconvolution or other data processing of the LPSIM image can yield an additional boost to the image quality.

Another test of resolving power is satisfaction of the Rayleigh criteria. These can be approximated by finding the distance at which two point particles are separated by a 30% dip in intensity. As shown in Fig. 4.17e, two point particles spaced 51 nm apart are totally unresolvable with a diffraction-limited setup. When LPSIM is applied, the two objects are resolved, with a 30% dip, as expected. This confirms the achievement of a threefold improvement over the standard image resolution.

Although point particles are useful test objects for specifying the FWHM, most practical objects of interest to biologists and others are significantly more complex. For this reason, we also examined results for a distribution of many quantum dots, including several that are clustered together. Figure 4.18a shows the standard image of such an object. In Fig. 4.18b, the reconstructed LPSIM image is shown. By a simple normalization of the collected image's frequency

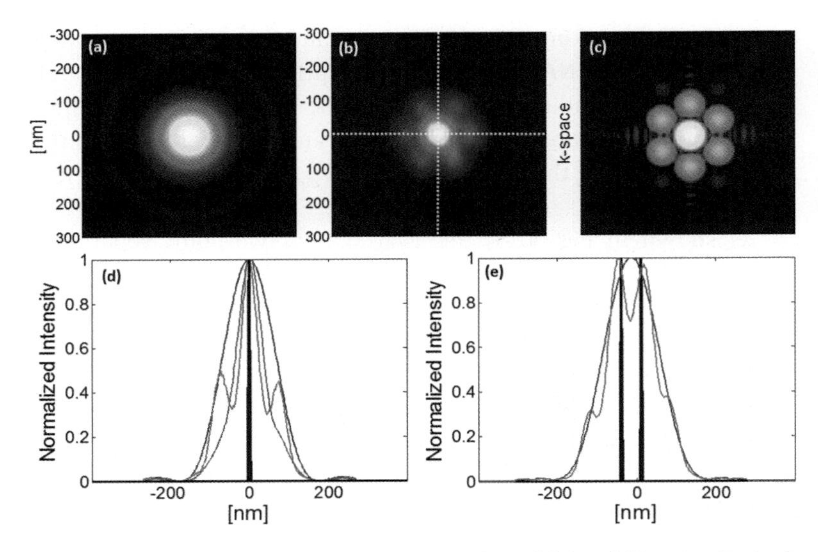

Figure 4.17 Resolution characterization. PSFs of (a) a diffraction-limited system and (b) the LPSIM. (c) Expanded OTF after LPSIM reconstruction. Note that the six lobes surrounding the central region are high-frequency information captured by LPSIM. (d) FWHM comparison of diffraction-limited (blue) and LPSIM (red) techniques for fluorescent emission at 430 nm. The green curve shows the FWHM taken along the vertical line in (b). (e) Two-point resolution. For two 5 nm quantum dots placed 51 nm apart, a standard setup (blue) will not resolve the points, but with LPSIM (red_, the typical Rayleigh criteria of a 30% intensity dip are satisfied.

domain magnitude to a generally expected magnitude envelope, a clean, highly resolved image is obtained, shown in Fig. 4.18c. By examination, it can easily be seen that several areas with unresolved clusters of quantum dots can now be resolved with LPSIM as separate individual objects.

The quantum dot distribution is more general than a single quantum dot, but it is still a sparse object. Another nonsparse converging stripe object (Fig. 4.18d) was tested. Figure 4.18e and Fig. 4.18f show the diffraction-limited and LPSIM imaging results for this object, respectively. There is a large unresolved area in the center of the diffraction-limited image, because the stripe features are too small to capture in the far field. However, with LPSIM, even without any additional postprocessing, we see the converging

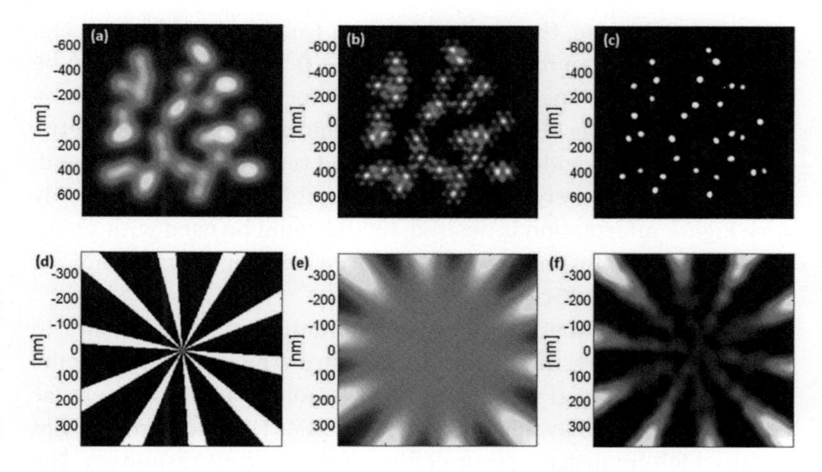

Figure 4.18 Imaging results. (a) Diffraction-limited image of a distribution of 5 nm wide fluorescent quantum dots. (b) LPSIM image of the same object. Previously unresolvable points are resolved. (c) Additional Fourier-based deconvolution combined with LPSIM yields a clean, highly resolved image. (d) A solid striped object. (e) Diffraction-limited image of (d). (f) LPSIM image of (d). The stripes are now resolvable much further in toward the center of the image.

stripes are resolved deep into the center of the object, once again in agreement with the expected threefold resolution enhancement.

As is the case with any imaging method, LPSIM has its relative strengths and weaknesses. Silver as a choice of plasmonic material has drawbacks due to oxidation issues. Gold will not oxidize, but it exhibits its strongest plasmonic behavior at longer wavelengths, making it an inefficient material to use with a 405 nm laser source. Although LPSIM images are recorded in the far field, they rely on the near-field evanescent excitations to induce fluorescent emission. Therefore, any object of interest must be within approximately 100 nm of the LPSIM substrate in order for the near field to have sufficient strength.

There are significant advantages to LPSIM that should be noted. The resonant nature of the localized plasmon fields means there will be enhancement of the fluorescent signal, which can improve visibility of faint objects or potentially be used to lower exposure times and increase imaging speed. Because LPSIM is not a point-scanning technique, the frame rate for a full, wide-field image can be much faster than methods such as STED. This technique can be scaled to different operating wavelengths for a given application by simply changing the fluorophore, plasmonic medium, and geometry, as necessary, to provide the same threefold resolution improvement. In certain cases, where there are constraints on an object and only very high-k information is desired, LPSIM could be used with a very small geometry to resolve features even smaller than 50 nm, as suggested in Fig. 4.15c.

Work on the experimental implementation of LPSIM is sure to yield interesting results in the near future. The numerical results shown here demonstrate a potential for resolution improvement far beyond what is achievable with standard SIM. This method does not require high-intensity light or nonlinear fluorescent responses. The example objects imaged in this section consistently show threefold resolution improvement over the diffraction limit. In the future, LPSIM may find useful application in the fields of biological and medical research, especially among researchers interested in taking live videos of subcellular dynamics.

4.6 Perspective and Outlook

The resolution enhancement capability of PSIM and LPSIM is ultimately limited by the achievable wavevectors of the system, which could reach several times of $\mathbf{k}_0$ through plasmonic mode engineering and fine plasmonic structure fabrication. However, it should also be noticed that multiple illumination pattern frequencies are needed if the desired resolving power of PSIM/LPSIM is more than three times as compared with that of conventional fluorescence microscopy. Otherwise, a missing band of information in the Fourier space is introduced, as schematically shown in Fig. 4.15c. Therefore, samples supporting multiple SP modes, or multiple interference angles, are needed to cover both the high- and the middle-range-spatial-frequency information of the object through PSIM. Or schemes that combine SIM/PSIM, or SIM/LPSIM, or PSIM/LPSIM, or SIM/PSIM/LPSIM to cover a large-frequency domain without any missing band. Therefore, by appropriate combination of SIM/PSIM/LPSIM techniques, the achievable resolution enhancement factor compared to conventional light microscopy could be way beyond 3X. As a consequence, more subimages should be acquired for extracting all the necessary spatial information of the object for proper super-resolution image reconstruction, resulting in longer image acquisition time. Although the basic super-resolution concept for PSIM and LPSIM has been proved, future experiments need to demonstrate its unique capability to obtain super-resolution and high imaging speed at the same time. In addition, the applications of these new tools in the field of biological and biomedical imaging should be further explored.

Acknowledgments

This work was supported by the Gordon and Betty Moore Foundation and the National Science Foundation (NSF) Electrical, Communications and Cyber Systems (ECCS; Grant No. 0969405). The authors thank Carl Zeiss for providing the loaned microscope and thank L. Wang for useful discussions on numerical reconstruction algorithms.

References

1. Murphy, D. B. (2001). *Fundamentals of Light Microscopy and Electronic Imaging* (John Wiley & Sons, Hoboken, NJ).
2. Lipson, A., Lipson, S. G., and Lipson, H. (2011). *Optical Physics* (Cambridge University Press, New York, NY).
3. Pohl, D. W., Denk, W., and Lanz, M. (1984). Optical stethoscopy: image recording with resolution $\lambda/20$, *Appl. Phys. Lett.*, **44**, pp. 651–653.
4. Hell, S. W., and Wichmann, J. (1994). Breaking the diffraction resolution limit by stimulated emission: stimulated-emission-depletion fluorescence microscopy, *Opt. Lett.*, **19**(11), pp. 780–782.
5. Dyba, M., and Hell, S. W. (2002). Focal spots of size lambda/23 open up farfield fluorescence microscopy at 33 nm axial resolution, *Phys. Rev. Lett.*, **88**(16), p. 163901.

6. Willig, K. I., Rizzoli, S. O., Westphal, V., Jahn, R., and Hell, S. W. (2006). STED microscopy reveals that synaptotagmin remains clustered after synaptic vesicle exocytosis, *Nature*, **440**(7086), pp. 935–939.
7. Westphal, V., Rizzoli, S. O., Lauterbach, M. A., Kamin, D., Jahn, R., and Hell, S. W. (2008). Video-rate far-field optical nanoscopy dissects synaptic vesicle movement, *Science*, **320**(5873), pp. 246–249.
8. Rust, M. J., Bates, M., and Zhuang, X. (2006). Sub-diffraction-limit imaging by stochastic optical reconstruction microscopy (STORM), *Nat. Methods*, **3**(10), pp. 793–795.
9. Huang, B., Wang, W., Bates, M., and Zhuang, X. (2008). Three-dimensional superresolution imaging by stochastic optical reconstruction microscopy, *Science*, **319**(5864), pp. 810–813.
10. Huang, B., Jones, S. A., Brandenburg, B., and Zhuang, X. (2008). Whole-cell 3D STORM reveals interactions between cellular structures with nanometerscale resolution, *Nat. Methods*, **5**(12), pp. 1047–1052.
11. Zhu, L., Zhang, W., Elnatan, D., and Huang, B. (2012). Faster STORM using compressed sensing, *Nat. Methods*, **9**(7), pp. 721–723.
12. Betzig, E., Patterson, G. H., Sougrat, R., Lindwasser, O. W., Olenych, S., Bonifacino, J. S., Davidson, M. W., Lippincott-Schwartz, J., and Hess, H. F. (2006). Imaging intracellular fluorescent proteins at nanometer resolution, *Science*, **313**(5793), pp. 1642–1645.
13. Manley, S., Gillette, J. M., Patterson, G. H., Shroff, H., Hess, H. F., Betzig, E., and Lippincott-Schwartz, J. (2008). High-density mapping of single-molecule trajectories with photoactivated localization microscopy, *Nat. Methods*, **5**(2), pp. 155–157.
14. Heintzmann, R., and Cremer, C. (1999). Laterally modulated excitation microscopy: improvement of resolution by using a diffraction grating, *Proc. SPIE*, **3568**, pp. 185–196.
15. Gustafsson, M. G. (2000). Surpassing the lateral resolution limit by a factor of two using structured illumination microscopy, *J. Microsc.*, **198**(Pt 2), pp. 82–87.
16. Schermelleh, L., Carlton, P. M., Haase, S., Shao, L., Winoto, L., Kner, P., Burke, B., Cardoso, M. C., Agard, D. A., Gustafsson, M. G., Leonhardt, H., and Sedat, J. W. (2008). Subdiffraction multicolor imaging of the nuclear periphery with 3D structured illumination microscopy, *Science*, **320**(5881), pp. 1332–1336.
17. Kner, P., Chhun, B. B., Griffis, E. R., Winoto, L., and Gustafsson, M. G. (2009). Super-resolution video microscopy of live cells by structured illumination, *Nat. Methods*, **6**(5), pp. 339–342.

18. Heintzmann, R., Jovin, T. M., and Cremer, C. (2002). Saturated patterned excitationmicroscopy–a concept for optical resolution improvement, *J. Opt. Soc. Am. A*, **19**(8), pp. 1599–1609.

19. Gustafsson, M. G. (2005). Nonlinear structured-illumination microscopy: wide-field fluorescence imaging with theoretically unlimited resolution, *Proc. Natl. Acad. Sci. U.S.A.*, **102**(37), pp. 13081–13086.

20. Raether, H. (1988). *Surface Plasmons on Smooth and Rough Surfaces and on Gratings* (Springer-Verlag, Berlin Heidelberg, Germany).

21. Luo, X., and Ishihara, T. (2004). Surface plasmon resonant interference nanolithography technique, *Appl. Phys. Lett.*, **84**, p. 4780.

22. Liu, Z.-W., Wei, Q.-H., and Zhang, X. (2005). Surface plasmon interference nanolithography, *Nano Lett.*, **5**, pp. 957–961.

23. Xiong, Y., Liu, Z., and Zhang, X. (2008). Projecting deep-subwavelength patterns from diffraction-limited masks using metal-dielectric multilayers, *Appl. Phys. Lett.*, **93**(11), p. 111116.

24. Liu, Z., Wang, Y., Yao, J., Lee, H., Srituravanich, W., and Zhang, X. (2009). Broad band two-dimensional manipulation of surface plasmons, *Nano. Lett.*, **9**, pp. 462–466.

25. Shao, D. B., and Chen, S. C. (2006). Direct patterning of three-dimensional periodic nanostructures by surface-plasmon-assisted nanolithography, *Nano Lett.*, **6**(10), pp. 2279–2283.

26. Liu, Z., Steele, J. M., Lee, H., and Zhang, X. (2006). Tuning the focus of a plasmonic lens by the incident angle, *Appl. Phys. Lett.*, **88**(17), p. 171108.

27. Durant, S., Liu, Z., Steele, J. M., and Zhang, X. (2006). Theory of the transmission properties of an optical far-field superlens for imaging beyond the diffraction limit, *J. Opt. Soc. Am. B*, **23**, pp. 2383–2392.

28. Liu, Z., Durant, S., Lee, H., Pikus, Y., Fang, N., Xiong, Y., Sun, C., and Zhang, X. (2007). Far-field optical superlens, *Nano Lett.*, **7**, pp. 403–408.

29. Xiong, Y., Liu, Z., Sun, C., and Zhang, X. (2007). Two-dimensional imaging by far-field superlens at visible wavelengths, *Nano Lett.*, **7**, pp. 3360–3365.

30. Salandrino, A., and Engheta, N. (2006). Far-field subdiffraction optical microscopy using metamaterial crystals: theory and simulations, *Phys. Rev. B*, **74**(7), p. 075103.

31. Jacob, Z., Alekseyev, L. V., and Narimanov, E. (2006). Optical hyperlens: far-field imaging beyond the diffraction limit, *Opt. Express*, **14**, pp. 8247–8256.

32. Liu, Z., Lee, H., Xiong, Y., Sun, C., and Zhang, X. (2007). Far-field optical hyperlens magnifying sub-diffraction-limited objects, *Science*, **315**, p. 1686.
33. Ma, C., and Liu, Z. (2010). Focusing light into deep subwavelength using metamaterial immersion lenses, *Opt. Express*, **18**, p. 4838.
34. Pile, D. F. P., and Gramotnev, D. K. (2005). Plasmonic subwavelength waveguides: next to zero losses at sharp bends, *Opt. Lett.*, **30**, pp. 1186–1188.
35. Bozhevolnyi, S. I., Volkov, V. S., Devaux, E., Laluet, J.-Y., and Ebbesen, T. W. (2006). Channel plasmon subwavelength waveguide components including interferometers and ring resonators, *Nature*, **440**, pp. 508–511.
36. Yao, J., Liu, Z., Liu, Y., Wang, Y., Sun, C., Bartal, G., Stacy, A. M., and Zhang, X. (2008). Optical negative refraction in bulk metamaterials of nanowires, *Science*, **321**, p. 930.
37. Lerosey, G., de Rosny, J., Tourin, A., Fink, M. (2007). Focusing beyond the diffraction limit with far-field time reversal, *Science*, **315**, 1120–1122.
38. Lemoult, F., Lerosey, G., de Rosny, J., and Fink, M. (2010). Resonant metalenses for breaking the diffraction barrier, *Phys. Rev. Lett.*, **104**(20), p. 203901.
39. Lemoult, F., Fink, M., and Lerosey, G. (2012). A polychromatic approach to far-field superlensing at visible wavelengths, *Nat. Commun.*, **3**, p. 889.
40. Lu, D., and Liu, Z. (2012). Hyperlenses and metalenses for far-field super-resolution imaging, *Nat. Commun.*, **3**, p. 1205.
41. Sentenac, A., Chaumet, P. C., and Belkebir, K. (2006). Beyond the Rayleigh criterion: grating assisted far-field optical diffraction tomography, *Phys. Rev. Lett.*, **97**(24), p. 243901.
42. Chung, E., Kim, Y.-H., Tang, W. T., Sheppard, C. J., and So, P. T. (2009). Wide-field extended-resolution fluorescence microscopy with standing surface-plasmon-resonance waves, *Opt. Lett.*, **34**, pp. 2366–2368.
43. Wei, F., and Liu, Z. (2010). Plasmonic structured illumination microscopy, *Nano Lett.*, **10**, pp. 2531–2536.
44. Wang, Q., Bu, J., Tan, P. S., Yuan, G. H., Teng, J. H., Wang, H., and Yuan, X. C. (2012). Subwavelength-sized plasmonic structures for wide-field optical microscopic imaging with super-resolution, *Plasmonics*, **7**, pp. 427–433.
45. Gjonaj, B., Aulbach, J., Johnson, P. M., Mosk, A. P., Kuipers, L., and Lagendijk, A. (2011). Active spatial control of plasmonic fields, *Nat. Photonics*, **5**, 360–363.

46. Gjonaj, B., Aulbach, J., Johnson, P. M., Mosk, A. P., Kuipers, L., and Lagendijk, A. (2013). Focusing and scanning microscopy with propagating surface plasmons, *Phys. Rev. Lett.*, **110**(26), p. 266804.
47. Frohn, J. T., Knapp, H. F., and Stemmer, A. (2000). True optical resolution beyond the Rayleigh limit achieved by standing wave illumination, *Proc. Natl. Acad. Sci. U.S.A.*, **97**, pp. 7232–7236.
48. Sentenac, A., Belkebir, K., Giovannini, H., and Chaumet, P. C. (2009). High-resolution total-internal-reflection fluorescence microscopy using periodically nanostructured glass slides, *J. Opt. Soc. Am. A*, **26**, p. 2550.
49. Girard, J., Scherrer, G., Cattoni, A., Le Moal, E., Talneau, A., Cluzel, B., de Fornel, F., and Sentenac, A. (2012). Far-field optical control of a movable subdiffraction light grid, *Phys. Rev. Lett.*, **109**(18), p. 187404.
50. Mudry, E., Belkebir, K., Girard, J., Savatier, J., Le Moal, E., Nicoletti, C., Allain, M., and Sentenac, A. (2012). Structured illumination microscopy using unknown speckle patterns, *Nat. Photonics*, **6**, pp. 312–315.
51. Yao, J., Liu, Y., Liu, Z., Sun, C., and Zhang, X. (2006). Surface plasmon beats formed on thin metal films, *Proc. SPIE*, **6323**, pp. 63231K.1–63231K.9.
52. Wei, F., Lu, D., Shen, H., Wan, W., Ponsetto, J. L., Huang, E., and Liu, Z. (2014). Wide field super-resolution surface imaging through plasmonic structured illumination microscopy, *Nano Lett.*, **14**, pp. 4634—4639.
53. Weber, W. H., and Eagen, C. F. (1979). Energy transfer from an excited dye molecule to the surface plasmons of an adjacent metal, *Opt. Lett.*, **4**, pp. 236–238.
54. Barnes, W. L. (1998). Fluorescence near interfaces: the role of photonic mode density, *J. Mod. Opt.*, **45**, pp. 661–699.
55. Ponsetto, J. L., Wei, F., and Liu, Z. (2014). Localized plasmon assisted structured illumination microscopy for wide-field high-speed dispersion-independent super resolution imaging, *Nanoscale*, **6**, pp. 5807–5812.

Chapter 5

Optical Super-Resolution Imaging Using Surface Plasmon Polaritons

Igor Smolyaninov
Department of Electrical and Computer Engineering, University of Maryland, College Park, MD 20742, USA
smoly@umd.edu

Over the past century the resolution of conventional optical microscopes, which rely on optical waves that propagate into the far field, has been limited because of diffraction to a value on the order of a half-wavelength ($\lambda_0/2$) of the light used. Although immersion microscopes had slightly improved resolution, on the order of $\lambda_0/2n$, the increased resolution was limited by the small range of refractive indices n of available transparent materials. We are experiencing quick demolition of the diffraction limit in optical microscopy. In the last few years numerous nonlinear optical microscopy techniques based on photoswitching and saturation of fluorescence have demonstrated far-field resolution of 20 to 30 nm. In a parallel development, recent progress in metamaterials has demonstrated that artificial optical media can be created, whose resolution is not determined by the conventional diffraction limit. The resolution of linear immersion microscopes based on such metamaterials is only limited by losses, which can be minimized

Plasmonics and Super-Resolution Imaging
Edited by Zhaowei Liu

ISBN 978-981-4669-91-7 (Hardcover), 978-1-315-20653-0 (eBook)
www.panstanford.com

by appropriate selection of the constituents of the metamaterials used and by the wavelength(s) used for imaging. It is also feasible to compensate for losses by adding gain to the structure. Thus, optical microscopy is quickly moving toward a 10 nm resolution scale, which should bring about numerous revolutionary advances in lithography and imaging.

5.1 Introduction

Optical microscopy is one of the oldest research tools. Its development began in about 1590 with the observation by the Dutch spectacle maker Zaccharias Janssen and his son Hans that a combination of lenses in a tube made small objects appear larger. In 1609 Galileo Galilei improved on their ideas and developed an occhiolino, or compound microscope, with a convex and a concave lens. The acknowledged father of microscopy is, however, Anton van Leeuwenhoek (1632–1723), who developed improved grinding and polishing techniques for making short-focal-length lenses and was the first person to consequently see bacteria, protozoa, and blood cells.

Although various electron and scanning probe microscopes have long surpassed the compound optical microscope in resolving power, optical microscopy remains invaluable in many fields of science. The practical limit to the resolution of an optical microscope is determined by diffraction: a wave cannot be localized to a region much smaller than half of its vacuum wavelength $\lambda_0/2$. Immersion microscopes introduced by Abbé in the 19th century have slightly improved resolution, on the order of $\lambda_0/2$, because of the shorter wavelength of light, λ_0/n, in a medium with refractive index n. However, immersion microscopes are limited by the small range of refractive indices n of available transparent materials. For a while it was believed that the only way to achieve nanometer-scale spatial resolution in an optical microscope was to detect evanescent optical waves in very close proximity to a studied sample using a near-field scanning optical microscope (NSOM) [1]. Although many fascinating results are being obtained with NSOM, such microscopes are not as versatile and convenient to use as regular far-field optical microscopes. For example, an image from a near-field optical microscope is obtained by point-by-point scanning, which is an indirect and a rather slow process, and can be affected by artifacts of the sample.

Over the past few years two major developments in optical microscopy have circumvented the resolution barrier of conventional diffraction. The first one makes use of nonlinear optics. A comprehensive review of developments in nonlinear microscopy has been published very recently by Hell [2]. Broadly speaking, these techniques rely on photoswitching and/or saturation of fluorescence from individual molecules. They demonstrate far-field resolution of 20 to 30 nm, which is limited by light collection. A parallel, yet revolutionary development in linear optical microscopy was inspired by a seminal paper by Veselago [3], who examined the properties of materials with relative permittivity (ε_r) and relative permeability (μ_r) that were simultaneously both negative—so-called double-negative materials. In such materials the refractive index is $n = \sqrt{\mu_r \varepsilon_r}$, where for double-negative materials the negative square root must be taken. Pendry pointed out that such negative-refractive-index materials could allow the fabrication of a perfect lens [4], and his discussion of the use of metal films as media for making a practical approximation of such media has stimulated extraordinary progress in the optics of metamaterials. In this context, metamaterials are composite structures of metals and dielectrics that mimic, at least partially, the double-negative characteristics of an ideal perfect lens. The way that they affect the propagation of waves can be controlled so that the dependence on μ_r is eliminated. In Pendry's implementation of a flat perfect lens made from an artificial negative-refractive-index (meta-) material, a high-resolution optical image could be obtained by amplifying the evanescent waves that decay away in conventional far-field imagery, but must be retained to obtain a high-resolution image. In essence the relative permittivity should be tuned as close as possible to $\varepsilon_r = -1$, which is the condition for surface plasmon polaritons (SPPs) to exist at the surface of the metal. In its simplest form an SPP is an electromagnetic excitation that propagates in a wave-like fashion along the planar interface between a metal and a dielectric medium, often vacuum, and whose amplitude decays exponentially with increasing distance into each medium from the interface. Thus, an SPP is a surface electromagnetic wave whose electromagnetic field is confined to the near vicinity of the dielectric–metal interface. This confinement leads to an enhancement of the electromagnetic field at the interface, resulting in an extraordinary sensitivity of SPPs to surface conditions. In practice, all metals, or metamaterials, exhibit some loss so that the best that can be achieved is $\varepsilon_r = -1 - j\varepsilon''$, where ε'' is the part of the relative permittivity that contributes to loss. This ultimately limits the resolving power of a microscope based on such SPP-based perfect lenses [5]. In Pendry's originally proposed implementation of a perfect lens the image appears in the image plane with subdiffraction-limited resolution, but this image is accessible only if it is recorded photolithographically in the image plane or examined with an auxiliary near-field microscope. Indeed, imaging of this kind was reported in 2005 in two independent experiments [6, 7]. Nevertheless, until recently this technique was limited by the fact that the magnification of the planar superlens is equal to 1.

5.2 Surface Plasmon Microscopy

An important early step to overcome this limitation was made in surface plasmon–assisted microscopy experiments [8], in which 2D image magnification was achieved. In this microscope design the dispersion behavior of SPPs propagating in the boundary between a thin metal film and a dielectric was exploited to use the 2D optics of SPPs with a short wavelength to produce a magnified local image of an object on the surface. If the dispersion curve of SPPs on gold is examined, an excitation wavelength that provides a small group velocity gives rise to a 2D SPP diffraction limit that is on the order of $\lambda_{SPP} = \lambda/n_g$. If a 2D mirror structure is fabricated in the surface, in the case of [8] using parabolic droplets on the surface (see Fig. 5.1b), then SPPs propagating in the surface reflect or scatter at the boundaries of an object placed on the surface. These reflected or scattered SPPs are then imaged by the surface structure to produce a magnified 2D image. This magnified image can be examined by a far-field microscope by using light scattered from surface roughness or from lithographically generated surface structures that scatter propagating SPPs in to far-field radiation. The increased spatial resolution of microscopy experiments performed with SPPs [9] is based on the hyperbolic dispersion law of such waves, which may be written in the form

$$\mathbf{k}_{xy}^2 - |\mathbf{k}_z|^2 = \frac{\varepsilon_d \omega^2}{c^2} \tag{5.1}$$

where ε_d is the dielectric constant of the medium bounding the metal surface, which for air is $\varepsilon_d = 1$, $\mathbf{k}_{xy} = \mathbf{k}_p$ is the wavevector component in the plane of propagation, and $\mathbf{k}_z$ is the wavevector component perpendicular to the plane. This form of the dispersion relation originates from the exponential decay of the surface wave field away from the propagation plane. The effective negative group refractive index (n_g) behavior of surface plasmons was also shown to play a very important role in these early experiments [10].

Various new geometries exhibiting image magnification beyond the usual diffraction limit have been proposed theoretically [11–13], which make use of newly developed optical metamaterials. For example, in the optical hyperlens design described by Jacob et al. [12] an optical metamaterial made of a concentric arrangement of metal and dielectric cylinders may be characterized by a strongly anisotropic dielectric permittivity tensor in which the tangential ε_θ and the radial ε_r components have opposite signs. The resulting hyperbolic dispersion relation

$$\frac{\mathbf{k}_r^2}{\varepsilon_\theta} - \frac{\mathbf{k}_\theta^2}{|\varepsilon_r|} = \frac{\omega^2}{c^2} \tag{5.2}$$

does not exhibit any lower limit on the wavelength of propagating light at a given frequency. Therefore, in a manner similar to the 2D optics of SPPs, there is no usual diffraction limit in this metamaterial medium. Abbe's resolution limit simply does not exist. Optical energy propagates through such a metamaterial in the form of radial rays. Moreover, it is easy to demonstrate that a pattern of polymethyl methacrylate (PMMA) stripes formed on a metal surface (as shown in Fig. 5.1a) behaves as a 2D plasmonic equivalent of 3D hyperbolic metamaterial.

Let us consider an SPP wave which propagates over a flat metal–dielectric interface. The electromagnetic field of an SPP at a dielectric–metal interface is obtained from the solution of Maxwell's equations in each medium, and the associated boundary conditions. The latter express the continuity of the tangential components of the electric and magnetic fields across the interface, and the vanishing of these fields infinitely far from the interface. If the metal film is thick, the SPP wavevector is defined by expression

$$\mathbf{k}_{\mathrm{p}} = \frac{\omega}{c}\left(\frac{\varepsilon_{\mathrm{d}}\varepsilon_{\mathrm{m}}}{\varepsilon_{\mathrm{d}} + \varepsilon_{\mathrm{m}}}\right)^{1/2} \tag{5.3}$$

where $\varepsilon_{\mathrm{m}}(\omega)$ and $\varepsilon_{\mathrm{d}}(\omega)$ are the frequency-dependent dielectric constants of the metal and the dielectric, respectively [14]. Let us introduce an effective 2D dielectric constant $\varepsilon_{2\mathrm{D}}$ such that $\mathbf{k}_{\mathrm{p}} = \varepsilon_{2\mathrm{D}}^{1/2}\omega/c$, and thus

$$\varepsilon_{2\mathrm{D}} = \left(\frac{\varepsilon_{\mathrm{d}}\varepsilon_{\mathrm{m}}}{\varepsilon_{\mathrm{d}} + \varepsilon_{\mathrm{m}}}\right) \tag{5.4}$$

Now it is easy to see that depending on the frequency, SPPs perceive the dielectric material bounding the metal surface in drastically different ways. At low frequencies $\varepsilon_{2\mathrm{D}} \approx \varepsilon_{\mathrm{d}}$. Therefore, plasmons perceive a PMMA stripe as a dielectric. On the other hand, at high enough frequencies around $\lambda_0 \approx 500$ nm, $\varepsilon_{2\mathrm{D}}$ changes sign and becomes negative since $\varepsilon_{\mathrm{d}}(\omega) > -\varepsilon_{\mathrm{m}}(\omega)$. As a result, around $\lambda_0 \approx 500$ nm plasmons perceive PMMA stripes on gold as if they are metallic layers, while gold/vacuum portions of the interface are perceived as dielectric layers. Thus, at these frequencies plasmons perceive a PMMA stripe pattern from Fig. 5.1b as a layered 3D hyperbolic metamaterial. This qualitative picture is supported by numerical calculations of real and imaginary parts of the surface plasmon wavevector at the gold–PMMA and gold–vacuum interfaces as a function of frequency presented in Fig. 5.2. In the frequency range marked by the box, PMMA has a negative effective dielectric constant, as perceived by plasmons, while the gold–vacuum interface looks like a medium with a positive dielectric constant. Moreover, a rigorous theoretical description of the PMMA-based plasmonic metamaterials developed in Ref. [15] produces a similar description.

Thus, two modes of operation of a 2D plasmonic microscope may be implemented, as shown in Fig. 5.1a,b. A plasmon microscope may be operated in the hyperlens mode (Fig. 5.1a) in which the plasmons

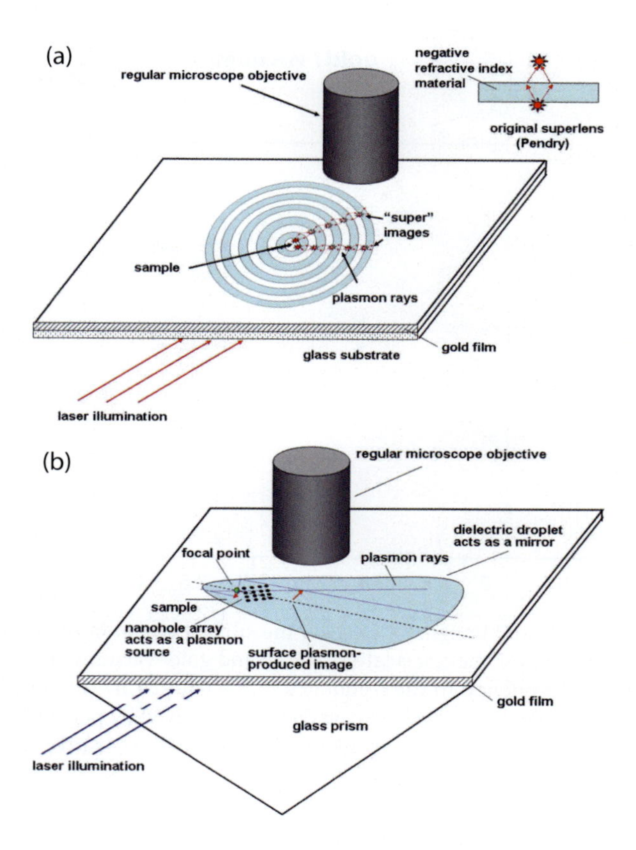

Figure 5.1 Two modes of operation of a 2D plasmonic microscope. (a) Plasmon microscope operating in the hyperlens mode: the plasmons generated by the sample located in the center of the hyperlens propagate in the radial direction. The lateral distance between plasmonic rays grows with distance along the radius. The images are viewed by a regular microscope. (b) Plasmon microscope operating in the geometrical optics mode: nanohole array illuminated by external laser light acts as a source of surface plasmons, which are emitted in all directions. Upon interaction with the sample positioned near the focal point of the parabolically shaped dielectric droplet, and reflection off the droplet edge, the plasmons form a magnified planar image of the sample. The image is viewed by a regular microscope. The droplet edge acts as an efficient plasmon mirror because of total internal reflection.

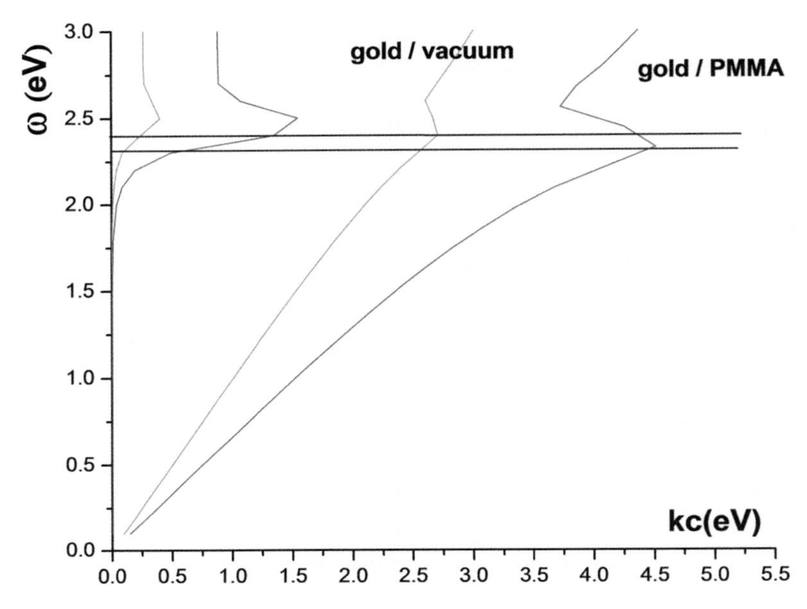

Figure 5.2 Real and imaginary parts of the surface plasmon wavevector at the gold–polymethyl methacrylate (PMMA) and gold–vacuum interfaces as a function of frequency. In the frequency range marked by the box, PMMA has a negative effective dielectric constant, as perceived by plasmons, while the gold–vacuum interface looks like a medium with a positive dielectric constant.

generated by the sample located in the center of the plasmonic hyperlens propagate in the radial direction. The lateral distance between plasmonic rays grows with distance along the radius. The images are viewed by a regular microscope. Alternatively, a 2D plasmon microscope may be operated in the geometrical optics mode, as shown in Fig. 5.1b. A nanohole array illuminated by external laser light may act as a source of surface plasmons, which are emitted in all directions. Upon interaction with the sample positioned near the focal point of the parabolically-shaped dielectric droplet, and reflection off the droplet edge, the plasmons form a magnified planar image of the sample. The image is viewed by a regular microscope. The droplet edge acts as an efficient plasmon mirror because of total internal reflection. It appears that

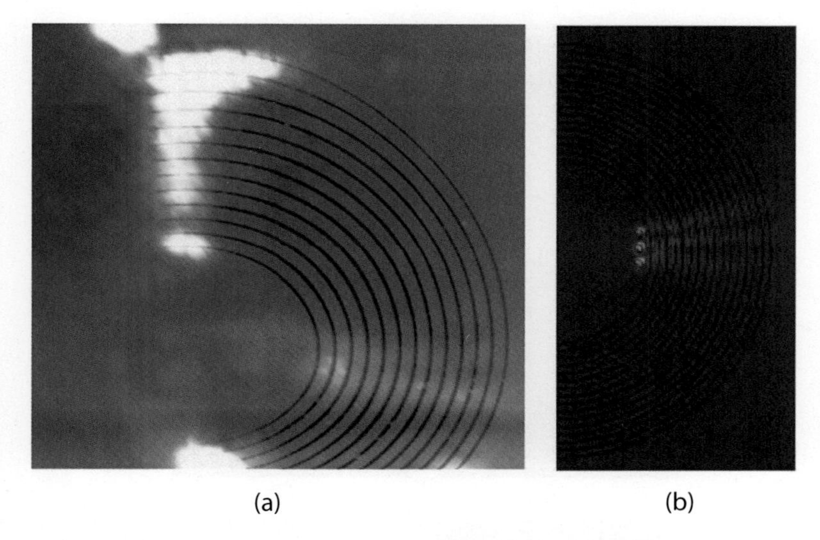

Figure 5.3 (a) Superposition image composed of an AFM image of PMMA on a gold plasmonic metamaterial structure superimposed onto the corresponding optical image obtained using a conventional optical microscope, illustrating the imaging mechanism of the magnifying hyperlens [16]. Near the edge of the hyperlens the separation of three rays is large enough to be resolved using a conventional optical microscope. (b) Theoretical simulation of ray propagation in the magnifying hyperlens microscope. From Ref. [16]. Reprinted with permission from AAAS.

both modes of operation exhibit strong evidence of optical super-resolution.

5.3 The Surface Plasmon Hyperlens

First, let us review imaging results obtained in the hyperlens mode of the 2D plasmonic microscope. The internal structure of the magnifying hyperlens (Fig. 5.4a) consists of concentric rings of PMMA deposited on a gold film surface. The required concentric structures were defined using a Raith e-line electron beam lithography (EBL) system with $\sim$70 nm spatial resolutions. The written structures were subsequently developed using a 3:1 IPA/MIBK solution (Microchem) as the developer and imaged using atomic

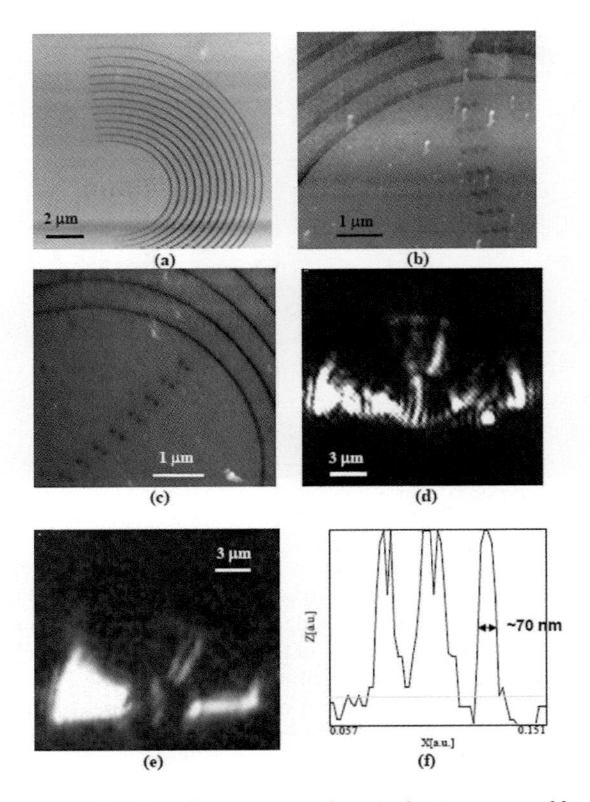

Figure 5.4 AFM (a–c) and conventional optical microscope (d, e) images of the resolution test samples composed of three (a, b) and two (c) rows of PMMA dots positioned near the center of the magnifying hyperlens. The conventional microscope images presented in (d) and (e) correspond to the samples shown in (b) and (c), respectively. The rows of PMMA dots give rise to either three or two divergent plasmon rays, which are visible in the conventional optical microscope images. (f) Cross section of the optical image along the line shown in (d) indicates a resolution of at least 70 nm or $\sim\lambda/7$. From Ref. [16]. Reprinted with permission from AAAS.

force microscopy (AFM) (see Fig. 5.4a). According to theoretical proposals in Refs. [12, 13], optical energy propagates through a hyperbolic metamaterial in the form of radial rays. This behavior is clearly demonstrated in Fig. 5.3b. If point sources are located near the inner rim of the concentric metamaterial structure, the lateral separation of the rays radiated from these sources increases upon propagation toward the outer rim. Therefore, resolution of an immersion microscope based on such a metamaterial structure is defined by the ratio of inner to outer radii. Resolution appears limited only by losses, which can be compensated by optical gain.

Following these theoretical ideas, magnifying superlenses (or hyperlenses) have been independently realized in two experiments [16, 17]. In particular, experimental data obtained using a 2D plasmonic hyperlens (shown in Fig. 5.3a) do indeed demonstrate ray-like propagation of subwavelength plasmonic beams emanated by test samples. A far-field optical resolution of at least 70 nm (see Fig. 5.4f) has been demonstrated using such a magnifying hyperlens based on a 2D plasmonic metamaterial design [16]. Rows of either two or three PMMA dots have been produced near the inner ring of the hyperlens (Fig. 5.4b,c). These rows of PMMA dots had 0.5 μm periodicity in the radial direction so that phase matching between the incident laser light and surface plasmons can be achieved. Upon illumination with an external laser, the three rows of PMMA dots in Fig. 5.4b gave rise to three divergent plasmon rays, which are clearly visible in the plasmon image in Fig. 5.4d obtained using a conventional optical microscope. The cross-sectional analysis of this image across the plasmon rays (Fig. 5.4f) indicates a resolution of at least 70 nm or $\sim\lambda/7$. The lateral separation between these rays increased by a factor of 10 as the rays reached the outer rim of the hyperlens. This increase allowed visualization of the triplet using a conventional microscope. In a similar fashion, two rows of PMMA dots shown in Fig. 5.4c gave rise to two plasmon rays, which are visualized in Fig. 5.4e.

The magnifying action and the imaging mechanism of the hyperlens have been further verified by control experiments presented in Fig. 5.5. The image shown in Fig. 5.5a presents results of two actual imaging experiments (top portion of Fig. 5.5a) performed simultaneously with four control experiments seen at the bottom of the same image. In these experiments, two rows of PMMA dots have been produced near the inner ring of the hyperlens structures seen at the top and at the bottom of Fig. 5.5a (the AFM image of the dots is seen in the inset). These rows of PMMA dots had 0.5 μm periodicity in the radial direction so that phase matching between the incident 515 nm laser light and surface plasmons can be achieved. On the

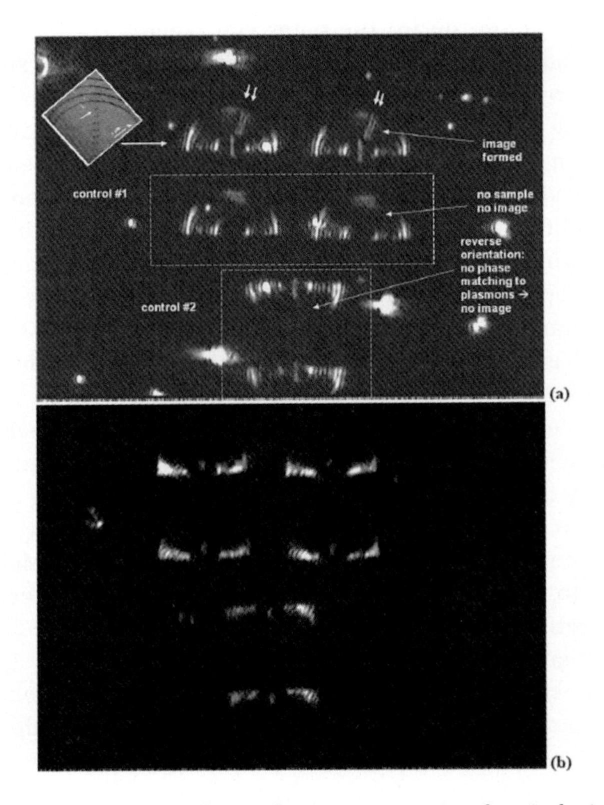

Figure 5.5 (a) This image obtained using a conventional optical microscope presents the results of two imaging experiments (top portion of the image) performed simultaneously with four control experiments seen at the bottom of the same image. The rows of PMMA dots shown in the inset AFM image were fabricated near the two top and two bottom hyperlenses. No such pattern was made near the two hyperlenses visible in the center of the image. Upon illumination with an external laser, the two rows of PMMA dots separated by a 130 nm gap gave rise to two divergent plasmon rays shown by the arrows, which are clearly visible in the top portion of the image. The four control hyperlenses visible at the bottom do not produce such rays, because there is no sample to image for the two hyperlenses in the center, and the two bottom hyperlenses are inverted. (b) The same pattern produced on an ITO instead of gold film demonstrates a pattern of ordinary light scattering by the structure without any hyperlens imaging effects.

other hand, no such PMMA dot structure was fabricated near the control hyperlenses seen in the center of Fig. 5.5a. Upon illumination with an external laser, the two rows of PMMA dots gave rise to the two divergent plasmon rays, which are clearly visible in the top portion of the image in Fig. 5.5a obtained using a conventional optical microscope. No such rays were observed in the four control hyperlenses visible in the bottom portion of the same image. There was no sample to image for the two hyperlenses located in the center of Fig. 5.5a. On the other hand, the PMMA dot structure was designed for phase-matched plasmon generation in the upward direction, as seen in the image. That is why no plasmon rays are visible when the hyperlens structures are inverted, as seen in the bottom of Fig. 5.5a. When the gold film was replaced with an ITO film in another control experiment performed using the same experimental geometry, no hyperlens imaging occurred since no surface plasmons are generated on ITO surface (see Fig. 5.5b). These experiments clearly verify the imaging mechanism and increased spatial resolution of the plasmonic hyperlens.

5.4 Surface Plasmon Microscope Operation in the Geometric Optics Mode

Operation of the 2D plasmonic microscope in the geometric optics mode (see Fig. 5.1b) leads to similar resolution. This result is natural since both modes of operation rely on the hyperbolic dispersion law—compare Eqs. 5.1 and 5.2. In our microscopy experiments the samples were immersed inside glycerin droplets on the gold film surface. The droplets were formed in desired locations by bringing a small probe (Fig. 5.6a) wetted in glycerin into close proximity to a sample. The probe was prepared from a tapered optical fiber, which has an epoxy microdroplet near its apex. Bringing the probe to a surface region covered with glycerin led to a glycerin microdroplet formation under the probe (Fig. 5.6b). The size of the glycerin droplet was determined by the size of the seed droplet of epoxy. The glycerin droplet under the probe can be moved to a desired location under the visual control, using a regular microscope. Our

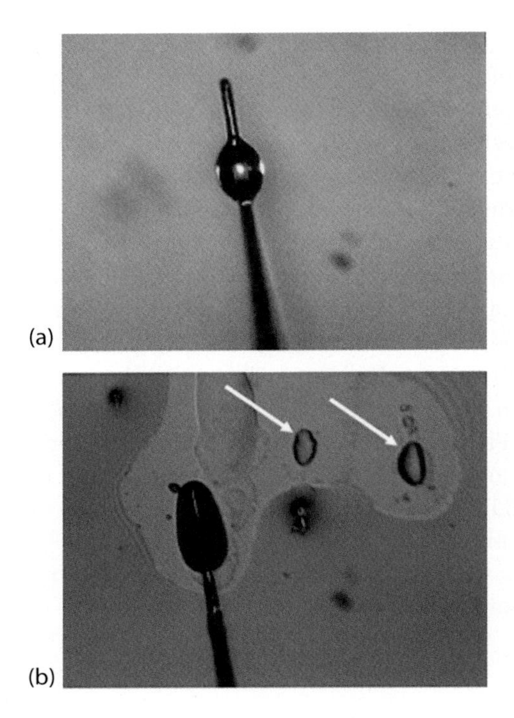

Figure 5.6 Glycerin droplets used in the geometric optics mode of an SPP microscope were formed in desired locations by bringing a small probe (a) wetted in glycerin into close proximity to a sample. The probe was prepared from a tapered optical fiber, which has an epoxy microdroplet near its apex. Bringing the probe to a surface region covered with glycerin led to glycerin microdroplet formation (b) under the probe in locations indicated by the arrows.

droplet deposition procedure allowed us to form droplet shapes, which were reasonably close to parabolic. In addition, the liquid droplet boundary may be expected to be rather smooth because of the surface tension, which is essential for the proper performance of the droplet boundary as a 2D plasmon mirror. Thus, the droplet boundary was used as an efficient 2D parabolic mirror for propagating surface plasmons excited inside the droplet by external laser illumination. Since the plasmon wavelength is much smaller than the droplet sizes, the image formation in such a mirror can be analyzed by simple geometrical optics in two dimensions.

The resolution test of the microscope has been performed using a 30 × 30 μm^2 array of triplet nanoholes (100 nm hole diameter with 40 nm distance between the hole edges) shown in Fig. 5.7c. This array was imaged using a glycerine droplet shown in Fig. 5.7a. Periodic nanohole arrays first studied by Ebbesen et al. [18] appear to be ideal test samples for the plasmon microscope. Illuminated by laser light, such arrays produce propagating surface waves, which explains the anomalous transmission of such arrays at optical frequencies. The image of the triplet array obtained at 515 nm using a 100× microscope objective is shown in Fig. 5.7b (compare it with an image in Fig. 5.7d calculated using 2D geometrical optics). Even though some discrepancy between the experimental and theoretical images can be seen (the image pattern observed in Fig. 5.7b looks convex looking from the left compared to the concave pattern observed in the calculation in Fig. 5.7d), the overall match between these images is impressive. The most probable reason for the observed convex/concave discrepancy is the fact that the droplet shape is not exactly parabolic, which produces some image aberrations. Although the expected resolution of the microscope at 515 nm is somewhat lower than at 502 nm, the 515 nm laser line is brighter, which allowed us to obtain more contrast in the 2D image. The least distorted part of the image, Fig. 5.7b (far from the droplet edge, yet close enough to the nanohole array, so surface plasmon decay does not affect resolution), is shown at higher digital zooms of the charge-coupled device (CCD) camera mounted onto our conventional optical microscope in Fig. 5.7e,f. These images clearly visualize the triplet nanohole structure of the sample. Moreover, using the experimentally measured point spread function (PSF) of the SPP microscope, resolution of 2D plasmon microscopy may be further improved to the ~30 nm scale (as shown in Fig. 5.10) by implementing digital resolution enhancement techniques [19].

The spatial resolution of the optical images (the PSF of the microscope) may be measured directly by calculating the cross-correlation $P{*}E$ between the optical image P and the scanning electron microscopy (SEM) image E of the same nanohole:

$$P^{*}E\,(r) = \int P(r_1)E(r_1 + r)dr_1 \tag{5.5}$$

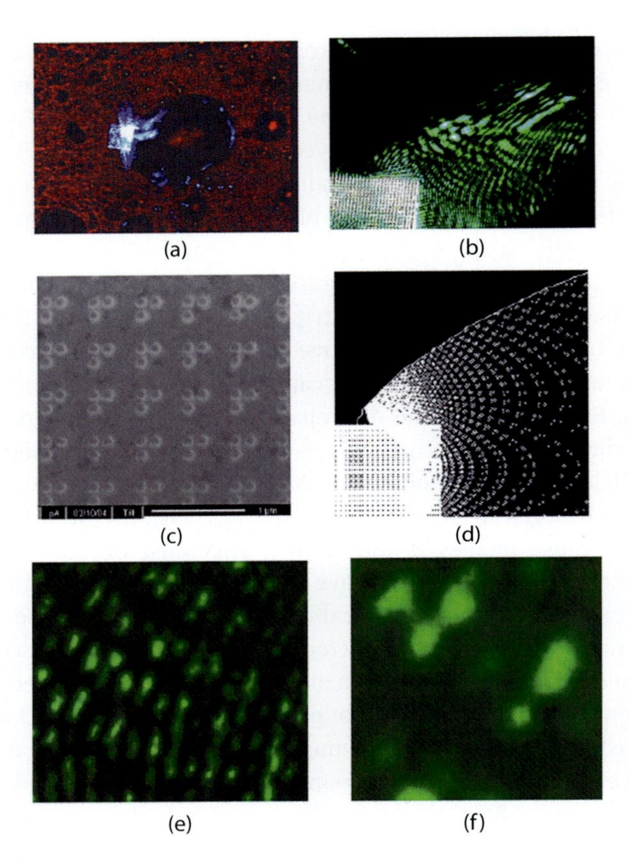

Figure 5.7 Resolution test of the 2D plasmonic microscope operated in the geometric optics mode. The array of triplet nanoholes (c) is imaged using a glycerine droplet shown in (a) using a 10× microscope objective. The image of the triplet array obtained using a 100× objective at 515 nm is shown in (b). The least distorted part of the image (b) is shown at higher digital zooms of the CCD camera mounted onto the microscope in (e) and (f). Comparison of the image (b) with the theoretically calculated image (d) clearly proves the resolving of the triplet structure. Reprinted (figure) with permission from Ref. [8]. Copyright (2005) by the American Physical Society.

The results of these calculations in the cases of triplet nanoholes from Figs. 5.7 and 5.8 demonstrate that a resolution of the order of PSF $\approx$ 70 nm or $\sim\lambda/8$ is achieved in these particular imaging experiments. Such an improved resolution in an SPP microscopy

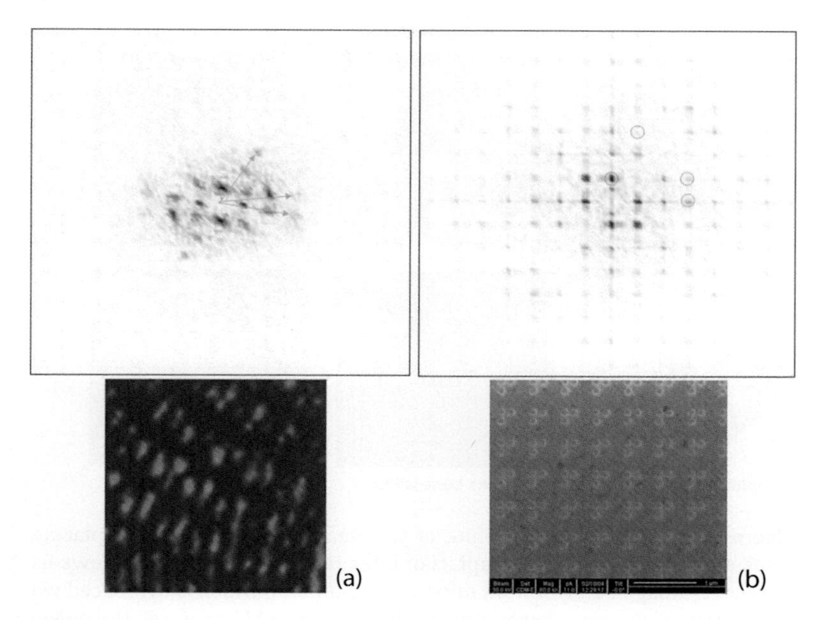

Figure 5.8 Comparison of the SPP-produced optical (a) and the SEM (b) images of the test array of triplet nanoholes. Comparison of the Fourier transforms of these images indicates spatial resolution in the optical image of ~78 nm. This conclusion may be reached from the apparent visibility of higher harmonics of the triplet structure (indicated by the arrows) in the optical image. Reproduced from Ref. [19]. With permission of Springer.

experiment is due to the fact that the SPP wavelength is shorter than the wavelength of guided modes at the same laser frequency. Photonic crystal effects and the effects of negative refraction also play some role in achieving better resolution, as described in Ref. [10].

Even though quite an improvement compared to a regular optical microscope, the ~70 nm resolution is not sufficient to achieve clear visibility of many nanoholes in the test pattern in Fig. 5.7c. While recognizable, most nanoholes appear quite fuzzy. However, the blurring of optical images at the limits of optical device resolution is a very old problem (one may recall the well-publicized recent problem of Hubble telescope repair). One solution of this problem is also well known. There exists a wide variety of image recovery techniques which successfully fight image blur based on

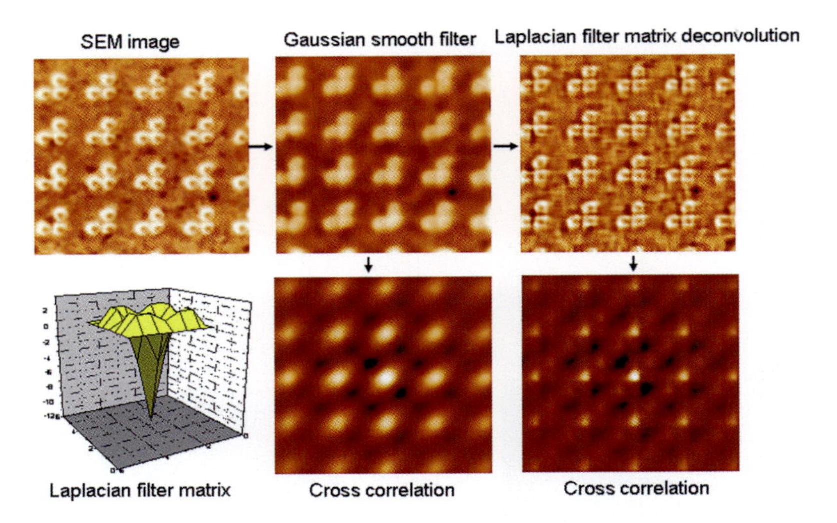

Figure 5.9 Theoretical modeling of the image recovery using Laplacian filter matrix deconvolution: Laplacian filter (shown in the inset) allows us to recover image deterioration due to Gaussian blur, which is evidenced via calculation of the cross-correlation of the original SEM image and the image recovered using the Laplacian matrix deconvolution method. Reproduced from Ref. [19]. With permission of Springer.

the known PSF of the optical system. One of such techniques is matrix deconvolution based on the Laplacian filter [20] (see Fig. 5.9 as an example). Utilization of such techniques is known to improve resolution by at least a factor of 2. However, precise knowledge of the PSF of the microscope in a given location in the image is absolutely essential for this technique to work, since it involves matrix convolution of the experimental image with a rapidly oscillating Laplacian filter matrix (an example of such 5×5 matrix is shown in Fig. 5.9). In our test experiments the PSF of the microscope was measured directly in some particular location of the optical image. This measured PSF was used to digitally enhance images of the neighboring nanohole arrays. A similar technique may be used to enhance resolution in the SPP-induced optical images of biological samples, which are measured using the nanohole array background, as described in Ref. [10].

Not surprisingly, the use of such digital filters led to approximately twofold improvement of resolution in the optical images

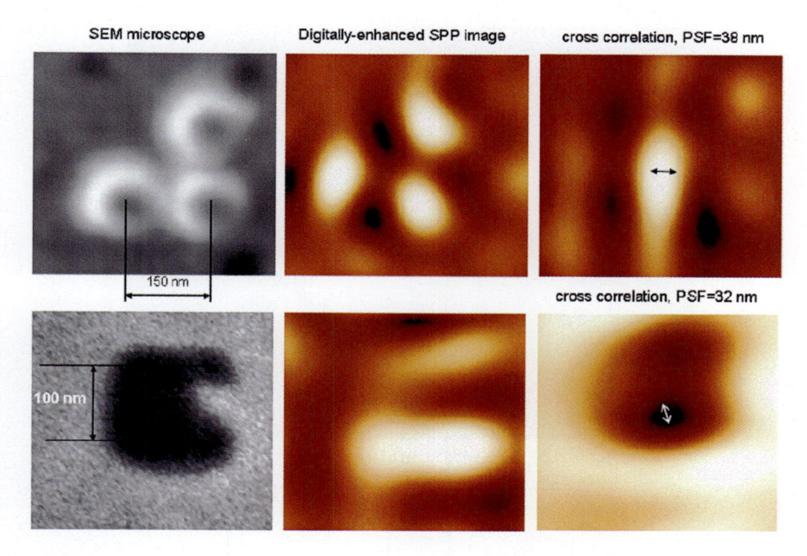

Figure 5.10 Calculated cross-correlation functions between SEM and digitally enhanced optical images of a triplet and a U-shaped arrangement of nanoholes in a gold film. The calculated point spread function of the digitally enhanced optical images appears to be on the order of 30 nm. Reproduced from Ref. [19]. With permission of Springer.

formed by the 2D plasmon microscope. This twofold improvement is demonstrated in Fig. 5.10 for both the triplet and the U-shaped nanoholes. The PSF measured as the cross-correlation between the digitally processed optical image and the corresponding SEM image appears to fall firmly into the 30 nm range, which represents improvement of resolution of the SPP-assisted optical microscope down to the $\sim\lambda/20$ range. This result may bring about direct optical visualization of many important biological systems.

5.5 Conventional Plasmon Focusing Devices

It is interesting to note that outside the hyperbolic-frequency-range plasmonic metamaterial devices described above may be used as efficient focusing elements for SPPs. Figure 5.11 demonstrates 2D focusing properties of the concentric ring structure illuminated by 532 nm external laser light. A 4×4 array of the focusing devices

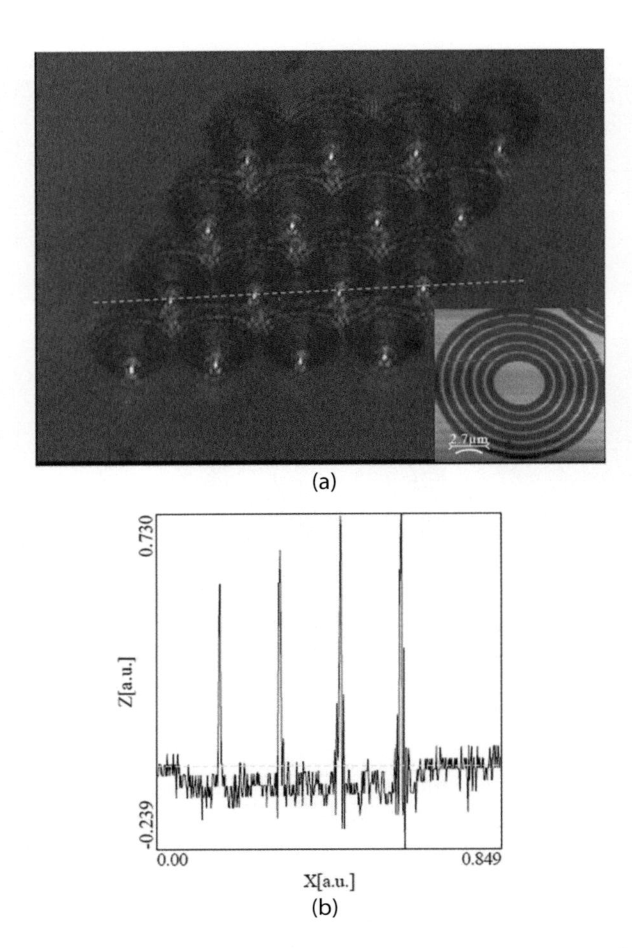

Figure 5.11 (a) Optical microscope image of the 4 × 4 array of the 2D focusing elements based on the concentric ring structure illuminated with 532 nm laser light. The inset shows the AFM image of an individual focusing structure. (b) Analysis of the image cross section along the dashed line shown in (a) indicates considerable enhancement of the field intensity in the focal spot, which is estimated to be of the order of 20.

is shown. Analysis of the image cross section along the dashed line shown in Fig. 5.11a indicates considerable enhancement of the field intensity in the focal spot, which is measured to be of the order of 20. The radius of the circular focusing area is limited by

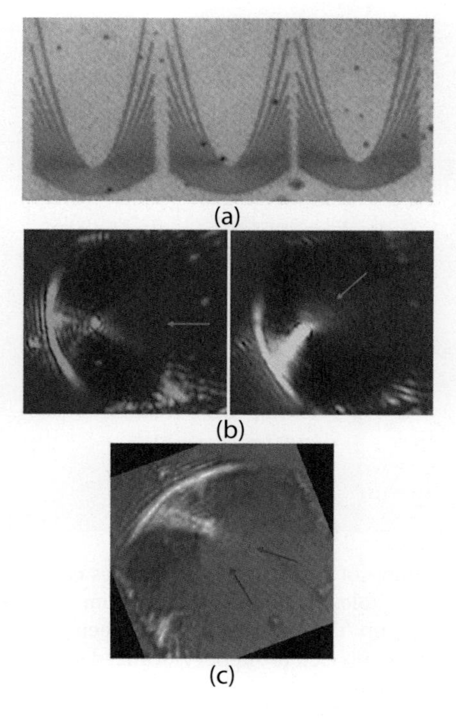

Figure 5.12 (a) Geometry of the plasmonic focusing devices based on parabolic gratings. Panels (b) and (c) demonstrate the possibility to scan the focal spot as a function of illumination angle.

the propagation length of SPP in a given frequency range. Near 500 nm wavelengths, the Re(**k**)/Im(**k**) ratio for the gold–vacuum interface is of the order of 30 [21], which means that plasmon energy may be collected over a $\sim 30\lambda/2\pi \approx 2.5$ μm radius. Ideal focusing of this energy into the diffraction-limited $(\lambda/2n_{\text{eff}})^2$ spot would produce the field intensity enhancement of the order of 100. However, surface plasmons are strongly scattered into 3D photons by surface defects. In reality, the reported field enhancement factors reach the ~20–50 range [22]. This design of the SPP 2D focusing device is very well suited for sensing applications using fluorescence or surface-enhanced Raman scattering (SERS). Fluorescence and SERS detection schemes benefit from exponential enhancement of the SPP field near the metal surface. Additional enhancement

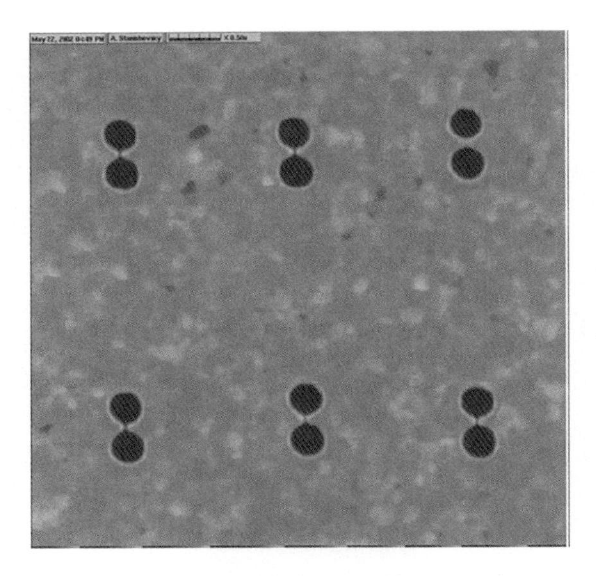

Figure 5.13 SEM image of an array of double holes through a 500 nm thick gold membrane. The holes have been milled from the back side using a 50 keV focused ion beam. The size of gaps in between the hole pairs is about 20 nm. Such gaps are supposed to support localized surface plasmon modes with about the same mode size.

due to SPP focusing is highly desirable in these applications. We anticipate that the SPP focusing devices may be combined with metal nanostructures, which support localized surface plasmon modes. This will lead to further enhancement of SERS signals. Other efficient geometries for SPP focusing include parabolic gratings shown in Fig. 5.12. The latter geometry permits scanning of the focal spot as a function of the illumination angle, which would be useful if the position of the focal spot needs to be matched with the location of the nanofocusing plasmonic structure, such as the one shown in Fig. 5.13 reproduced from Ref. [23].

5.6 Conclusion

Thus, it appears that both major recent developments in far-field optical microscopy—nonlinear super-resolution techniques [2] and linear techniques based on plasmonic and optical metamaterials—are quickly moving the resolution scale of far-field optical microscopy down toward the 10 nm level. While fabrication of 3D photonic metamaterials faces numerous technological challenges, many concepts and ideas in the optics of metamaterials may be tested much easier in two spatial dimensions using planar optics of SPPs. The 2D plasmonic metamaterials and devices may be used in various super-resolution microscopy, waveguiding, and laser cavity schemes. We have reviewed various examples of 2D plasmonic super-resolution imaging devices, which are reasonably easy to fabricate and study. These devices exhibit spatial resolution of the order of 70 nm. Moreover, utilization of well-known digital image recovery techniques enables further improvement of resolution of far-field optical microscopy down to the $\sim$30 nm scale. As a result, spatial resolution of far-field optical microscopy approaches resolution of SEM. Unlike more time-consuming near-field optical techniques, far-field imaging allows very simple, fast, robust, and straightforward image acquisition. Widespread availability of these techniques to the research community should bring about numerous revolutionary advances in super-resolution imaging, lithography, and sensing. However, metamaterial losses remain an important performance-limiting issue. It remains to be seen if loss compensation using gain media [24, 25] will be able to overcome this problem.

References

1. Pohl, D. W., and Courjon, D. (eds.), (1993). *Near Field Optics. NATO ASI-E Series* (Kluwer Academic, the Netherlands).
2. Hell, S. W. (2007). Far field optical nanoscopy, *Science*, **316**, pp. 1153–1158.
3. Veselago, V. G. (1968). The electrodynamics of substances with simultaneously negative values of ε and μ, *Sov. Phys. Usp.*, **10**, pp. 509–514.
4. Pendry, J. B. (2000). Negative refraction makes a perfect lens, *Phys. Rev. Lett.*, **85**, pp. 3966–3969.
5. Efros, A. L. (2008). The problem of subwavelength imaging by Veselago lens, *Solid State Commun.*, **146**, pp. 198–201.

6. Fang, N., Lee, H., Sun, C., and Zhang, X. (2005). Sub–diffraction-limited optical imaging with a silver superlens, *Science*, **308**, pp. 534–537.
7. Melville, D. O. S., and Blaikie, R. J. (2005). Super-resolution imaging through a planar silver layer, *Opt. Express*, **13**, pp. 2127–2134.
8. Smolyaninov, I. I., Elliott, J., Zayats, A. V., and Davis, C. C. (2005). Far-field optical microscopy with a nanometer-scale resolution based on the in-plane image magnification by surface plasmon polaritons, *Phys. Rev. Lett.*, **94**, p. 057401.
9. Zayats, A. V., and Smolyaninov, I. I. (2003). Near-field photonics: surface plasmon polaritons and localized surface plasmons, *J. Opt. A: Pure Appl. Opt.*, **5**, pp. S16–S50.
10. Smolyaninov, I. I., Davis, C. C., Elliott, J., Wurtz, G., and Zayats, A. V. (2005). Super-resolution optical microscopy based on photonic crystal materials, *Phys. Rev. B*, **72**, p. 085442.
11. Ramakrishna, S. A., and Pendry, J. B. (2004). Spherical perfect lens: solutions of Maxwell's equations for spherical geometry, *Phys. Rev. B*, **69**, p. 115115.
12. Jakob, Z., Alekseyev, L. V., and Narimanov, E. (2006). Optical hyperlens: far-field imaging beyond the diffraction limit, *Opt. Express*, **14**, pp. 8247–8256.
13. Salandrino, A., and Engheta, N. (2006). Far-field subdiffraction optical microscopy using metamaterial crystals: theory and simulations, *Phys. Rev. B*, **74**, p. 075103.
14. Zayats, A. V., Smolyaninov, I. I., and Maradudin, A. (2005). Nano-optics of surface plasmon-polaritons, *Phys. Rep.*, **408**, pp. 131–314.
15. Jacob, Z., and Narimanov, E. E. (2008). Optical hyperspace for plasmons: Dyakonov states in metamaterials, *Appl. Phys. Lett.*, **93**, p. 221109.
16. Smolyaninov, I. I., Hung, Y. J., and Davis, C. C. (2007). Magnifying superlens in the visible frequency range, *Science*, **315**, pp. 1699–1702.
17. Liu, Z., Lee, H., Xiong, Y., Sun, C., and Zhang, X. (2007). Far-field optical hyperlens magnifying sub-diffraction-limited objects, *Science*, **315**, p. 1686.
18. Ebbesen, T. W., Lezec, H. J., Ghaemi, H. F., Thio, T., and Wolff, P. A. (1998). Extraordinary optical transmission through sub-wavelength hole arrays, *Nature*, **391**, pp. 667–669.
19. Smolyaninov, I. I., Davis, C. C., Elliott, J., Wurtz, G., and Zayats, A. V. (2006). Digital resolution enhancement in surface plasmon microscopy, *Appl. Phys. B*, **84**, pp. 253–256.

20. Poon, T.-C., and Banerjee, P. P. (2001). *Contemporary Optical Image Processing with MATLAB* (Elsevier Science, New York).
21. Smolyaninov, I. I., Hung, Y. J., and Davis, C. C. (2007). Imaging and focusing properties of plasmonic metamaterial devices, *Phys. Rev. B*, **76**, p. 205424.
22. Bonod, N., Popov, E., Gérard, D., Wenger, J., and Rigneault, H. (2008). Field enhancement in a circular aperture surrounded by a single channel groove, *Opt. Express*, **16**, pp. 2276–2287.
23. Smolyaninov, I. I., Zayats, A. V., Stanishevsky, A., and Davis, C. C. (2002). Optical control of photon tunneling through an array of nanometer scale cylindrical channels, *Phys. Rev. B*, **66**, p. 205414.
24. Xiao, S., Drachev, V. P., Kildishev, A. V., Ni, X., Chettiar, U. K., Yuan, H.-K., and Shalaev, V. M. (2010). Loss-free and active optical negative-index metamaterials, *Nature*, **466**, pp. 735–738.
25. Burke, J. E. (1964). Scattering of surface waves on an infinitely deep fluid, Scattering of surface waves on an infinitely deep fluid, *J. Math. Phys.*, **6**, pp. 805–819.

Chapter 6

Hyperlenses and Metalenses

Dylan Lu[a] **and Zhaowei Liu**[a,b,c]

[a]*Department of Electrical and Computer Engineering, University of California, San Diego, 9500 Gilman Drive, La Jolla, CA 92093, USA*
[b]*Center for Magnetic Recording Research, University of California, San Diego, 9500 Gilman Drive, La Jolla, CA 92093, USA*
[c]*Materials Science and Engineering, University of California, San Diego, 9500 Gilman Drive, La Jolla, CA 92093, USA*
zhaowei@ucsd.edu

The imaging resolution of conventional optical lenses is known as being hampered by the diffraction limit. To break such a diffraction limit, research efforts on artificial metamaterials provide a new platform for building new optical imaging systems based on hyperlenses and metalenses. Hyperlenses project super-resolution images to the far field through a magnification mechanism, whereas metalenses not only resolve subwavelength details but also enable optical Fourier transforms. Recently, there have been emerging numerous designs for hyperlenses and metalenses, advancing both theoretical and experimental investigations, though challenges remain to be overcome for future development.

Plasmonics and Super-Resolution Imaging
Edited by Zhaowei Liu

ISBN 978-981-4669-91-7 (Hardcover), 978-1-315-20653-0 (eBook)
www.panstanford.com

6.1 Introduction

The invention of optical microscopy has enabled revolutions in many fields, including microelectronics, biology, and medicine. However, the resolution of a conventional optical lens system is always limited by diffraction to about half the wavelength of light. This diffraction barrier arises because subwavelength information from an object, carried by high-spatial-frequency evanescent waves, exists only in the near field. To overcome this diffraction limit, pioneering approaches have been carried out, including near-field scanning microscopy [1], as well as various fluorescence-based imaging methods, such as stimulated emission depletion (STED) and stochastic optical reconstruction microscopy imaging (STORM) [2, 3]. These scanning- or random-sampling-based nonprojection techniques achieve super-resolving power by sacrificing imaging speed, making them uncompetitive for dynamic imaging.

Lens-based projection imaging remains the best option for high-speed microscopy. Although having been widely used to enhance resolution, immersion techniques are limited by the low refractive indices of natural materials [4]. Since the last decade, tremendous development in the fields of plasmonics and metamaterials provide solutions for engineering extraordinary material properties not existing in nature, such as a negative index of refraction [5, 6] or strongly anisotropic materials [7–10]. This has brought new opportunities for novel lens designs with unprecedented resolution [11, 12]. Inspired by the concept of the perfect lens [13], a series of superlenses were demonstrated with resolving powers beyond the diffraction limit [14–20]. The first optical superlens with subdiffraction-limited resolution was achieved by enhancing evanescent waves through a slab of silver [14]. As the evanescent field enhancement is associated with surface plasmon excitation, the subwavelength image is typically limited to the near-field of the metal slab. However, the hyperlens—a metamaterial-based lens—can send super-resolution images to the far field by incorporating a magnification mechanism. It has become one of the most promising candidates for practical applications since its first demonstration in 2007 [16, 17]. Various other metamaterial-based lenses, called metalenses, have also been developed combining both super-resolving power and Fourier transform function, making them more like conventional lenses but with superior capabilities [21–24].

In this chapter, we will start with the underlying physics of the hyperlens, introduce its experimental demonstrations, and then describe the working mechanism of the metalens, its design principles, and extraordinary imaging properties. We will also discuss major limitations and practical challenges, as well as suggestions for future development of super-resolution hyperlenses. The extension of the hyperlens concept from electromagnetic waves to acoustic waves is also included.

6.2 Physics of the Hyperlens

Light emitted or scattered from objects comprises propagating and evanescent components corresponding to low and high wavevectors, respectively. The propagating waves can reach the far field, preserving large-feature information, whereas the evanescent waves are nonpropagating in a natural material environment, and thus the detailed information is confined in the near field. The far-field imaging becomes diffraction limited because deep-subwavelength feature information carried by the evanescent waves cannot reach the far field and has no contribution to the formation of the final image.

To realize far-field super-resolution imaging, the invention of the hyperlens builds on two basic requirements, (i) a new material that supports high wavevector propagation and (ii) a magnification mechanism that converts high-wavevector waves to low-wavevector waves to send the super-resolution information to the far field.

Anisotropic plasmonic metamaterials provide one of the most practical solutions to fulfill the material requirement. Because only the direction-dependent permittivities need to be considered and engineered in anisotropic plasmonic metamaterials, the overall material loss is significantly reduced [8, 25]. The simplest anisotropic plasmonic metamaterials can be constructed by the deposition of alternating metal/dielectric multilayers (Fig. 6.1a). When the layer thickness is much smaller than the probing wavelength, an effective medium approximation is commonly used to describe the

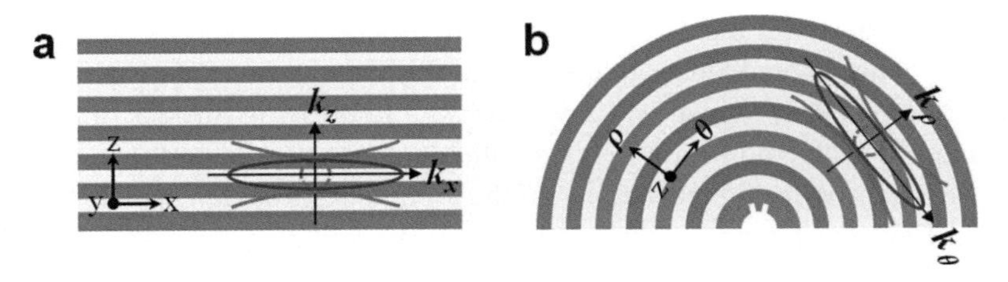

Figure 6.1 (a) Schematic of multilayer metamaterials with different colors representing two constituent materials. The inset shows hyperbolic (red) or eccentric elliptic (blue) dispersions compared with the isotropic spherical dispersion (green). (b) Multilayer metamaterials in a curved geometry result in hyperlenses with corresponding dispersions represented in cylindrical coordinates. The green blocks in the inner boundary define the subwavelength objects.

permittivities along different directions, as follows [26]:

$$\begin{cases} \varepsilon_x = \varepsilon_y = p\varepsilon_{\mathrm{m}} + (1-p)\,\varepsilon_{\mathrm{d}} \\ \varepsilon_z = \dfrac{\varepsilon_{\mathrm{m}}\varepsilon_{\mathrm{d}}}{p\varepsilon_{\mathrm{d}} + (1-p)\,\varepsilon_{\mathrm{m}}} \end{cases} \tag{6.1}$$

where ε_x, ε_y, and ε_z are the effective permittivities along x, y, and z directions; ε_{m} and ε_{d} are the permittivities for the constituent metal and dielectric; and p corresponds to the volumetric filling ratio of the metallic layer. In contrast to normal isotropic media with a spherical dispersion, the dispersion properties for multilayer metamaterials can be designed to have $\varepsilon'_x > 0$ and $\varepsilon'_z < 0$ for hyperbolic dispersions, where ε'_x and ε'_z represent the real part of the permittivity along x and z directions, as shown in Fig. 6.1a. Anisotropic metamaterials support the propagation of high-wavevector waves due to unbounded wavevector values or very large wavevector cutoff.

The magnification mechanism that compresses the wavevector is realized by bending the flat multilayer metamaterials into co-centrically curved layers, as shown in Fig. 6.1b [11, 18], which can be explained by transformation optics [28]. It makes hyperlenses with corresponding dispersions represented in cylindrical coordinates. For a subwavelength object placed at the inner boundary, wave propagation from the object across the radial direction gradually compresses its tangential wavevectors, owing to the conservation

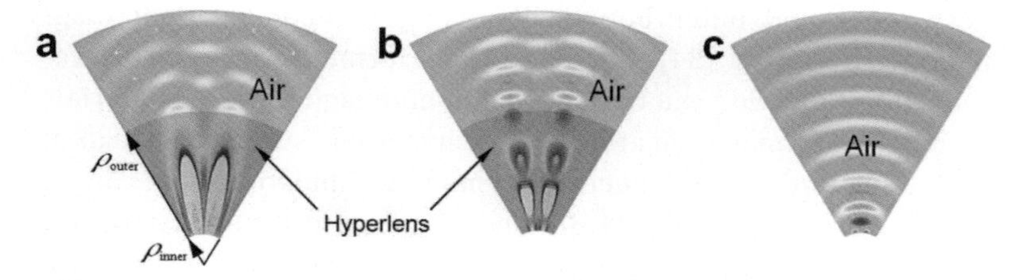

Figure 6.2 Numerical simulation of multilayer-metamaterial-based lens with eccentric elliptic (a) and hyperbolic (b) dispersions. In the simulation, two subwavelength line objects were separated by a distance of 80 nm. The radii of the hyperlens of either dispersion at the inner and outer boundaries are 240 nm and 1200 nm, respectively. The permittivity for metal and dielectric in the hyperbolic (elliptic) hyperlens is $-2.3-0.3i$ ($-2.3-0.3i$) and 2.7 (2.25), respectively. The filling ratio of metal in both hyperlenses is 50%. The same objects cannot be resolved in air (c).

of angular momentum, resulting in a magnified image at the outer boundary. The magnified image, once larger than the diffraction limit, will be resolved in the far field. The magnification at the output surface is simply given by the ratio of the radii at the two boundaries.

Although the term "hyperlens" was initially referring to a superlens made of metamaterials with a hyperbolic dispersion relation [18], elliptically dispersive metamaterials, that is, $\varepsilon'_z \gg \varepsilon'_x > 0$, can also be used to build a more generalized hyperlens as long as the coverage for lateral wavevectors is large enough [27]. Figure 6.2 shows that two subdiffraction-limited line sources separated by a distance of $\sim$80 nm can be clearly resolved by using either eccentric elliptic (Fig. 6.2a) or hyperbolic (Fig. 6.2b) metamaterials but not by air alone (Fig. 6.2c).

Various practical configurations of the hyperlens have been proposed by different groups, including a cylindrical geometry [18, 28, 29], tapered metallic wire arrays [30, 31], and specifically designed material dispersions [32, 33]. To circumvent the problem of absorption in real materials and thereby avoid image distortions in the far field, proper configurational design must ensure that rays originating from the object travel equal path lengths through the metamaterial. Besides, scattering at surfaces of the hyperlens has proven detrimental for real imaging applications. It has been proposed to impedance match the hyperlens metamaterial with both inner and outer boundaries using radius-dependent magnetic permeability [34]. Comparable performance by the surface-reflection reduction can be achieved in nonmagnetic metamaterials by impedance matching at one of the hyperlens surfaces instead of both. This trade-off has been confirmed by simulations and can be practically implemented [34, 35]. Owing to its use of curved surfaces, the cylindrical hyperlens design may not be convenient for some real applications. Design of the planar hyperlens has been shown to be theoretically feasible via specific material dispersion based on transformation optics [32, 33]. In these designs, the metamaterial properties are configured to bend light rays from subwavelength features in such a way as to form a resolvable image on a flat output plane. By using different material combinations, the working wavelength for multilayer metamaterials can be tuned across a broad band of wavelengths. Metallic nanowire-based metamaterials can also be used to shift toward much longer operational wavelengths. The wavelength range for either hyperbolic or elliptic dispersion is determined by not only the constituents' permittivity but also their filling ratios.

6.3 Experimental Demonstration of the Hyperlens

Theoretical predictions on hyperlenses made of anisotropic curved multilayer-based metamaterials [18, 29] have inspired experimental investigations into the hyperlens concept. The first 1D optical hyperlens was demonstrated experimentally at UV frequencies [16, 36]. The fabrication process of the hyperlens is shown in Fig. 6.3. A thin Cr layer (150 nm) was firstly deposited on a quartz wafer (150 μm thick) by an electron beam evaporation process (EB3, BOC Edwards). A 50 nm wide etch slit was then made by focused ion beam (FIB) milling (STRATA 201XP, FEI) before the half-cylindrical groove was defined through an isotropic wet etching of quartz in buffered oxide etch (BOE) solution of 10:1 dilution ratio. After 12 minutes, groove diameter of approximately 1.6 μm was achieved. The root-mean-square (RMS) surface roughness of the groove was measured to be 1.3 nm by atomic force microscopy (Dimension

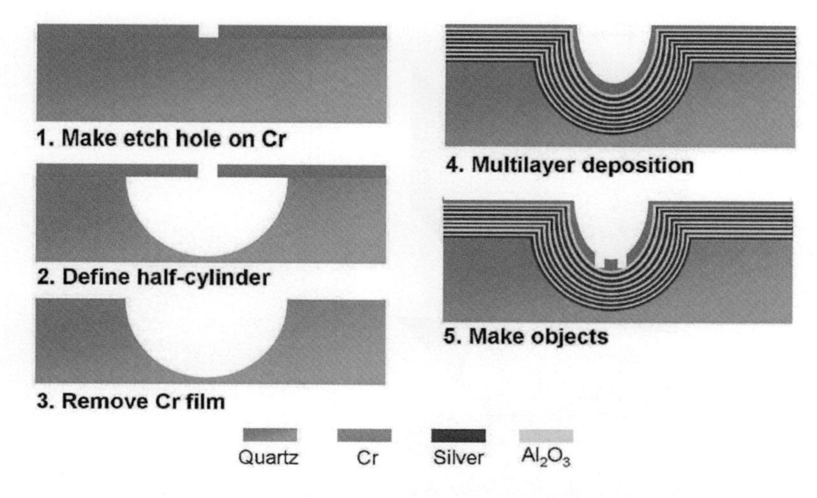

Figure 6.3 Fabrication process flow of a cylindrical hyperlens. Isotropic wet etching makes a cylindrical groove in quartz (2) through a hole etched in a Cr film (1). After the Cr film is removed (3), a multilayer hyperlens structure is fabricated using alternate deposition of Ag and Al_2O_3 (4). A Cr film caps the hyperlens structure for object fabrication (5).

3100, Veeco). Since multiple Ag and Al_2O_3 thin films were to be deposited, the initial surface quality was a crucial factor for tolerable smoothness of subsequent layers. The Cr layer was later removed by another wet-etching step (CR-7 Cr etchant). The multilayer structure was deposited at a low predeposition pressure of 3 μTorr to ensure the best possible surface quality of each layer. Starting from the Ag layer, Al_2O_3 and Ag films were deposited alternatively at 0.3 nm/s, and a 50 nm of Cr film was deposited on top of the 16th layer (Al_2O_3). Despite the directional nature of electron beam evaporation deposition, reasonably good side-wall coverage was achieved. Finally, various objects were inscribed on the Cr film by FIB.

The fabricated optical hyperlens was therefore made of 8 pairs of Ag (35 nm) and Al_2O_3 (35 nm) multilayer films on a cylindrical quartz cavity, with an inner cavity of 950 nm wide (Fig. 6.4a). Subwavelength objects were inscribed into a 50-nm-thick chrome layer at the inner boundary. The sample was illuminated from the top air side by a mercury lamp passing through a band-pass filter

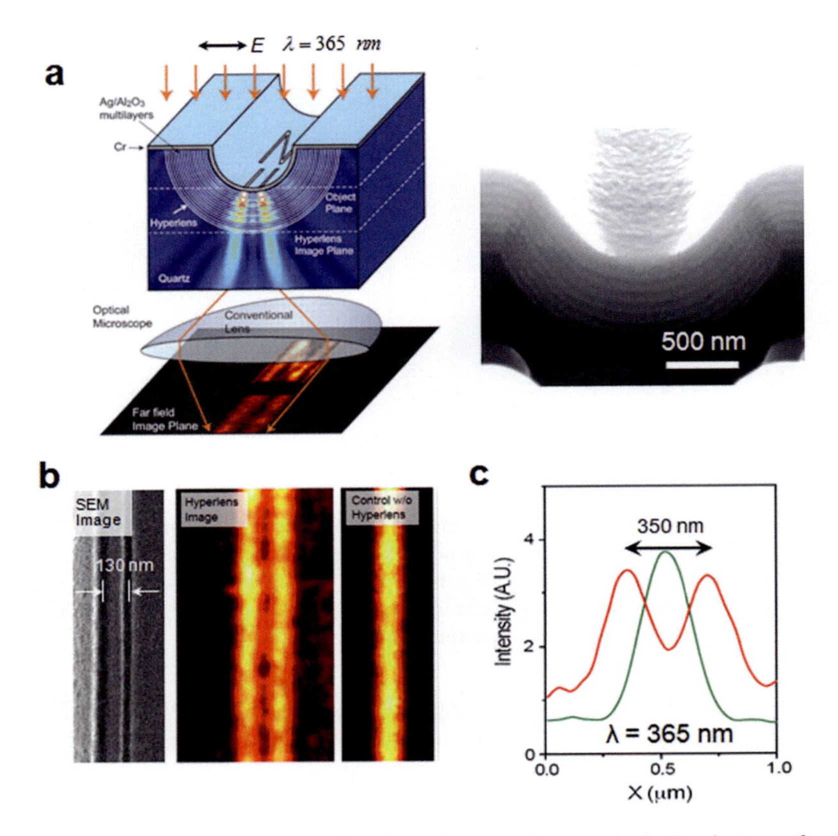

Figure 6.4 (a) Schematic of a hyperlens and numerical simulation of imaging of subwavelength objects. Right: Scanning electron microscope (SEM) image of the cross section of a cylindrical hyperlens structure with 16 Ag (bright) and Al_2O_3 (dark) layers. The thick bright layer at the top is Cr. (b) Hyperlens imaging of a pair of line objects with a line width of 35 nm and a center-to-center distance of 130 nm. From left to right: SEM image of the line-pair object, magnified hyperlens image, and the resulting diffraction-limited image in a control experiment. (c) The cross-sectional profile for the hyperlens image of the line-pair object (red) and for a diffraction-limited image obtained without the hyperlens (green). A.U., arbitrary units.

(center wavelength: 365 nm; bandwidth: 10 nm) and a UV polarizer on an optical microscope system. The multilayer metamaterial was designed to have different signs in the radial and tangential permittivities. The resulting hyperbolic dispersion supports the propagation of evanescent waves from the subwavelength objects along the radial direction in the metamaterial. As the tangential wavevectors are progressively compressed, a magnified image will be formed at the outer boundary of the hyperlens and captured by an oil immersion objective (100×, NA = 1.4) before recorded by a UV-sensitive charge-coupled device (CCD) camera. To minimize the possible focusing error that could affect the image resolution, a set of images were taken at different Z-focusing positions and the one with highest contrast was chosen. Figure 6.4b,c shows that a subdiffraction-limited object with a 130-nm center-to-center distance was observed directly in the far field. The magnified hyperlens image shows the object clearly resolved, as compared with a diffraction-limited image in a control experiment. So far, hyperlens experiments have achieved 125 nm pitch resolution ($\lambda/2.92$) working at a wavelength of 365 nm [36], although some simulations provided designs with much higher resolution around $\lambda/9$ [37].

The cylindrical hyperlens demonstrates 1D resolution improvement, which limits its practical applications. A 2D super-resolution capability is far more desirable, and it has recently been experimentally demonstrated in the visible spectral region by using a spherical hyperlens [20]. This 2D hyperlens was created by deposition of alternating layers of Ag and Ti_3O_5 thin films in a hemispherical geometry, as shown in Fig. 6.5a. To work at visible wavelength, a relative high-index dielectric material, titanium oxide, was chosen to match the magnitude of metal permittivity [35]. Titanium oxide has a dielectric constant of 5.83 at 410 nm, whereas the permittivity of silver is $-4.99–0.22i$. The filling ratio is chosen to be 1:1, with each layer thickness of 30 nm. According to the effective medium approximation, the multilayer metamaterial thus has a small positive tangential permittivity $\varepsilon_\theta = 0.42–0.11i$ together with a large negative radial permittivity $\varepsilon_r = -64–19.83i$, resulting in an ultraflat hyperbolic dispersion that allows the propagation of waves with very high spatial frequencies. The arbitrary subwavelength features are inscribed on the top Cr layer by FIB milling to serve as objects for imaging. To overcome the restriction of magnification along one dimension for transverse magnetic illumination, the sample was illuminated with unpolarized

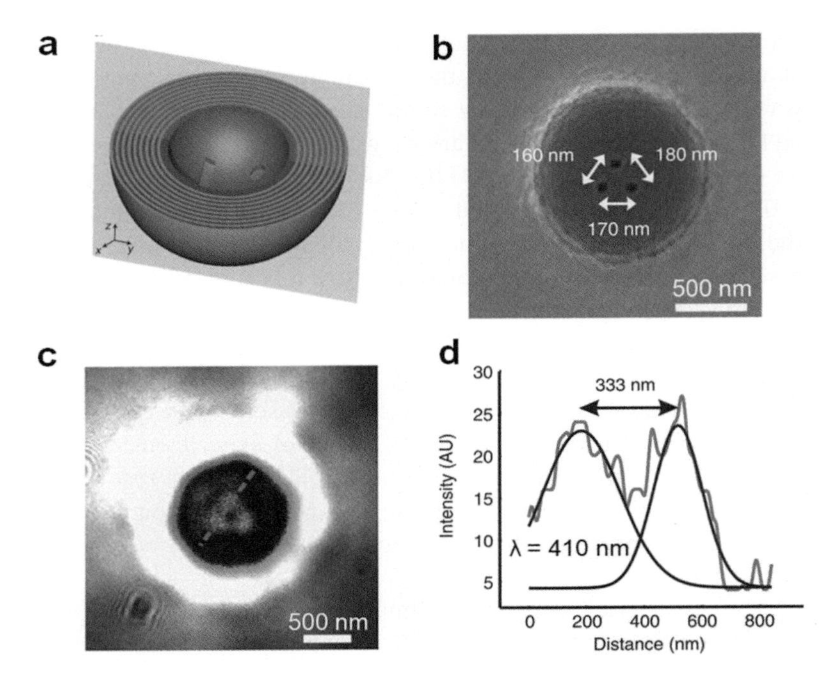

Figure 6.5 (a) Schematic of a spherical hyperlens comprising nine pairs of Ag and Ti_3O_5 layers. The subwavelength object is carved in the top Cr layer. (b) SEM image of three dots positioned triangularly with gaps of 180 nm, 170 nm, and 160 nm on a spherical hyperlens comprising nine pairs of Ag and Ti_3O_5 layers. (c, d) Image of the object after being magnified and its cross section along the red dashed line. The black curves in (d) correspond to the cross-sectional analysis for identifying the separation of the two apertures measured to be 333 nm. A.U., arbitrary units.

light, which contains transverse magnetic components spanning the whole 2D reciprocal space, thus supporting subwavelength features in both dimensions.

Figure 6.5b shows one example of subwavelength objects consisting of three apertures located in triangular configuration with 160 nm, 170 nm, and 180 nm spacing in three sides. The magnified images were collected by an objective lens in a transmission optical microscope as shown in Fig. 6.5c. The result indicates that the separation down to 160 nm is clearly resolved in the far field, beyond the diffraction limit, 205 nm, at a wavelength of 410 nm. The cross section along the red line is redrawn in Fig. 6.5d, showing the corresponding separation of the image to be 333 nm, which results in an averaged magnification factor of 2.08. Although this ratio is smaller than the designed value, it can be improved by better conformal film deposition and thickness control. Further increase is also expected by using larger ratio of inner and outer radius of the hyperlens, while keeping good light transmission.

These proof-of-concept experiments have investigated hyperlenses with hyperbolic dispersions working at UV and visible wavelengths. A broader working wavelength range across the visible or near-infrared becomes attainable with appropriate material combinations for both hyperbolic dispersions and highly elliptical dispersions [35]. The resolution of hyperlenses is essentially determined by the geometry, material loss, and film quality. In addition to increasing the magnification with a higher ratio of the outer-to-inner radii, the resolution could be further improved through the choice of appropriate filling ratios of metal and dielectric components because it results in the propagation of higher-spatial-frequency information as the dispersion approaches flatness [35]. Although the intrinsic metal loss is unavoidable, the hyperlens can be optimized by increasing the amount of the dielectric component or working at longer wavelengths. A higher optical transmission can also be realized with resonant structures [38]. Moreover, instead of conformal film deposition, fabrication based on rolled-up technology provides an alternative method for controlling film conformality and roughness [39].

6.4 Working Mechanism of the Metalens

Although capable of super-resolution for far-field detection, the hyperlens lacks the ability to focus plane waves. For conventional optical lenses, refraction and focusing of light rely on curved interfaces, fundamental for Fourier transforms and imaging. However, hyperlenses cannot realize the Fourier transform function due to the lack of a phase compensation mechanism. In this and the

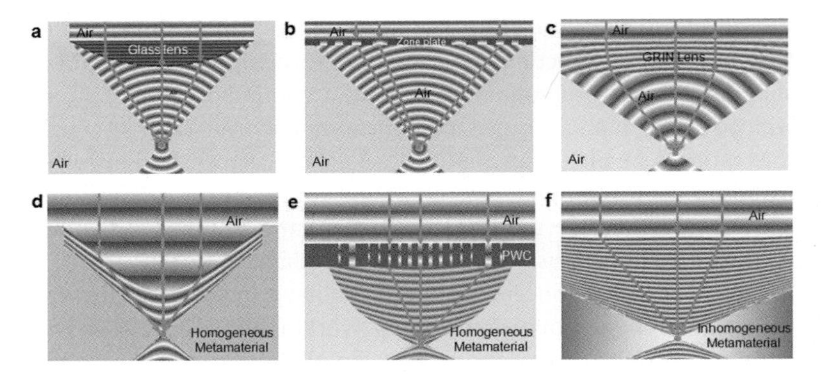

Figure 6.6 Schematics of plane wave focusing by a glass lens with a convex surface (a), Fresnel zone plate (b), and GRIN lens (c), in comparison with homogeneous metamaterials with shaped interfaces (d), metamaterials with a planar plasmonic metal-insulator-metal waveguide coupler (e), and GRIN metamaterials (f). Green arrows are a guide to the eye for wave propagation.

following sections, we will introduce the metalens—a metamaterial-based lens that provides not only the advantages of super-resolving capabilities but also the Fourier transform function, making it an exceptional combination of super-resolution and desirable functions of conventional lenses.

As shown in Fig. 6.6a–c, the general methods to introduce a phase compensation mechanism into a lens design include shaping the surface of a piece of homogeneous material, such as polishing a piece of glass to have a convex surface; using diffractive structures to engineer the wavefront, as in the Fresnel zone plate; and using material inhomogeneity to modify the phase change in space with, for instance, a gradient-index (GRIN) lens. A metalens is created by incorporating similar phase compensation mechanisms into a metamaterial slab to bring a plane wave to a focus (Fig. 6.6d–f). The anisotropic metamaterial slab with either a hyperbolic or eccentric elliptic dispersion is essential to achieve subdiffraction-limited resolution since it supports the propagation of super-resolution information carried by high wavevectors. The hyperlens and the metalens thus share the same material requirement, with practical realizations including multilayers [16] and nanowire metamaterials [25]. On the other hand, as high-wavevector waves are totally reflected at the metamaterial–air interface, a bidirectional coupler is needed to convert waves from high wavevectors in the metamaterial to low wavevectors in air, and vice versa. To make a focusing lens, this coupler must also function as a phase compensation component in contrast with the far-field superlens that functions as an evanescent-to-propagating convertor through the use of a grating [19].

Various types of phase compensation mechanisms have been proposed for focusing plane waves in metalenses [21–23]. The underlying design principle is to satisfy the phase-matching condition to constructively bring plane waves from air into deep-subwavelength focus inside the metamaterial. This can be accomplished by either geometric variations, such as plasmonic waveguide couplers (PWCs) [21] and shaped metamaterial–air interfaces [22], or material refractive index variations like GRIN metamaterials [23]. A PWC, which is made of an array of specially designed nonperiodic metal-insulator-metal (MIM) waveguides, turns nonflat wavefronts at the metamaterial–PWC interface into flat ones at the PWC–air interface, thus matching the phase condition for realizing a deep-subwavelength focus inside the metamaterial slab [21]. Instead of relying on MIM waveguides, a metamaterial with a hyperbolic dispersion can also achieve low and high wavevector conversion and phase compensation after being shaped to a concave surface. In contrast, a convex surface is needed for metamaterials with an elliptic dispersion. The curvature of the metalens is determined by a lens maker's equation that is similar to the conventional one [22]. Using transformation optics [40], such geometry variations can be transformed to refractive index variations within the material space, resulting in a GRIN metalens. Analogous to a conventional GRIN lens [41], the designed permittivity profiles for a GRIN metalens gradually bring plane waves to a super-resolution focus inside the metamaterials. As the focal length is defined, the imaging magnification is obtained by placing an object at an appropriate distance from the metalens, which is similar to the case of a conventional lens.

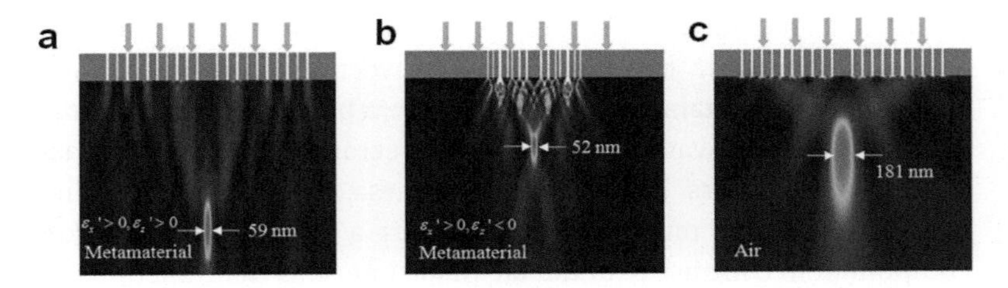

Figure 6.7 (a) Deep-subwavelength focusing of a metalens consisting of a Ag PWC and a metamaterial with an elliptic dispersion ($\varepsilon_x = 3.4 + 0.05i$; $\varepsilon_z = 11.3 + 0.9i$). A focus with FWHM = 59 nm is formed at a focal length $f = 1.0$ μm under a normal transverse magnetic (TM) incidence. (b) Deep-subwavelength focusing of a metalens consisting of an Al PWC and a metamaterial with a hyperbolic dispersion ($\varepsilon_x = 3.7 + 0.06i$; $\varepsilon_z = -8.9 + 1.8i$). A 52 nm focus is formed at $f = 0.5$ μm under a normal TM incidence. (c) An Ag waveguide plate in free space with a 181 nm focus at $f = 0.5$ μm. The incident wavelength is 365 nm. Metalens dimensions: $x = 1.8$ μm; $z = 1.4$ μm. PWC height: 200 nm. Slit widths: 10–20 nm. Material in slits: air.

6.5 Metalens Demonstration

6.5.1 Design of the Metalens: Plane Wave Focusing for Optical Fourier Transform

One demonstration of the metalens was first proposed by utilizing a planar PWC at UV wavelengths [21]. Due to the surface plasmonic waveguiding in the MIM structure, a large range of propagation constants can be obtained by varying the waveguide geometry and constituent materials. A carefully designed nonperiodically distributed PWC thus is able to tailor the wavefront of light, achieving phase compensation. As shown in Fig. 6.7a,b, based on an Ag (Al) PWC, the metalens with an elliptic (hyperbolic) dispersion has a deep-subwavelength focal spot with a full-width at half-maximum (FWHM) of 59 nm, $\sim\lambda/6.2$, (52 nm, $\sim\lambda/7.0$) in the metamaterials. Compared to the diffraction-limited focal spot in air (Fig. 6.7c), about threefold resolution improvement was demonstrated. Under tilted illumination, the focal spot of the metalens with the elliptic dispersion shifts along the incident direction similar to a conventional optical lens, whereas it shifts to the opposite side for the hyperbolic dispersive metalens [21].

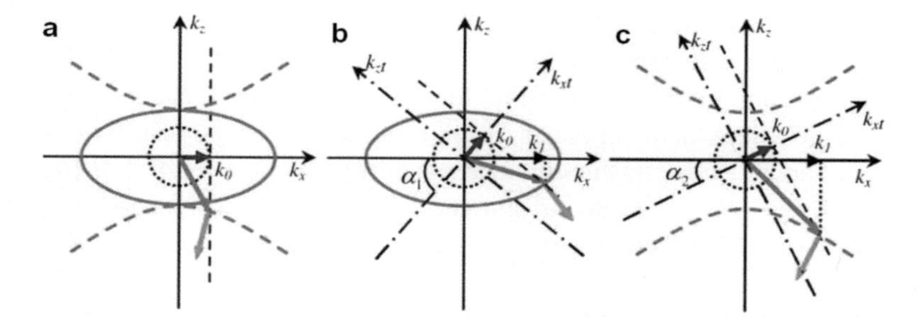

Figure 6.8 Equifrequency curves of air (dotted blue circle) and the metamaterials with an elliptic (solid red ellipse) and a hyperbolic (dashed red hyperbola) dispersion, respectively. The interface of the metamaterials, $\mathbf{k}_{xt}$, is a straight line along the $\mathbf{k}_x$ direction (a), an angle α_1 with respect to the $\mathbf{k}_x$ axis of elliptic dispersion (b), and an angle α_2 with respect to the $\mathbf{k}_x$ axis of hyperbolic dispersion (c).

This counterintuitive behavior arises from the negative refraction at the hyperbolic metamaterials interfaces and will be discussed in detail in the next section. According to the same design principle, an alternative PWC satisfies the phase condition by tuning the waveguide height instead of width. Improved resolution ($\sim\lambda/8.6$) was verified at longer wavelengths, due to smaller material loss [42].

Although the PWC-based metalens is capable of exceptional imaging resolution, it may demand a sophisticated fabrication process. As a metamaterial counterpart to the solid immersion lens [4], metalenses formed by shaping metamaterial surfaces have recently been proposed for super-resolution imaging with the Fourier transform function [22]. Compared to the isotropic spherical dispersion in air, the dispersion for metamaterials has either elliptic or hyperbolic dispersion, as shown in Fig. 6.8a. When the interface of the metamaterial is along the $\mathbf{k}_x$ axis, only the incident light within a small light cone indicated by the dashed blue circle can transmit in and out of the metamaterial. However, the transverse $\mathbf{k}$-vector coverage can be extended to $\mathbf{k}_l$ when the metamaterial surface is along $\mathbf{k}_{xt}$ axis, in an angle with respect to the material principle axes, as shown in Fig. 6.8b,c. The enlarged wavevector is not achieved in natural solid immersion materials, and therefore super-resolution is expected. To demonstrate this concept, an elliptic metamaterial made of multilayers of silver and gallium phosphide laying in the x direction after shaped into a convex geometry can satisfy the phase compensation condition for subwavelength focusing. The volume filling factor of silver in the multilayer is 20%, resulting in a metamaterial with the effective permittivities $\varepsilon_x = 5.05 + 0.10i$ and $\varepsilon_z = 15.99 + 0.08i$. As shown in Fig. 6.9a, the corresponding eccentric elliptic dispersion enables a deep-subwavelength focal spot with an FWHM of 70 nm ($\sim\lambda/9$) formed at a focal length of 3 μm when the plane wave is incident normally at a red wavelength of 633 nm. This metalens achieved an effective numerical aperture of 4.5, which is much larger than that found in naturally occurring materials [43]. When the incident light is tilted, the focus shifts accordingly, as estimated by $x = \mathrm{Re}\left(f\sqrt{\varepsilon_x}/\varepsilon_z \sin\theta\right)$. A shift of $x =$ 310 nm is observed in Fig. 6.9b for an incident angle of 14°, which is comparable to the calculated shift. In addition, a hyperbolic metamaterial designed using silver nanowires aligned in the z axis in an air background with a volume filling factor of 70% achieves a focus with an FWHM of 66 nm ($\sim\lambda/9.6$) for normal incidence at a wavelength of 633 nm (Fig. 6.9c). Owing to negative refraction, a hyperbolic metamaterial must be made into a concave geometry to focus light, and the focus shifts to the opposite side to the elliptic metalens with $x = -108$ nm, which is exceptionally different from conventional lenses.

Such geometry manipulation can alternatively be converted to the design of a spatially variant material property. A GRIN metalens, with appropriately designed permittivity profiles, was first investigated for its super-resolution imaging as well as its Fourier transform capabilities at optical wavelengths [23]. It is well known that a conventional GRIN lens requires a symmetric refractive index profile with a maximum in the center. The permittivity profiles for a GRIN metalens can also be designed through a ray model. Figure 6.10a shows the schematic ray model of focusing inside the GRIN metamaterial. It can be modeled by discretizing the material into infinitesimally thin layers in the x direction so that each layer has a uniform permittivity. The converging trajectory AC is simplified to a series of refraction processes between two adjacent anisotropic layers. All possible combinations of the symmetric gradient profiles can be obtained for the GRIN metalens to achieve focusing [23].

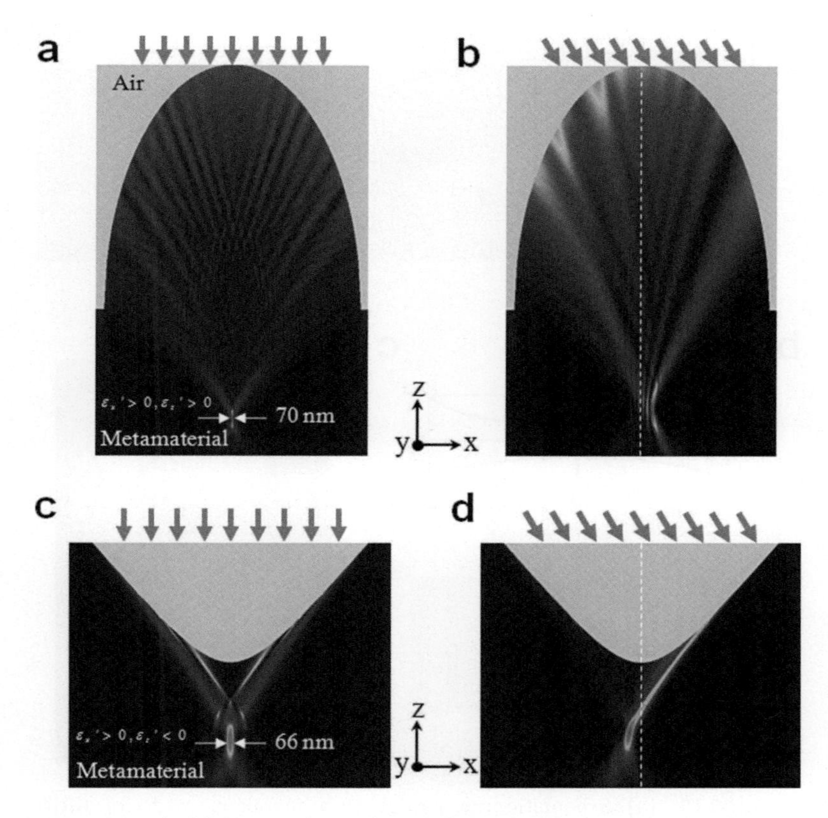

Figure 6.9 (a, b) Deep-subwavelength focusing of a metalens by shaping a metamaterial with an elliptic dispersion ($\varepsilon_x = 5.05 + 0.10i$; $\varepsilon_z = 15.99 + 0.08i$). A focus with FWHM = 70 nm is formed at a focal length f = 3.0 μm under a normal (a) and an angle of 14° (b) TM incidence. (c, d) Deep-subwavelength focusing of a metalens by shaping a metamaterial with a hyperbolic dispersion ($\varepsilon_x = 8.13 + 0.09i$; $\varepsilon_z = -12.51 + 0.35i$). A 66 nm focus is formed at f = 2 μm under a normal (c) and an angle of 37.8° (d) TM incidence. The incident wavelength is 633 nm.

As an example, Fig. 6.10b shows one design of permittivity profile for a hyperbolic GRIN metalens, which is realized by using silver nanowires grown vertically in an alumina template with a volume filling ratio of silver varying symmetrically from 30% at the center to 60% on the edges. Numerical simulation indicates such a hyperbolic metalens gradually converges a normally incident wave

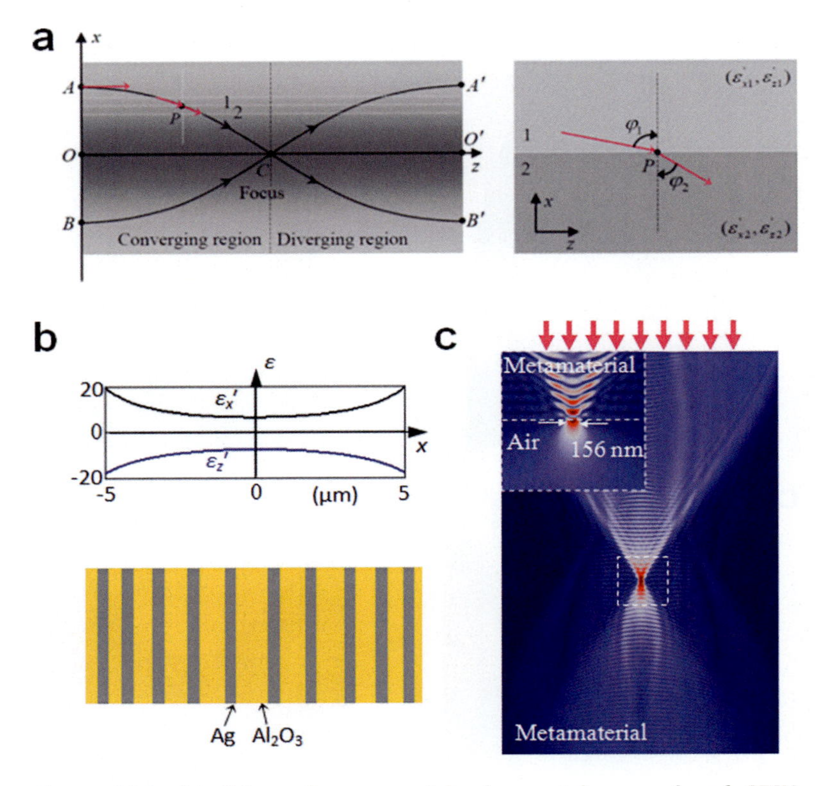

Figure 6.10 (a) Schematic ray model of a metalens made of GRIN metamaterials and the discretization in the x direction. Zoom-in view at a point P is on the right panel. (b) Permittivity profiles of a hyperbolic GRIN metalens using silver nanowires (gray) in an alumina (yellow). The volume filling ratio of silver varies symmetrically from 30% at the center to 60% at the edges. (c) Simulated electrical field intensity showing a deep-subwavelength focusing inside the metamaterial for a normal incident light at a wavelength of 830 nm. The inset shows the zoom-in focus area corresponding to the dashed box when the metalens is truncated at the focal plane.

to a subdiffraction-limited focus inside the metamaterial (Fig. 6.10c). The super-resolution focus with FWHM $= \lambda/5.3$ can be accessed externally on the metamaterials' surface when the metalens is truncated at a plane close to the internal focal plane (Fig. 6.10c, inset).

All the metalenses introduced so far were designed based on real material properties. In addition to material and coupling loss, nonmonochromatic illumination may reduce the focusing performance, as the metamaterials are made of dispersive metal and dielectric composites. Nonetheless, they work fairly well in a relatively wide range of wavelengths, and can always be optimized for further resolution improvement by careful choice of material combinations for desired working wavelengths. Although we mainly focused on super-resolution focusing in 1D at optical frequencies, these principles can be easily extended to 2D, as well as to other frequencies, such as the infrared, terahertz, or microwave regions. The same principles can also be applied to acoustics due to the nature of waves.

6.5.2 Extraordinary Imaging Properties of the Hyperbolic Metalens

The metalens exhibits deep-subwavelength plane-wave focusing with a defined focal length, which is in stark contrast to the hyperlens. The imaging properties of a lens, that is, the image characteristics of an object placed at different locations with respect to the lens focal length f, are well understood, being governed by the imaging equation $1/s_{\text{o}} + 1/s_{\text{i}} = 1/f$, where s_{o} and s_{i} are the object and image distances to the lens, respectively. A metalens achieves subwavelength focusing by combining either hyperbolic or elliptically dispersive metamaterials with a phase compensation mechanism. Extraordinarily, the hyperbolic metalens has negative refraction at the metalens–air interface, leading to an opposite focal shift with respect to the tilted incident plane waves [21–23]. Unprecedented imaging properties are expected from such lenses with anomalous Fourier transform properties [24].

Consider a PWC-based hyperbolic metalens comprising a PWC and a metamaterial slab with a hyperbolic dispersion ($\varepsilon'_x > 0$ and $\varepsilon'_z < 0$). A plane wave incident from the air side converges to a focus F_{m} at a focal length of f_{m} in the hyperbolic metamaterial side (Fig. 6.11a,c). Meanwhile, a plane wave from the metamaterial side diverges after passing through the metalens, resulting in a virtual focus at F_{d} at a focal length of f_{d} also in the metamaterial

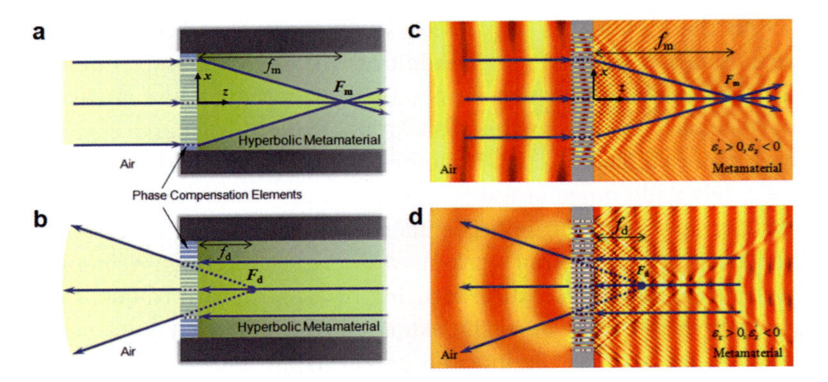

Figure 6.11 Extraordinary focusing properties of a Janus lens by schematics (a, b) and numerical simulations (c, d). (a, c) Plane waves from air converge through a phase-compensating element to a focus (F_{m}) in the metamaterial. (b, d) Plane waves from the metamaterial diverge in air, resulting in a virtual focus (F_{d}) also in the metamaterial. The arrows are eye-guiding rays.

(Fig. 6.11b,d). In contrast to a converging lens with one focus on either side, this distinctive focusing behavior makes the hyperbolic metalens into a Janus lens, having two different focusing behaviors in opposite directions.

For a Janus lens, the imaging behavior is governed by a new metalens imaging equation [24]:

$$\frac{1}{v_{\mathrm{d}}} + \frac{\varepsilon_z'/\sqrt{\varepsilon_x'}}{v_{\mathrm{m}}} = \frac{\varepsilon_z'/\sqrt{\varepsilon_x'}}{f_{\mathrm{m}}}, \tag{6.2}$$

or

$$\frac{1}{v_{\mathrm{d}}} + \frac{\varepsilon_z'/\sqrt{\varepsilon_x'}}{v_{\mathrm{m}}} = \frac{1}{f_{\mathrm{d}}}, \tag{6.3}$$

where v_{d} and v_{m} are the image and object distances in air and the metamaterial, respectively; the focal lengths satisfy $f_{\mathrm{d}} = f_{\mathrm{m}}\sqrt{\varepsilon_x'}/\varepsilon_z'$. In contrast to the conventional imaging equations, the material properties for a metalens also have an important role in its imaging characteristics. Therefore, the Janus lens imaging properties with respect to the focal lengths are completely different from those of a conventional glass lens. Take a hyperbolic metalens with $f_{\mathrm{m}} = 2.0\ \mu\mathrm{m}$ at a wavelength of 690 nm, for instance. The metamaterial is constructed by silver nanowires oriented in the z direction in a

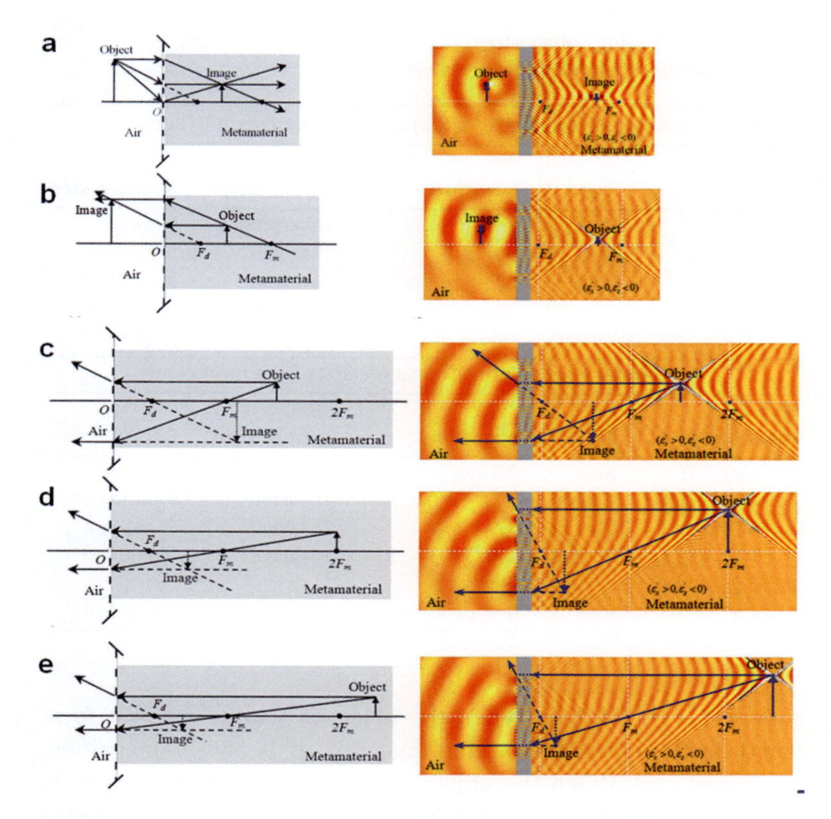

Figure 6.12 Imaging behavior and corresponding simulations of a hyperbolic metalens for an object in the air (a), an object inside point F_m (b), an object between points F_m and $2F_\mathrm{m}$ (c), an object at point $2F_\mathrm{m}$ (d), and an object outside point $2F_\mathrm{m}$ (e).

dielectric host (a refractive index of 1.3) with a filling ratio of 50%, resulting an effective permittivities $\varepsilon_x = 6.4 + 0.03i$ and $\varepsilon_z = -10.3 + 0.2i$. Figure 6.12a shows schematics and numerical simulations of the extraordinary imaging behavior in such a hyperbolic metalens. For an object in air, the image in the metamaterial is always minified, erect, real, and within the first focal length, regardless of the distance of the object in air to the metalens, which is completely different from a conventional lens. Similarly, the image of an object in the metamaterial can also be determined using characteristic rays as

Table 6.1 Imaging properties of a conventional converging lens in comparison with a hyperbolic metalens

Imaging Properties of a Conventional Converging Lens				
Object location	**Type**	**Location**	**Image orientation**	**Relative size**
$\infty > s > 2f$	Real	$f < p < 2f$	Inverted	Minified
$s = 2f$	Real	$p = 2f$	Inverted	Same size
$f < s < 2f$	Real	$2f < p < \infty$	Inverted	Magnified
$s = f$		$\pm\infty$		
$s < f$	Virtual	$-\infty < p < 0$	Erect	Magnified
A Summary of Imaging Characteristics for a Janus Lens				
Object location	**Type**	**Location**	**Image orientation**	**Relative size**
$\infty > v_d > 0$	Real	$0 < v_m < f_m$	Erect	Minified
$\infty > v_m > 2f_m$	Virtual	$2f_d < v_d < f_d$	Inverted	Minified
$v_m > 2f_m$	Virtual	$v_d = 2f_d$	Inverted	Same size
$f_m > v_m < 2f_m$	Virtual	$-\infty < v_d < 2f_d$	Inverted	Magnified
$v_m < f_m$		$\pm\infty$		
$v_m < f_m$	Real	$0 < v_d < \infty$	Erect	Magnified

shown in Fig. 6.12b–e. When the object is within the first focal length in the metamaterial, the image in air is always magnified, erect, and real. When the object is outside the first focal length, the image is always virtual and inverted. Depending on the relative distance of the object, the image is magnified, the same size, and minified for the object distance $f_{\mathrm{m}} < v_{\mathrm{m}} < 2f_{\mathrm{m}}$, $v_{\mathrm{m}} = 2f_{\mathrm{m}}$, and $v_{\mathrm{m}} > 2f_{\mathrm{m}}$, respectively. All the cases are also numerically verified and summarized in Table 6.1 in comparison to imaging properties in a conventional lens.

When the metalens is truncated to a plane within F_{m}, a magnified real image is formed in air for a subwavelength object placed on the truncated surface of the metamaterials [24]. Therefore, a super-resolution image for far-field detection becomes possible. In addition to super-resolution imaging, studies on hyperbolic metalenses with this type of new imaging property pave the way for novel optical devices and systems that could be extremely hard to achieve with conventional lenses.

6.6 Hyperlenses for Acoustic Waves

As in the case for electromagnetic waves, the resolution for conventional acoustic imaging is also limited by the wavelength of acoustic waves due to loss of spatial details carried by evanescent waves. The concept of anisotropic metamaterials has therefore been suggested for super-resolution imaging of acoustic waves for potential applications in sonar sensing, ultrasonic imaging, and nondestructive material testing [44–46]. In 2009, a 2D acoustic hyperlens was experimentally demonstrated for deep-subwavelength imaging over a broadband frequency range and with low loss [47]. Similar to the optical hyperlens, it consists of a nonresonant radially symmetric layered structure but with tapered metallic wires that gradually convert evanescent waves into propagating waves in the far field (Fig. 6.13). Two sound sources with a subwavelength separation of 1.2 cm were well resolved at the outer surface of the acoustic hyperlens when the frequency was swept from 4.2 kHz to 7 kHz, corresponding to a $\lambda/6.8$–$\lambda/4.1$ resolution. Owing to their longitudinal nature, the dispersion relation for acoustic waves is described by the effective mass density together with the effective bulk modulus. Instead of using a hyperbolic dispersion, the acoustic hyperlens in this demonstration has a highly eccentric elliptic dispersion with positive effective mass density that supports a broadband working frequency and a large range of wavevectors. With an elliptic rather than a hyperbolic dispersion, it avoids the use of resonating elements in constructing the metamaterial, resulting in a low-loss hyperlens with very large propagation distances and, thus, a very large magnification. This kind of anisotropic acoustic metamaterial was later utilized in the fabrication of a 3D holey-structured metamaterial for acoustic deep-subwavelength imaging [48]. A 2D object with a line width of 3.18 mm can be clearly resolved at the operating frequency of 2.18 kHz, demonstrating a subwavelength resolution of $\lambda/50$ [48]. This was achieved by the effective transmission of evanescent wave components carrying deep-subwavelength details to the output side through the Fabry–Perot resonances excited inside the holey structure. Although the nearly perfect imaging was confined to the near field of the

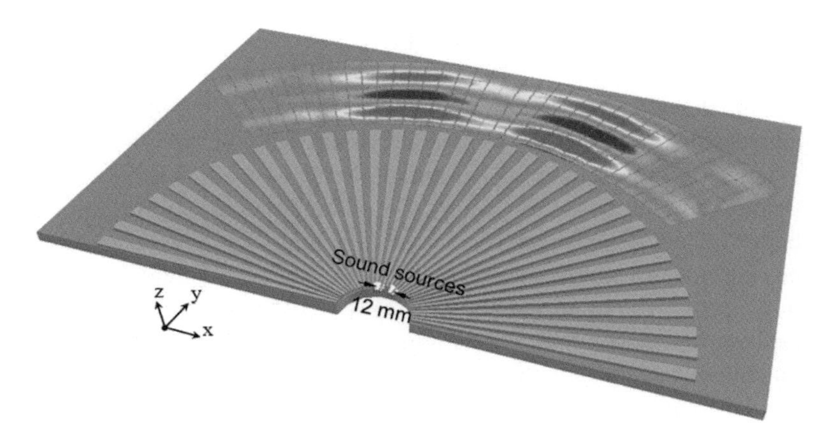

Figure 6.13 Super-resolution imaging of acoustic waves. A 2D acoustic hyperlens made of 36 brass fins (running radially from a radius of 2.7–21.8 cm) in air embedded on a brass substrate, spanning 180° in the angular direction. The cover aluminum sheet has been removed to expose the microstructure for this view. The intensity profile in the green background represents experimental pressure measurements in the propagation region that show a magnified image of a subdiffraction-limited object consisting of two in-phase sound sources separated by 1.2 cm at 6.6 kHz. The pressure is measured at various points ranging from 0.4 cm to 9.9 cm away from the lens.

acoustic metamaterial, it points to a possibility for constructing a 3D acoustic hyperlens for far-field subwavelength imaging by perforating a hemispherical solid with tapered holes in the radial direction.

Despite all the success achieved by acoustic metamaterials, other strategies such as time-reversal techniques [49] have also been extended to overcome acoustic imaging limitations. Based on a sonic analogue to the resonant metalens proposed for electromagnetic waves [50, 51], one group recently proved by experiment that broadband sounds can be controlled and focused on a subwavelength scale by using acoustic resonators [49]. The demonstration of foci with a $\lambda/25$ size for audible sound testifies to the generality of the time reversal principle. Nevertheless, experimental proof of the capability of the time reversal metalens for subwavelength imaging is still needed.

6.7 Perspectives and Outlook

Over the past few years, the hyperlens, built on anisotropic metamaterials with magnifying power, has been demonstrated to achieve subdiffraction-limited resolution for far-field detection across the optical regime. On the other hand, the metalens has recently been proposed as a new member of the superlens family, possessing not only super-resolution power but also Fourier transform function by the incorporation of a phase compensation mechanism. The combined functionality makes the metalens a more suitable candidate for integration with conventional optical systems. Although its deep-subwavelength focusing performance has been proven by theoretical analysis and numerical simulations, experimental demonstration of the metalens is expected to prove its practical feasibility.

There are two major limiting factors on future applications of both hyperlenses and metalenses. First, material loss is an intrinsic property of metamaterials at optical wavelengths, restricting the ultimate resolution and signal transmission efficiency of the lens especially when resonant elements are involved. To mitigate the loss issue, several approaches may provide solutions, including compensating for loss in the metal by addition of a gain medium in dielectrics [52], searching for better plasmonic materials among existing elements, band structure engineering, or material doping or alloying [53–55]. Second, the object has to be placed in the near field of the lens to make use of the evanescent waves that normally decay away from the object, although the image can be projected and detected in the far field. Super-resolution is thus limited to surface imaging, which may encounter difficulties in extension to the third dimension. New imaging schemes may be needed to solve this issue.

Besides, current designs for both hyperlenses and metalenses have a limited field of view due to the limited physical size of the lens. Better designs, such as periodic array geometries with an improved field-of-view-to-lens-diameter ratio, may solve the problem, although it remains wide open to solutions. Image distortion and corresponding compensation methods are also important issues that need further investigation for practical applications of these lenses. In the case of hyperlenses, imaging at the back focal plane may result in a great reduction in distortions.

Despite existing limitations, super-resolution achieved by hyperlenses and metalenses has already shown exciting potential applications in various fields of science and technology. The implementation of these lenses in conventional optical microscopes may extend the resolution to the nanometer scale for real-time observations, which would have a great impact on modern biological imaging. Further improvement is expected by combining hyperlenses or metalenses with other imaging techniques, such as dark-field microscopy for high-contrast microscopy [56]. Super-resolution imaging and its reversed process for lithography may find applications in optical nanolithography and ultrahigh-density optical data storage [37]. The extension of the hyperlens and metalens concept to acoustic waves will also lead to super-resolution in applications such as sonar sensing and ultrasonic imaging.

Acknowledgments

We acknowledge financial support from NSF-ECCS under Grant No. 0969405 and NSF-CMMI under Grant No. 1120795.

References

1. Betzig, E., and Trautman, J. K. (1992). Near-field optics: microscopy, spectroscopy, and surface modification beyond the diffraction limit, *Science*, **257**, pp. 189–195.
2. Hell, S. W. (2007). Far-field optical nanoscopy, *Science*, **316**, pp. 1153–1158.
3. Huang, B., Wang, W., Bates, M., and Zhuang, X. (2008). Three-dimensional super-resolution imaging by stochastic optical reconstruction microscopy, *Science*, **319**, pp. 810–813.
4. Mansfield, S. M., and Kino, G. S. (1990). Solid immersion microscope, *Appl. Phys. Lett.*, **57**, pp. 2615–2616.
5. Soukoulis, C. M., Linden, S., and Wegener, M. (2007). Negative refractive index at optical wavelengths, *Science*, **315**, pp. 47–49.

6. Shelby, R. A., Smith, D. R., and Schultz, S. (2001). Experimental verification of a negative index of refrraction, *Science*, **292**, pp. 77–79.
7. Podolskiy, V. A., and Narimanov, E. E. (2005). Strongly anisotropic waveguide as a nonmagnetic left-handed system, *Phys. Rev. B*, **71**, p. 201101(R).
8. Podolskiy, V. A., Alekseyev, L. V., and Narimanov, E. E. (2005). Strongly anisotropic media: the THz perspectives of left-handed materials, *J. Mod. Opt.*, **52**, pp. 2343–2349.
9. Lu, D., Kan, J. J., Fullerton, E. E., and Liu, Z. (2014). Enhancing spontaneous emission rates of molecules using nanopatterned multilayer hyperbolic metamaterials, *Nat. Nanotechnol.*, **9**, pp. 48–53.
10. Ferrari, L., Lu, D., Lepage, D., and Liu, Z. W. (2014). Enhanced spontaneous emission inside hyperbolic metamaterials, *Opt. Express*, **22**, pp. 4301–4306.
11. Zhang, X., and Liu, Z. (2008). Superlenses to overcome the diffraction limit, *Nat. Mater.*, **7**, pp. 435–441.
12. Lu, D., and Liu, Z. (2012). Hyperlenses and metalenses for far-field super-resolution imaging, *Nat. Commun.*, **3**, p. 1205.
13. Pendry, J. B. (2000). Negative refraction makes a perfect lens, *Phys. Rev. Lett.*, **85**, p. 3966.
14. Fang, N., Lee, H., Sun, C., and Zhang, X. (2005). Sub-diffraction-limited optical imaging with a silver superlens, *Science*, **308**, pp. 534–537.
15. Taubner, T., Korobkin, D., Urzhumov, Y., Shvets, G., and Hillenbrand, R. (2006). Near-field microscopy through a SiC superlens, *Science*, **313**, p. 1595.
16. Liu, Z., Lee, H., Xiong, Y., Sun, C., and Zhang, X. (2007). Far-field optical hyperlens magnifying sub-diffraction-limited objects, *Science*, **315**, pp. 1686–1701.
17. Smolyaninov, I. I., Hung, Y.-J., and Davis, C. C. (2007). Magnifying superlens in the visible frequency range, *Science*, **315**, p. 1699.
18. Jacob, Z., Alekseyev, L. V., and Narimanov, E. (2006). Optical hyperlens: far-field imaging beyond the diffraction limit, *Opt. Express*, **14**, pp. 8247–8256.
19. Liu, Z., et al. (2007). Far-field optical superlens, *Nano Lett.*, **7**, pp. 403–408.
20. Rho, J., et al. (2010). Spherical hyperlens for two-dimensional sub-diffractional imaging at visible frequencies, *Nat. Commun.*, **1**, p. 143.
21. Ma, C. B., and Liu, Z. W. (2010). A super resolution metalens with phase compensation mechanism, *Appl. Phys. Lett.*, **96**, p. 183103.

22. Ma, C. B., and Liu, Z. W. (2010). Focusing light into deep subwavelength using metamaterial immersion lenses, *Opt. Express*, **18**, pp. 4838–4844.

23. Ma, C. B., Escobar, M. A., and Liu, Z. W. (2011). Extraordinary light focusing and Fourier transform properties of gradient-index metalenses, *Phys. Rev. B*, **84**, p. 195142.

24. Ma, C., and Liu, Z. (2012). Breaking the imaging symmetry in negative refraction lenses, *Opt. Express*, **20**, pp. 2581–2586.

25. Yao, J., et al. (2008). Optical negative refraction in bulk metamaterials of nanowires, *Science*, **321**, p. 930.

26. Ramakrishna, S. A., Pendry, J. B., Wiltshire, M. C. K., and Stewart, W. J. (2003). Imaging the near field, *J. Mod. Opt.*, **50**, pp. 1419–1430.

27. Ma, C., Aguinaldo, R., and Liu, Z. (2010). Advances in the hyperlens, *Chin. Sci. Bull.*, **55**, pp. 2618–2624.

28. Pendry, J. B., and Ramakrishna, S. A. (2002). Near-field lenses in two dimensions, *J. Phys.: Condens. Matter*, **14**, pp. 8463–8479.

29. Salandrino, A., and Engheta, N. (2006). Far-field subdiffraction optical microscopy using metamaterial crystals: theory and simulations, *Phys. Rev. B*, **74**, p. 075103.

30. Shvets, G., Trendafilov, S., Pendry, J. B., and Sarychev, A. (2007). Guiding, focusing, and sensing on the subwavelength scale using metallic wire arrays, *Phys. Rev. Lett.*, **99**, p. 053903.

31. Ono, A., Kato, J., and Kawata, S. (2005). Subwavelength optical imaging through a metallic nanorod array, *Phys. Rev. Lett.*, **95**, p. 267407.

32. Han, S., et al. (2008). Ray optics at a deep-subwavelength scale: a transformation optics approach, *Nano Lett.*, **8**, pp. 4243–4247.

33. Kildishev, A. V., and Shalaev, V. M. (2008). Engineering space for light via transformation optics, *Opt. Lett.*, **33**, pp. 43–45.

34. Kildishev, A. V., and Narimanov, E. E. (2007). Impedance-matched hyperlens, *Opt. Lett.*, **32**, pp. 3432–3434.

35. Smith, E. J., Liu, Z., Mei, Y. F., and Schmidt, O. G. (2009). System investigation of a rolled-up metamaterial optical hyperlens structure, *Appl. Phys. Lett.*, **95**, p. 083104.

36. Lee, H., Liu, Z., Xiong, Y., Sun, C., and Zhang, X. (2007). Development of optical hyperlens for imaging below the diffraction limit, *Opt. Express*, **15**, pp. 15886–15891.

37. Xiong, Y., Liu, Z. W., and Zhang, X. (2009). A simple design of flat hyperlens for lithography and imaging with half-pitch resolution down to 20 nm, *Appl. Phys. Lett.*, **94**, p. 203108.

38. Li, G. X., Tam, H. L., Wang, F. Y., and Cheah, K. W. (2008). Half-cylindrical far field superlens with coupled Fabry-Perot cavities, *J. Appl. Phys.*, **104**, p. 096103.
39. Schwaiger, S., et al. (2009). Rolled-up three-dimensional metamaterials with a tunable plasma frequency in the visible regime, *Phys. Rev. Lett.*, **102**, p. 163903.
40. Pendry, J. B., Schurig, D., and Smith, D. R. (2006). Controlling electromagnetic fields, *Science*, **312**, pp. 1780–1782.
41. Gómez-Reino, C., Pérez, M., and Bao, C. (2002). *Gradient-Index Optics: Fundamentals and Applications* (Springer Verlag, New York).
42. Ma, C. B., and Liu, Z. W. (2011). Designing super-resolution metalenses by the combination of metamaterials and nanoscale plasmonic waveguide couplers, *J. Nanophotonics*, **5**, p. 051604.
43. Wu, Q., Feke, G. D., Grober, R. D., and Ghislain, L. P. (1999). Realization of numerical aperture 2.0 using a gallium phosphide solid immersion lens, *Appl. Phys. Lett.*, **75**, p. 4064.
44. Guenneau, S., Movchan, A., Petursson, G., and Ramakrishna, S. A. (2007). Acoustic metamaterials for sound focusing and confinement, *New J. Phys.*, **9**, p. 399.
45. Ao, X. Y., and Chan, C. T. (2008). Far-field image magnification for acoustic waves using anisotropic acoustic metamaterials, *Phys. Rev. E*, **77**, p. 025601(R).
46. Lee, H. J., Kim, H. W., and Kim, Y. Y. (2011). Far-field subwavelength imaging for ultrasonic elastic waves in a plate using an elastic hyperlens, *Appl. Phys. Lett.*, **98**, p. 241912.
47. Li, J. S., Fok, L., Yin, X. B., Bartal, G., and Zhang, X. (2009). Experimental demonstration of an acoustic magnifying hyperlens, *Nat. Mater.*, **8**, pp. 931–934.
48. Zhu, J., et al. (2011). A holey-structured metamaterial for acoustic deep-subwavelength imaging, *Nat. Phys.*, **7**, pp. 52–55.
49. Lemoult, F., Fink, M., and Lerosey, G. (2011). Acoustic resonators for far-field control of sound on a subwavelength scale, *Phys. Rev. Lett.*, **107**, p. 064301.
50. Lerosey, G., De Rosny, J., Tourin, A., and Fink, M. (2007). Focusing beyond the diffraction limit with far-field time reversal, *Science*, **315**, pp. 1120–1122.
51. Lemoult, F., Lerosey, G., de Rosny, J., and Fink, M. (2010). Resonant metalenses for breaking the diffraction barrier, *Phys. Rev. Lett.*, **104**, p. 203901.

52. Xiao, S. M., et al. (2010). Loss-free and active optical negative-index metamaterials, *Nature*, **466**, pp. 735–738.
53. Boltasseva, A., and Atwater, H. A. (2011). Low-loss plasmonic metamaterials, *Science*, **331**, pp. 290–291.
54. Naik, G. V., Liu, J., Kildishev, A. V., Shalaev, V. M., and Boltasseva, A. (2012). Demonstration of Al:ZnO as a plasmonic component for near-infrared metamaterials, *Proc. Natl. Acad. Sci. U.S.A.*, **109**, pp. 8834–8838.
55. Lu, D., Kan, J., Fullerton, E. E., and Liu, Z. (2011). Tunable surface plasmon polaritons in Ag composite films by adding dielectrics or semiconductors, *Appl. Phys. Lett.*, **98**, p. 243114.
56. Hu, H., Ma, C., and Liu, Z. (2010). Plasmonic dark field microscopy, *Appl. Phys. Lett.*, **96**, p. 113107.

Chapter 7

Modeling Linear and Nonlinear Hyperlens Structures

Daniel Aronovich and Guy Bartal
Department of Electrical Engineering, Technion–Israel Institute of Technology, Haifa 32000, Israel
guy@ee.technion.ac.il

7.1 Motivation

Subwavelength imaging [1–5] has been studied extensively in the past few years due to its great importance in many fields, ranging from the fundamental understanding of light propagation at the nanoscale to applications such as sensing, nanolithography, and microscopy. Moreover, studying and manipulating light on the nanoscale is of great importance to many different disciplines other than subwavelength imaging, with already a huge influence on communication [6, 7], security [8], and biosensing applications [9]. Conventional imaging systems are generally restricted by the diffraction limit [10] in free space, acting as a low-pass filter that inhibits transmission of spatial frequencies larger than $1/\lambda$. These spatial frequencies are evanescent in a homogeneous medium and

Plasmonics and Super-Resolution Imaging
Edited by Zhaowei Liu

ISBN 978-981-4669-91-7 (Hardcover), 978-1-315-20653-0 (eBook)
www.panstanford.com

therefore cannot contribute to the reconstruction of the image at the system's output. Current modern methods overcome this limit using techniques such as near-field scanning with a subwavelength tip [11] or by randomly distributing fluorescent molecules on the sample and averaging over multiple exposures [12]. These solutions require scanning or repetitive experiments, thus limiting real-time applications.

In recent years, a different approach emerged with the introduction of *metamaterials*, which are artificially engineered media with designed properties beyond those available in nature. Metamaterials are composed of nanoinclusions with a unit cell much smaller than the wavelength of light, recently made possible with the advance in nanofabrication. Due to their periodic nature, metamaterials have much in common with photonic crystals [13]. However, photonic crystals generally have a periodicity of the same order as the wavelength and hence cannot be described by homogenization methods such as effective medium theory (EMT) [14]. The field of metamaterials paved the way toward achieving real-time subwavelength imaging, starting from the search for negative dielectric permittivity and magnetic permeability [15] and vastly evolved with the introduction of the negative-index [16] "perfect lens" and its subsequent realization by a metallic slab [17]. Following this single-slab device, a more advanced design of a structure made of alternating metal-dielectric layers was proposed [18], where, for periodicity much smaller than the wavelength, effective anisotropic properties were assigned utilizing the EMT. Such stacks are commonly named hyperbolic metamaterials (HMMs) or indefinite media [19, 20], which can be considered as a polaritonic crystal where the coupled states of light and matter give rise to a larger bulk density of states [21] and the support of high **k**-vectors [19, 22, 23]. The operating mechanism of these structures is attributed to surface plasmon polariton (SPP) coupling between adjacent layers, giving rise to Bloch-like modes that carry the subwavelength information [24]. While being capable of transferring subdiffraction-limited spatial features, the subwavelength information remains confined to the near field of the structures. This restriction was removed with the introduction of the hyperlens [1, 25–27], which utilizes cylindrical geometry to adiabatically magnify the subwavelength features of the imaged objects, pushing them above the diffraction limit at the device output such that they could be conveyed by conventional microscopy.

7.2 Background

7.2.1 The Hyperlens

The hyperlens was introduced in 2006 in back-to-back theoretical publications [25, 26], proposing adiabatic stretching of a subwavelength image to convert evanescent information into a propagating one, such that it can be processed by conventional optics at the far field. Experimental realization of the hyperlens was carried out in 2007 by two experimental groups [1, 27] utilizing cylindrical metal-dielectric multilayers (MDMLs). A schematic of the experimental scheme, performed in Ref. [1], is shown in Fig. 7.1a. This process can be also interpreted in the wavevector space (*k*-space); Fig. 7.1b

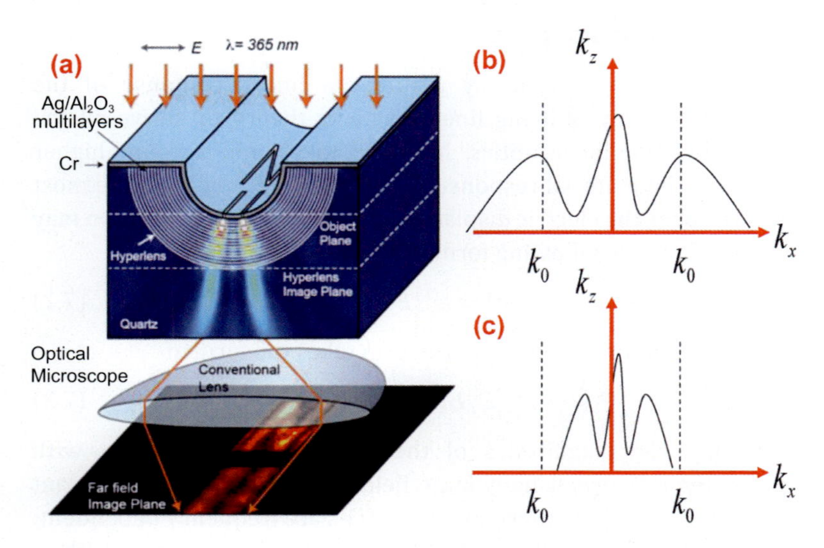

Figure 7.1 (a) The experimental hyperlens. (b) Fourier transform of the image at the input facet. (c) Fourier transform of the image at the output facet.

depicts the Fourier transform of the input image and clearly shows that part of it lies outside the propagation window. "Stretching" the image in real space corresponds to "squeezing" it in k-space (Fig. 7.1c).

The stretching is proportional to the magnification ratio of the hyperlens, which can be expressed as

$$M = \frac{r_{\text{out}}}{r_{\text{in}}} \tag{7.1}$$

Ideally, by constructing thick devices, a large magnification ratio can be achieved. In practice, however, the hyperlens relies on SPPs, which suffer high ohmic losses, posing a limit on the values of M in order to maintain a reasonable signal-to-noise ratio (SNR). Furthermore, the condition under which the diffraction is completely inhibited, which is necessary for perfect imaging, exhibits the highest losses. We shall refer to this condition as the "perfect imaging" condition. This is one of the major drawbacks of the hyperlens—a drawback we aim to tackle by introducing the nonlinear hyperlens.

7.2.2 Nonlinear Optics

Hyperlens analysis typically assumes a linear response of the medium, thereby utilizing linear systems theory to derive all the equations. Nonlinear optics, however, takes into account higher orders in the material response (susceptibility), where, in the most general form, the electric displacement in the frequency domain may be written in the following form:

$$\vec{D} = \varepsilon^{(\mathrm{L})}\vec{E} + \vec{P}_{\mathrm{NL}} \tag{7.2}$$

where the nonlinear polarization term takes the form of

$$P_{\mathrm{NL}} = \chi^{(2)}_{ijk}E_jE_k + \chi^{(3)}_{ijkl}E_jE_kE_l + \cdots \tag{7.3}$$

Generally, the magnitudes of the $\chi(i)$-s decrease rapidly with increasing i; hence, usually large fields are required for significant nonlinear effects. Furthermore, the $\chi(i)$-s are frequency dependent; however, most problems require a particular process with a given frequency and a well-defined polarization direction. In these scenarios, one can drop all frequency arguments and indices and use an effective scalar susceptibility $\chi(i)$. In this work we have assumed the presence of the third-order susceptibility $\chi(3)$, which is responsible for the optical (AC) Kerr effect; this nonlinearity is expressed as a local change in the refractive index, or permittivity, as a function of the local intensity of the illumination:

$$P^{(3)}_{\mathrm{NL},\omega} = \frac{1}{2}\varepsilon_0 \chi^{(3)}_{iiii}(-\omega;\omega,\omega,-\omega)\left(\left|\vec{E}_\omega\right|^2 \vec{E}_\omega + \frac{1}{2}\left(\vec{E}_\omega \cdot \vec{E}_\omega\right)\vec{E}^*_\omega\right) \tag{7.4}$$

Due to the fact that the nonlinear polarization is at the same frequency as the source, a phasor representation of the field is still valid and the total permittivity of the medium is the sum of the linear and the nonlinear ones:

$$\vec{\varepsilon}^{\,\mathrm{tot}} = \varepsilon^{(\mathrm{L})} + \varepsilon^{(\mathrm{NL})} \tag{7.5}$$

7.3 Numerical Techniques and Algorithms

This section focuses on the main numerical methods that were utilized in this work. The known techniques include the beam propagation method (BPM) and the transfer matrix method (TMM). The first is a split-step Fourier algorithm, typically used for solving the initial condition problem for the wave equation under the paraxial approximation. In this work, it is extended to tackle the nonlinear, nonparaxial problem in a cylindrically symmetric system for propagation along the radial direction. The TMM is extensively used for solving propagation problems in layered media. This tool was also extended to cylindrical layers rather than flat layers.

7.3.1 The Beam Propagation Method

The BPM was studied extensively under the paraxial approximation, first during the early 1980s for solving the Schrödinger equation (Eq. 7.11) and the paraxial wave equation (Eq. 7.12), whereas later it was developed to handle reflections (Eq. 7.13). In what follows, we introduce the cylindrical BPM which can simulate subwavelength propagation in the radial direction without assuming the paraxial approximation.

7.3.2 The Cylindrical Beam Propagation Method

The cylindrical BPM is a novel algorithm that was developed to analyze wave propagation along the *radial* direction in a cylindrically symmetric system. For transverse magnetic (TM) polarization ($E\rho$, $E\phi$, Hz), Maxwell's equations in cylindrical coordinates acquire the form as

$$\vec{\nabla} \times \vec{H} = -i\frac{\omega}{c}\overline{\varepsilon}\overrightarrow{E} \quad \Rightarrow \quad \frac{1}{\rho}\begin{vmatrix} \hat{\rho} & \rho\hat{\phi} & \widehat{z} \\ \partial_\rho & \partial_\phi & 0 \\ 0 & 0 & H_z \end{vmatrix}$$

$$= \left[\frac{1}{\rho}\frac{\partial}{\partial\phi}H_z\right]\widehat{\rho} - \left[\frac{\partial}{\partial\rho}H_z\right]\widehat{\phi} = -ik_0\left(\varepsilon_\phi E_\phi\widehat{\phi} + \varepsilon_\rho E_\rho\widehat{\rho}\right)$$

$$\vec{\nabla} \times \vec{E} = i\frac{\omega}{c}\overrightarrow{H} \quad \Rightarrow \quad \frac{1}{\rho}\begin{vmatrix} \widehat{\rho} & \rho\widehat{\phi} & \widehat{z} \\ \partial_\rho & \partial_\phi & 0 \\ E_\rho & \rho E_\phi & 0 \end{vmatrix}$$

$$= \frac{1}{\rho}\frac{\partial}{\partial\rho}\left(\rho E_\phi\right) - \frac{1}{\rho}\frac{\partial}{\partial\phi}E_\rho = ik_0 H_z \tag{7.6}$$

The rotor equations yield three coupled equations:

$$\begin{cases} \dfrac{1}{\rho}\dfrac{\partial}{\partial\rho}\left(\rho E_\phi\right) - \dfrac{1}{\rho}\dfrac{\partial}{\partial\phi}E_\rho = ik_0 H_z \\ E_\rho = \dfrac{i}{\rho\varepsilon_\rho k_0}\dfrac{\partial}{\partial\phi}H_z \\ E_\phi = \dfrac{-i}{\varepsilon_\phi k_0}\dfrac{\partial}{\partial\rho}H_z \end{cases} \tag{7.7}$$

Substituting the second and third equations into the first one, we obtain

$$\frac{\rho}{\varepsilon_\phi}\frac{\partial}{\partial\rho}\left(\rho\frac{\partial}{\partial\rho}H_z\right) + \frac{1}{\varepsilon_\rho}\frac{\partial}{\partial^2\phi}H_z = -k_0^2\rho^2 H_z \tag{7.8}$$

We next use variable separation to find solutions to this equation:

$$H_z = R\left(\rho\right)\Phi\left(\phi\right) \tag{7.9}$$

This results in the exponential (phase) solution in the azimuthal direction and the generalized Bessel equation in the radial one:

$$\begin{cases} \Phi\left(\phi\right) = \Phi_0 e^{im\phi} \\ \dfrac{\partial^2}{\partial\rho^2}R\left(\rho\right) + \dfrac{1}{\rho}\dfrac{\partial}{\partial\rho}R\left(\rho\right) + \left(k_0^2\varepsilon_\phi - \dfrac{\varepsilon_\phi}{\varepsilon_\rho}\left(\dfrac{m}{\rho}\right)^2\right)R\left(\rho\right) = 0 \end{cases} \tag{7.10}$$

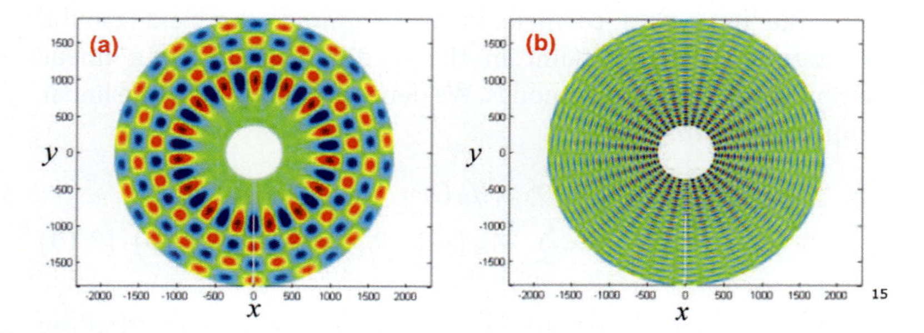

Figure 7.2 A comparison between an isotropic medium linear mode (a), which exhibits a cutoff at some radius, and the hyperbolic medium mode at the same order (b), which can penetrate to radii smaller than the wavelength.

The final solution of the linear mode in such a medium is

$$H_z(\rho, \phi) = R(\rho)\,\Phi(\phi) \propto J_{m\sqrt{\frac{\varepsilon_\phi}{\varepsilon_\rho}}}\left(k_0\sqrt{\varepsilon_\phi}\rho\right) e^{im\phi} \tag{7.11}$$

Given that a hyperlens possesses a hyperbolic dispersion relation, that is, $\varepsilon\phi > 0$, $\varepsilon\rho < 0$, the Bessel orders become purely imaginary if no loss is present or complex when loss is taken into account. These modes can penetrate to radii smaller than the wavelength, even at high orders, unlike their conventional counterparts, hence forming the basis for the hyperlens operation. A calculation of the complex Bessel mode is presented (Fig. 7.2b), showing no diffraction limit. The lack of diffraction limit is manifested in the absence of the cutoff for the high-order azimuthal mode. The isotropic linear mode (regular Bessel), on the other hand, is subjected to the diffraction limit and thus exhibits a cutoff (Fig. 7.2a). This comparison implies that the hyperbolic modes can support subwavelength imaging (a conclusion supported by the amplitude transfer function analysis which we carried out in Fig. 7.8).

The problem of propagation in the cylindrical hyperbolic media can be addressed as follows: An arbitrary input field distribution H_0 at a specific radius ρ_0 can be projected on each of the linear modes. The projection over the mth mode is defined by the overlap integral:

$$\begin{aligned} A_m(\rho_0) &= \int_{-\pi}^{\pi} H_0(\rho=\rho_0)\cdot R_m(\rho=\rho_0)\,\Phi_m(\phi)\,d\phi \\ &= \int_{-\pi}^{\pi} H_0(\rho=\rho_0) J_{m\sqrt{\frac{\varepsilon_\phi}{\varepsilon_\rho}}}\left(ik_0\sqrt{\varepsilon_\phi}\rho_0\right) e^{im\phi}\,d\phi \end{aligned} \tag{7.12}$$

Once the projection coefficients are acquired, the field can be reconstructed at any point in the medium simply by a linear combination of the linear modes. We define this process as the linear propagator $D\Delta\rho$:

$$\begin{aligned} H\left(\rho_0+\Delta\rho,\phi\right) &= D_{\Delta\rho}\left\{H_0\left(\rho=\rho_0,\phi\right)\right\} \\ &= \sum_m A_m\left(\rho_0\right)\cdot R_m\left(\rho_0+\Delta\rho\right)\Phi_m\left(\phi\right) \end{aligned} \quad (7.13)$$

This process is analogous to the Fourier transform, except that the input function is spanned in the Bessel mode base rather than the plane-wave base.

7.3.3 The Nonlinear Beam Propagation Method

Consider a nonlinear medium possessing Kerr-type nonlinearity characterized by its third-order susceptibility $\chi(3)$. The nonlinear polarization is given by

$$\begin{aligned} \overrightarrow{P}_{\mathrm{NL}}\left(\rho,\phi\right) &= \frac{1}{2}\varepsilon_0\chi^{(3)}\left[|E|^2\overrightarrow{E}+\frac{1}{2}\left(\overrightarrow{E}\cdot\overrightarrow{E}\right)\overrightarrow{E}^*\right] \\ &= \varepsilon_\phi^{(\mathrm{NL})}\left(\rho,\phi\right)E_\phi\widehat{\phi}+\varepsilon_\rho^{(\mathrm{NL})}\left(\rho,\phi\right)E_\rho\widehat{\rho} \end{aligned} \quad (7.14)$$

For TM polarization, the electric field has two components, $E\phi$ and $E\rho$, providing the ability to compute the total effective permittivity vector ε tot, which is the sum of the linear and the nonlinear permittivity:

$$\begin{aligned} \vec{\varepsilon}^{\mathrm{tot}}\left(\rho,\phi\right) &= \left(\varepsilon_\phi^{(\mathrm{L})}+\varepsilon_\phi^{(\mathrm{NL})}\left(\rho,\phi\right)\right)\widehat{\phi}+\left(\varepsilon_\rho^{(\mathrm{L})}+\varepsilon_\rho^{(\mathrm{NL})}\left(\rho,\phi\right)\right)\widehat{\rho} \\ &= \varepsilon_\phi^{\mathrm{tot}}\left(\rho,\phi\right)\widehat{\phi}+\varepsilon_\rho^{\mathrm{tot}}\left(\rho,\phi\right)\widehat{\rho} \end{aligned} \quad (7.15)$$

Care should be taken while using the total effective permittivity tensor, due to its inhomogeneous nature. We rewrite Eq. 7.7 by substituting one equation into the other two to obtain a set of two coupled equations, where we use the total effective permittivity tensor components similar to the one introduced for the Cartesian symmetry:

$$\frac{\partial}{\partial\rho}\Psi = M^{\mathrm{tot}}\Psi \quad (7.16)$$

This is a first-order partial differential equation (PDE), where we define a generalized vector Ψ and a propagation matrix M^{tot}:

$$M^{\text{tot}} = \begin{pmatrix} 0 & i\varepsilon_{\phi}^{\text{tot}}k_0 \\ ik_0 + \dfrac{i}{\rho^2\varepsilon_{\rho}^{\text{tot}}k_0}\dfrac{\partial^2}{\partial\phi^2} & -\frac{1}{\rho} \end{pmatrix}, \quad \Psi = \begin{pmatrix} H_z \\ E_{\phi} \end{pmatrix} \tag{7.17}$$

While M^{tot} contains the information for both linear and nonlinear contributions, for a sufficiently small $\Delta\rho$ it can be expressed as the sum of the linear and nonlinear propagation matrices: $M^{\text{tot}} = M^{\text{L}} + M^{\text{NL}}$. Hence, the extraction of the nonlinear propagation matrix M^{NL} can be achieved by subtracting M^{L} from M^{tot}, where

$$M^{\text{L}} = \begin{pmatrix} 0 & i\varepsilon_{\phi}^{L}k_0 \\ ik_0 + \dfrac{i}{\rho^2\varepsilon_{\rho}^{L}k_0}\dfrac{\partial^2}{\partial\phi^2} & -\frac{1}{\rho} \end{pmatrix}, \quad \Psi = \begin{pmatrix} H_z \\ E_{\phi} \end{pmatrix} \tag{7.18}$$

Assuming a small propagation step $\Delta\rho$, the PDE (Eq. 7.16) can be solved as

$$\begin{aligned} \Psi(\rho + \Delta\rho) &= e^{M^{\text{tot}}\Delta\rho}\Psi(\rho) + o\left[f(\Delta z)\right] \\ &= e^{M^{\text{NL}}\Delta\rho}e^{M^{\text{L}}\Delta\rho}\Psi(\rho) + o\left[f(\Delta z)\right] \end{aligned} \tag{7.19}$$

For a given input field distribution, this equation provides the ability to calculate the field at any point in space. The propagation operators are split to the linear propagator, $\exp(M^{\text{L}}\Delta\rho)$, which is responsible for all diffraction phenomena during propagation and, the nonlinear propagator, $\exp(M^{\text{NL}}\ \Delta\rho)$, manifesting the nonlinear effects. Both propagators are evaluated using finite difference methods; the differential term can be written as

$$\frac{\partial^2}{\partial\phi^2} = \frac{1}{\Delta\phi}\left(D_{\phi} - \frac{1}{12}D_{\phi}^2 + \frac{1}{90}D_{\phi}^4 - \ldots\right) \tag{7.20}$$

where D_{ϕ} is the central difference operator:

$$D_{\phi} = \begin{pmatrix} -2 & 1 & 0 & . & . & 0 \\ 1 & -2 & 1 & . & . & 0 \\ 0 & 1 & . & . & . & . \\ . & . & . & . & . & . \\ . & . & . & . & -2 & 1 \\ 0 & 0 & . & . & 1 & -2 \end{pmatrix} \tag{7.21}$$

Each term in Eq. 7.20 corresponds to a different truncation error; taking higher terms will decrease the error on one hand but increase the computation time on the other. One can reduce the error by replacing the linear propagator in Eq. 7.19 with the linear propagator $D\Delta\rho$ which was derived earlier, leading to the final cylindrical nonlinear propagation scheme:

$$\Psi(\rho + \Delta\rho) = e^{M^{NL}\Delta\rho} D_{\Delta}\{\Psi(\rho)\} + o[f(\Delta z)] \tag{7.22}$$

7.3.4 The Finite Difference Method

The finite difference method is a widely used mathematical method for representing differential equations as a set of algebraic equations, which may be packed into a matrix and solved numerically. The basic idea is to define a set of points in space and approximate the differential equation on these points. The derivative in Eq. (37) can be calculated as

$$\frac{1}{\varepsilon(x_i)}\frac{\partial}{\partial x}H(x_i) = \frac{1}{\varepsilon(x_i)}\frac{H(x_{i+1}) - H(x_i)}{\Delta x} \tag{7.23}$$

where

$$\Delta x = \frac{D}{N} \quad N = \text{number of samples.}$$

The second derivative can be written in the same way:

$$\frac{\partial}{\partial x}\left(\frac{1}{\varepsilon(i)}\frac{\partial}{\partial x}H(x_i)\right) = \frac{\frac{1}{\varepsilon(x_{i+1})}\frac{H(x_{i+2})-H(x_{i+1})}{\Delta x} - \frac{1}{\varepsilon(x_i)}\frac{H(x_{i+1})-H(x_i)}{\Delta x}}{\Delta x}$$

$$= \frac{1}{\varepsilon(x_i)\Delta x^2}H(x_i) - \frac{1}{\Delta x^2}\left(\frac{1}{\varepsilon(x_{i+1})} - \frac{1}{\varepsilon(x_i)}\right)$$

$$H(x_{i+1}) + \frac{1}{\varepsilon(x_{i+1})\Delta x^2}H(x_{i+2}) \tag{7.24}$$

and in matrix form:

$$\begin{pmatrix} \cdot & \cdot & \cdot & \cdot & \cdot & \cdot \\ \cdot & 0 & 0 & 0 & 0 & \cdot \\ \cdot & 0 & \frac{1}{\varepsilon(x_i)\Delta x^2} & -\frac{1}{\Delta x^2}\left(\frac{1}{\varepsilon(x_{i+1})} - \frac{1}{\varepsilon(x_i)}\right) & \frac{1}{\varepsilon(x_{i+1})\Delta x^2} & \cdot \\ \cdot & 0 & 0 & 0 & 0 & \cdot \\ \cdot & 0 & 0 & 0 & 0 & \cdot \\ \cdot & \cdot & \cdot & \cdot & \cdot & \cdot \end{pmatrix}$$

$$\begin{pmatrix} \cdot \\ H(i-1) \\ H(i) \\ H(i+1) \\ H(i+2) \\ \cdot \end{pmatrix} = \frac{k_z}{k_0} \begin{pmatrix} \cdot \\ H(i-1) \\ H(i) \\ H(i+1) \\ H(i+2) \\ \cdot \end{pmatrix}$$

7.3.5 The Cylindrical Transfer Matrix Method

A similar methodology can be applied to cylindrical symmetry. Solving the homogeneous Maxwell's equation in each layer yields backward and forward cylindrical waves, which are Bessel functions (Fig. 7.3).

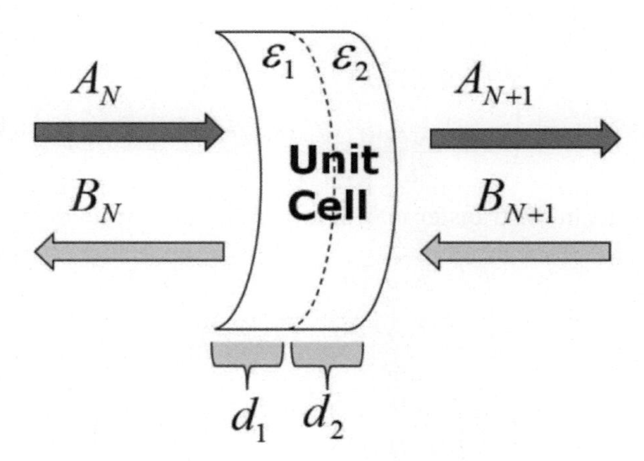

Figure 7.3 Cylindrical configuration.

The magnetic field in each of the slabs can be written as

$$H_z(r,\phi) = \begin{cases} A_0 J_n\left(k_0\sqrt{\varepsilon_1} r\right) + B_0 Y_n\left(k_0\sqrt{\varepsilon_1} r\right) & 0 < r < d_1 \\ A_1 J_n\left(k_o\sqrt{\varepsilon_2} r\right) + B_1 Y_n\left(k_o\sqrt{\varepsilon_2} r\right) & d_1 < r < D \end{cases} \tag{7.25}$$

where the electric field is derived by Eq. 7.7:

$$E_\phi(r) = \begin{cases} \dfrac{-i}{\varepsilon_1 k_0}\left[A_0 J_n'\left(k_0\sqrt{\varepsilon_1} r\right) + B_0 Y_n'\left(k_0\sqrt{\varepsilon_1} r\right)\right] & 0 < r < d_1 \\ \dfrac{-i}{\varepsilon_2 k_0}\left[A_1 J_n'\left(k_o\sqrt{\varepsilon_2} r\right) + B_1 Y_n'\left(k_o\sqrt{\varepsilon_2} r\right)\right] & d_1 < r < D \end{cases} \tag{7.26}$$

Imposing boundary conditions, manifesting the continuity of the fields on the metal/dielectric boundary yields

$$\begin{aligned} & A_0 J_n\left(k_0\sqrt{\varepsilon_1} d_1\right) + B_0 Y_n\left(k_0\sqrt{\varepsilon_1} d_1\right) \\ & \quad = A_1 J_n\left(k_o\sqrt{\varepsilon_2} d_1\right) + B_1 Y_n\left(k_o\sqrt{\varepsilon_2} d_1\right) \\ & \frac{1}{\varepsilon_1}\left[A_0 J_n'\left(k_0\sqrt{\varepsilon_1} d_1\right) + B_0 Y_n'\left(k_0\sqrt{\varepsilon_1} d_1\right)\right] \\ & \quad = \frac{1}{\varepsilon_2}\left[A_1 J_n'\left(k_o\sqrt{\varepsilon_2} d_1\right) + B_1 Y_n'\left(k_o\sqrt{\varepsilon_2} d_1\right)\right] \end{aligned} \tag{7.27}$$

and in matrix form:

$$\begin{aligned} & \begin{pmatrix} J_n\left(k_0\sqrt{\varepsilon_1} d_1\right) & Y_n\left(k_0\sqrt{\varepsilon_1} d_1\right) \\ \dfrac{J_n'\left(k_0\sqrt{\varepsilon_1} d_1\right)}{\varepsilon_1} & \dfrac{Y_n'\left(k_0\sqrt{\varepsilon_1} d_1\right)}{\varepsilon_1} \end{pmatrix} \begin{pmatrix} A_0 \\ B_0 \end{pmatrix} \\ & \quad = \begin{pmatrix} J_n\left(k_o\sqrt{\varepsilon_2} d_1\right) & Y_n\left(k_o\sqrt{\varepsilon_2} d_1\right) \\ \dfrac{J_n'\left(k_o\sqrt{\varepsilon_2} d_1\right)}{\varepsilon_2} & \dfrac{Y_n'\left(k_o\sqrt{\varepsilon_2} d_1\right)}{\varepsilon_2} \end{pmatrix} \begin{pmatrix} A_1 \\ B_1 \end{pmatrix} \end{aligned} \tag{7.28}$$

This results in the transfer matrix in cylindrical coordinates:

$$\begin{aligned} M_{\text{cell}} = & \begin{pmatrix} J_n\left(k_o\sqrt{\varepsilon_2} d_1\right) & Y_n\left(k_o\sqrt{\varepsilon_2} d_1\right) \\ \dfrac{J_n'\left(k_o\sqrt{\varepsilon_2} d_1\right)}{\varepsilon_2} & \dfrac{Y_n'\left(k_o\sqrt{\varepsilon_2} d_1\right)}{\varepsilon_2} \end{pmatrix}^{-1} \\ & \begin{pmatrix} J_n\left(k_0\sqrt{\varepsilon_1} d_1\right) & Y_n\left(k_0\sqrt{\varepsilon_1} d_1\right) \\ \dfrac{J_n'\left(k_0\sqrt{\varepsilon_1} d_1\right)}{\varepsilon_1} & \dfrac{Y_n'\left(k_0\sqrt{\varepsilon_1} d_1\right)}{\varepsilon_1} \end{pmatrix} \end{aligned} \tag{7.29}$$

7.4 The Nonlinear Hyperlens

This section introduces the concept of the nonlinear hyperlens. The performance of the linear hyperlens can be vastly improved by relaxing the perfect imaging condition and finding alternative ways to suppress the resultant diffraction. A promising direction is by utilizing self-focusing nonlinearity, for example, using dielectric layers with high Kerr-type nonlinearity. The nonlinearity induces a local change in the electrical permittivity that corresponds to the intensity pattern, which, in turn, alters the diffraction in the structure. This phenomenon was already used to predict subwavelength solitons in metal-dielectric nanolayers [46, 47]. In this section, we present a nonlinear hyperlens made of alternating layers of metal and nonlinear dielectric. We show that by introducing nonlinearity the diffraction is suppressed and the operation frequency range of the device can be extended where a longer propagation distance can be achieved. The subwavelength propagation in such a nonlinear hyperbolic medium is simulated using a self-developed NL-BPM (see Section 7.3.3) in cylindrical coordinates that is applicable for any nonparaxial propagation along the radial direction.

Before diving into the development and simulation of the nonlinear hyperlens we shall address some basic properties and drawbacks of hyperbolic metamaterials.

7.4.1 The Perfect Imaging Condition

To convey the main concept, we start by introducing a simplified system made of a metal-dielectric composite in Cartesian coordinates. Given the subwavelength period of the nanolayers we assume validity of the EMT; hence, for dielectric and metallic layers of the same thickness, the effective permittivities can be expressed via the homogenization principle:

$$\begin{aligned} \varepsilon_x &= \frac{1}{2}\left(\varepsilon_{\mathrm{m}} + \varepsilon_{\mathrm{d}}\right) \\ \varepsilon_z &= \frac{2}{\varepsilon_{\mathrm{m}}^{-1} + \varepsilon_{\mathrm{d}}^{-1}} \end{aligned} \qquad (7.30)$$

The linear modes of this system remain plane-wave, where each mode is characterized by its spatial frequency (transverse wavevector) $\mathbf{k}_x$. The phase it acquires during propagation is determined by its propagation constant (longitudinal wavevector):

$$\mathbf{k}_z = \sqrt{\varepsilon_x \mathbf{k}_0^2 - \frac{\varepsilon_x}{\varepsilon_z}\mathbf{k}_x^2} \tag{7.31}$$

Perfect imaging, where an object in the input plane is perfectly imaged to the output plane, demands that all the linear modes acquire the same phase during propagation. This condition is met for $\varepsilon_x = 0$, where $\varepsilon_x > 0$ results in diffraction that will blur the image.

7.4.2 Diffraction Loss Trade-Off

The system under investigation consists of metal-dielectric multilayers where the permittivity of the dielectric layers is assumed constant at $\varepsilon_\mathrm{d} = 3.2$. The perfect imaging case is investigated at a wavelength of $\lambda = 385$ nm with a metal permittivity of $\varepsilon_\mathrm{m} =$ -3.1–0.28i [48], while the effect of diffraction on the output image is studied at $\lambda = 365$ nm with $\varepsilon_\mathrm{m} = -2.4$–$0.24i$. The $\mathbf{k}_z$–$\mathbf{k}_x$ relations, which characterize the diffraction in the medium, are displayed in Fig. 7.4.

Namely, the curve becomes flat owing to the fact that ε_x approaches zero. This indicates that the diffraction is arrested as

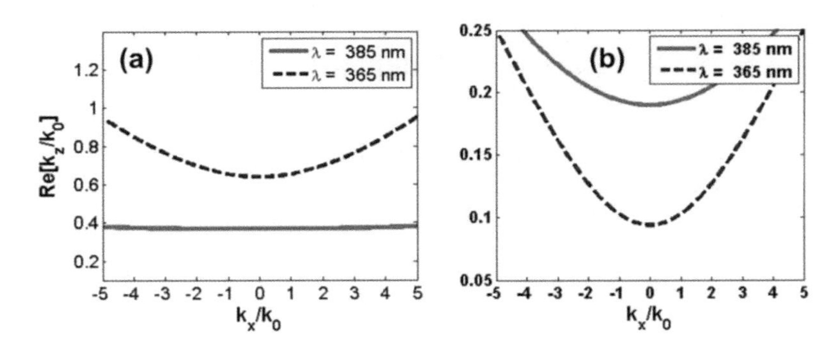

Figure 7.4 Dispersion relation: real (a) and imaginary (b) parts of the propagation constant at $\lambda = 385$ nm (red) and $\lambda = 365$ nm (blue). In the perfect imaging condition (red lines) the dispersion is flat, indicating diffractionless propagation.

all the plane wave constituents acquire a similar phase during propagation. Farther from the perfect imaging condition, the curve attains a hyperbolic shape, distinctive to indefinite media [19], resulting in anomalous diffraction characteristics [49, 50]. The propagation losses, associated with the absorption due to ohmic losses in the metal, can be obtained from the imaginary part of $\mathbf{k}_z$, shown in Fig. 7.4. Far from it, one encounters much higher loss at a broad range of wavevectors. This example demonstrates the inherent trade-off between diffraction and loss in a hyperbolic layered medium. This is evident from the linear propagation simulation shown in Fig. 7.5.

The propagation of a subwavelength object consisting of two 40 nm slits separated by 90 nm is simulated through a 500 nm thick hyperbolic medium, similar to what was in Fig. 7.4. Indeed, the propagation at $\lambda = 385$ nm suffers high loss in the perfect imaging regime and shows a propagation distance of 1.5λ (Fig. 7.5a; red curve in Fig. 7.5d). At the same time, a similar wave packet at 365 nm wavelength suffers lower loss but experiences strong diffraction that smears the two-slit object (Fig. 7.5b, green curve in Fig. 7.5d). Figure 7.5c depicts the nonlinear propagation where the dielectric layers possess Kerr nonlinearity. At the maximum index change of $\Delta n = 0.035$ the diffraction is arrested and the two-slit image is retained at the output, displaying higher efficiency (lower attenuation) compared to the linear resonant imaging.

7.4.3 Nonlinear Hyperlens Simulations

To analyze such hyperbolic structure in a cylindrically symmetric system, we utilize this radial BPM, which is applicable to a cylindrical coordinate system, to simulate an input field comprised of two 40 nm full-width at half-maximum (FWHM) Gaussian beams with 75 nm separation. The input radius is $r_{\text{in}} = 250$ nm and the output radius is $r_{\text{out}} = 1000$ nm, indicating a fourfold magnification at the hyperlens output. The hyperlens is made of dielectric layers with a dielectric constant of $\varepsilon_{\text{d}} = 3.2$ and silver layers, as can be seen in Fig. 7.6.

We examine again two operation wavelengths: Close to the perfect imaging condition at $\lambda = 385$ nm the metal permittivity is

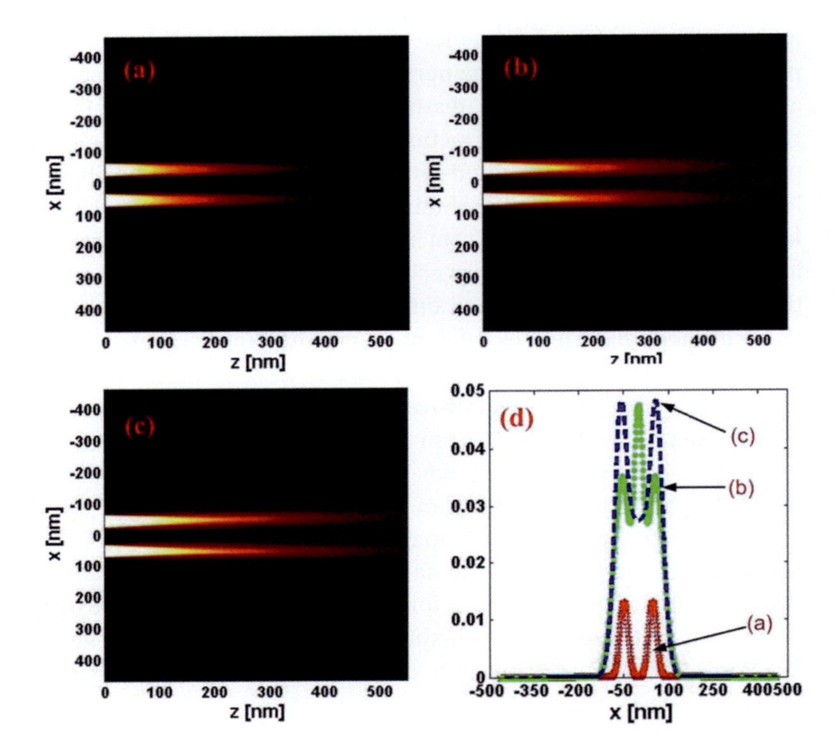

Figure 7.5 Imaging of two super-Gaussian sources 90 nm apart. (a) Linear propagation with perfect imaging evident at λ = 385 nm. (b) Linear propagation far from the condition at λ = 365 nm. (c) Nonlinear propagation at λ = 365 nm. (d) Output intensities: Close to the perfect imaging condition (red) the image is resolved but suffers high loss; far from the condition (green) the image is smeared due to strong diffraction; the nonlinear propagation in the medium (blue) resolves the image with better efficiency.

$\varepsilon_{\text{m}} = -3.1\text{–}0.28i$, while farther from the condition at $\lambda = 360$ nm, $\varepsilon_{\text{m}} = -2.2\text{–}0.23i$. Figure 7.6a shows the propagation simulation of the subwavelength object close to the perfect imaging condition. Just like its Cartesian counterpart, the loss in the metal layers reduces the power significantly such that the signal is almost undetectable at the output. Moving away from the condition at 360 nm the wavelength shows reduced loss but strong diffraction that smears the object (Fig. 7.6b; green curve in Fig. 7.6d). When incorporating the Kerr

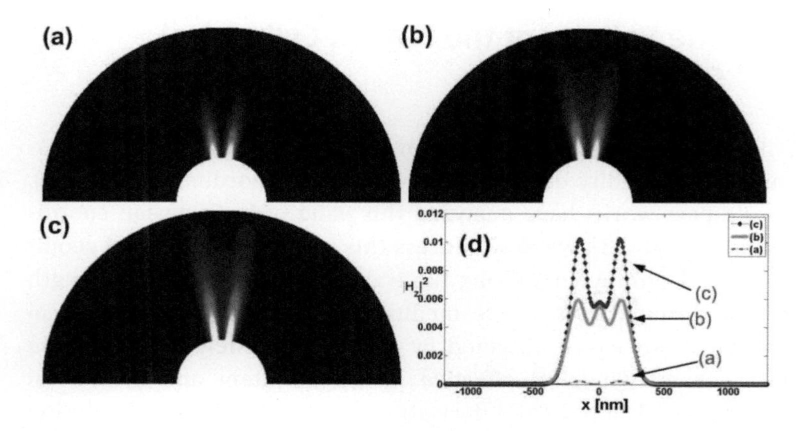

Figure 7.6 Imaging of two Gaussian sources 75 nm apart in a hyperlens with inner and outer radii of 240 nm and 1000 nm, respectively. Panels (a–d) are similar to Fig. 7.5.

nonlinearity, with a maximal refractive index change of $\Delta n = 0.013$ at this wavelength, the diffraction is suppressed and the output is again resolved with 50X better efficiency (Fig. 7.6; blue curve in Fig. 7.6d) and with 300 nm separation, which can be resolved with conventional diffraction-limited optics.

Note that the diffraction arrest can be achieved with a smaller nonlinear index change than in a Cartesian system; this is due to the different diffraction mechanism that dominates hyperbolic Bessel functions compared to plane waves.

7.4.4 Conclusions

In conclusion, we have shown that by incorporating self-defocusing Kerr nonlinearity in the dielectric constituents of a hyperbolic metamaterial, one can suppress diffraction and achieve broadband operation and longer propagation distances. The longer propagation distance enables larger devices and hence improves spatial resolution: a 75 nm separation at a 360 nm wavelength was demonstrated at moderate nonlinearity. The next section shows an in-depth investigation of the applicativity of the EMT in the hyperlens configuration.

7.5 The Validity of the EMT in Cylindrical Coordinates

The previous NL-BPM derivation and hyperlens simulations assumed the validity of the EMT in cylindrical coordinates (Fig. 7.7). While past works have analyzed this validity in Cartesian coordinates [20, 51–53], we now address this issue in a cylindrical geometry. We begin by comparing the transmission of a subwavelength object through an effective medium to that in a true cylindrical multilayer structure obtained by the TMM, namely a comparison between a cylindrical effective medium system and cylindrical multilayers. The full TMM derivation that provides the full solution of Maxwell's equations is presented in Chapter 3 and relies on previous works [54, 55]. It enables a deeper investigation of the linear modes of the system, including the subwavelength imaging capabilities. Figure 7.7 presents the calculation of two point sources 80 nm apart, propagating through a hyperlens with an inner radius of 250 nm and an outer radius of 750 nm. The device unit cell is 20 mm thick with a filling fraction of 0.5. The illumination wavelength is 365 nm, corresponding to a metal permittivity of $\varepsilon_{\mathrm{m}} = -2.4+0.23i$ [48]. The dielectric permittivity is chosen to be

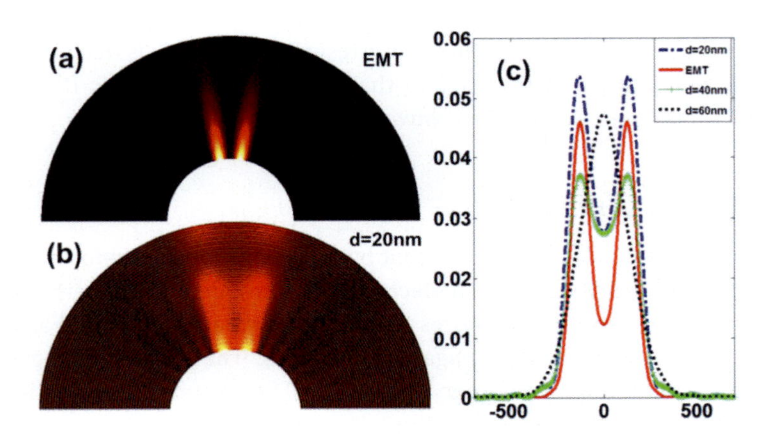

Figure 7.7 Diffraction in the EMT case (a) compared to the TMM simulation with a 20 nm unit cell period (b). A comparison of the amplitude at the output faced is depicted in (c).

$\varepsilon_d = 3.2$. The result is presented in Fig. 7.7a and compared with the EMT approximation in Fig. 7.7b. Figure 7.7c shows a comparison between the field intensity at the output of each one of the systems. The EMT (solid red curve) and the 20 nm unit cell (dashed blue curve) show the same magnification of 3 at the output, which corresponds to the ratio between the input and the output radius. To investigate the applicability of the EMT we have broadened the unit cell to 40 and 60 nm, respectively. While the 40 nm unit cell can still resolve the image at the output (green curve), the 60 nm unit cell hinders such a resolution (dashed black curve) and, therefore, proves to be outside the approximation limits.

7.5.1 The Cylindrical Amplitude Transfer Function

To better quantify the validity of the EMT in cylindrically symmetric systems we investigate its fundamental transmission property—the amplitude transfer function [56] (ATF)—which quantifies the dependence of the system's transmittance in the transverse wavevector. Note that the ATF has no analytical closed solution in cylindrical geometry, as neither of the dispersion relations possesses such a solution. Hence, we extract the ATF numerically by calculating the azimuthal fields at the input and output of the device:

$$\mathrm{ATF}^{\mathrm{Cyl}}(m) = \frac{A_m(\rho_{\mathrm{out}})}{A_m(\rho_{\mathrm{in}})} \tag{7.32}$$

The modes Am for the EMT are defined by Eq. 7.12 and the projection upon them is analogous to the Fourier transform in a Cartesian system. The same projection is carried out for the exact TMM solutions. We first examine the super-resolution capabilities of the hyperlens at the same wavelength of 365 nm. The green circles in Fig. 7.8a represent the reference case of using only a dielectric slab with permittivity $\varepsilon_d = 3.2$, which exhibits a maximal transmission of 60%, which approaches the cutoff around the 8th mode. This cutoff is attributed to the diffraction limit which inhibits transmission of features smaller than half the wavelength. The theoretical cutoff in a homogeneous system can be calculated using the relation [25]

$$m_{\mathrm{cutoff}} = \left\lceil r_{\mathrm{in}} k_0 \sqrt{\varepsilon_d} \right\rceil \tag{7.33}$$

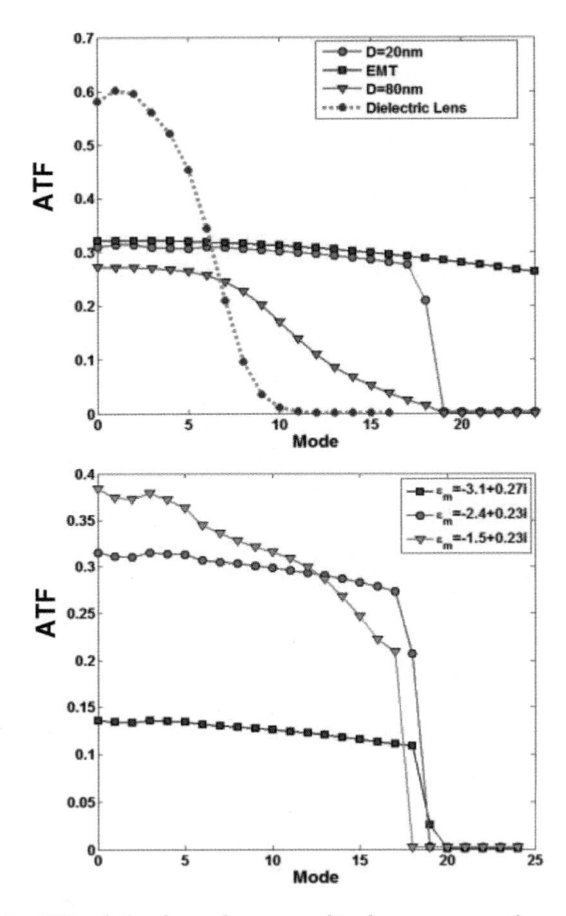

Figure 7.8 ATF of the hyperlens: amplitude versus mode number. (a) ATF of different lenses. Green solid circles: dielectric lens; blue solid rectangles: effective medium; red solid circles and pink solid triangles: TMM calculations of a hyperlens with a 20 and 80 nm unit cell, respectively. (b) ATF of a 20 nm unit cell hyperlens close to the perfect imaging condition (blue rectangles) and far from it (red solid circles and green solid triangles).

which results in a cutoff at $m = 8$, in good agreement with the simulation results. The blue rectangles represent the transmission of a hyperlens made of an effective medium. While the overall transmission is lower (due to the metallic losses embedded in the effective permittivities), there is almost no degradation in the

ATF for a higher mode order, even for those corresponding to wavevectors beyond the diffraction limit. The red circles represent the TMM results for a unit cell of 20 nm (filling fraction 0.5) which shows almost the same behavior as the EMT up to the 19th mode, where it undergoes a cutoff. The purple triangles represent the TMM simulation of an 80 nm unit cell which shows almost the same cutoff but at a nonflat ATF, indicating low suitability for imaging. Figure 7.8b depicts the ATF of circular multilayers at different frequencies with a 20 nm unit cell and examines the device's capability for broadband imaging. Close to the perfect imaging condition at a wavelength of 380, where $\varepsilon_{\mathrm{m}} = -3.1+0.27i$, the ATF is flat (blue rectangles), but as expected, the losses are highest and hence the transmittance is low. Away from the perfect imaging frequency (red circles) at a wavelength of 365 nm the curve is still flat, but with reduced loss, making this frequency optimal for imaging. At a farther frequency, for example, at a wavelength of 350 nm (green triangles) the perfect imaging capability is lost due to the coarse ATF.

7.5.2 The Mean Square Approximation Error

The ability to calculate the exact electromagnetic solution at any point allows quantification of the inaccuracy associated with the use of the EMT. The error figure can be defined as the mean square approximation error (MSAE), which is calculated by summing the difference between the exact TMM and the approximated EMT solutions to an impulse response (point-source excitation) over the area of interest:

$$\begin{aligned}\mathrm{MSAE} &= \int_{r_{\mathrm{in}}}^{r_{\mathrm{out}}} \left| h^{\mathrm{TMM}}\left(r'\right) - h^{\mathrm{EMT}}\left(\mathbf{r}'\right)\right|^2 d^2r' \\ &= \int_{r_{\mathrm{in}}}^{r_{\mathrm{out}}} \left| \sum_n \left[A_n^{\mathrm{TMM}}\left(r'\right) J_n^*\left(k_0\sqrt{\varepsilon}r'\right) \right.\right. \\ &\qquad \left.\left. -A_n^{\mathrm{EMT}}\left(r'\right) J^*_{n\sqrt{\frac{\varepsilon_\phi}{\varepsilon_\rho}}}\left(ik_0\sqrt{\varepsilon_\phi}r'\right)\right] e^{-in\phi}\right|^2 d^2r' \qquad (7.34)\end{aligned}$$

The MSAE is studied by varying two parameters, the unit cell size and the inner (input) radius of the device. The dependence of the MSAE on these two parameters is plotted in Fig. 7.9 for a 365 nm wavelength. While the error grows with the increase of the unit

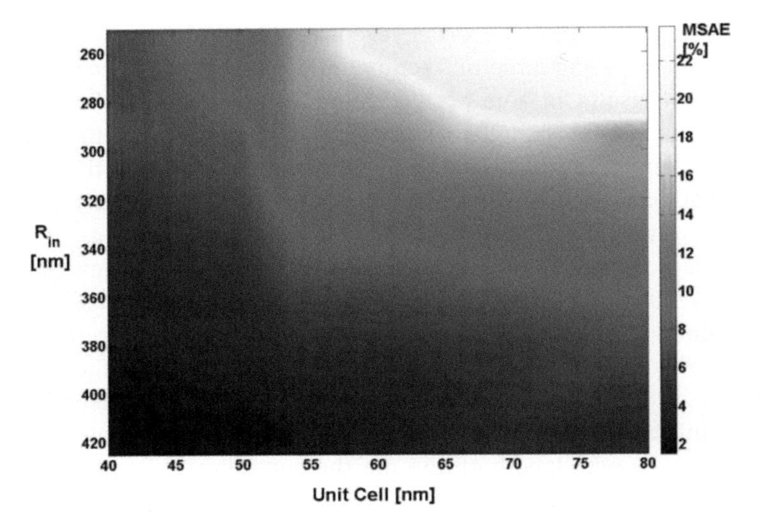

Figure 7.9 MSAE calculation. Where we varied the unit cell of the device and the input radius as well, the result error is in percentage of power between the TMM and EMT solutions.

cell similarly to Cartesian systems [52], the ratio between the input radius of the device and the unit cell plays an important role in the accumulated error. A larger input radius can compensate for the error accumulated due to a large unit cell and reduce it to below 6% in power; this degree of freedom is unique to cylindrical systems and not present in Cartesian ones. This can be explained by the lack of shift invariance along the radial direction, providing the role played by the inner radius of the device in the EMT applicability.

7.6 Conclusions

In conclusion, we applied the TMM to cylindrical coordinates and showed that it can be used to accurately calculate wave propagation in cylindrically symmetric metamaterials such as the hyperlens. We extracted the ATF of a cylindrically symmetric system and accurately quantified the super-resolution capabilities of cylindrical metal dielectric multilayers. Furthermore, we assessed the applicability of the EMT by calculating the MSAE and found that as long as the inner radius of the hyperlens is much larger than the unit cell period, the EMT coincides well with the exact electromagnetic solution. This conclusion is unique to the cylindrical geometry due to the fact that the system is no longer shift-invariant in the radial direction. Both ATF and MSAE analyses agree that the unit cell period should be at least five times smaller than the inner radius for the EMT to be valid (MSAE $<6\%$ error in power).

References

1. Liu, Z., Lee, H., Xiong, Y., Sun, C., and Zhang, X. (2007). Far-field optical hyperlens magnifying sub-diffraction-limited objects, *Science*, **315**, p. 1686.
2. Szameit, A., Shechtman, Y., Osherovich, E., Bullkich, E., Sidorenko, P., Dana, H., Steiner, S., et al. (2012). Sparsity-based single-shot subwavelength coherent diffractive imaging, *Nat. Mater.*, **11**(5), p.p. 455–459.
3. Rogers, E. T. F., et al. (2012). A super-oscillatory lens optical microscope for subwavelength imaging, *Nat. Mater.*, **11**, pp. 432–435.
4. Wood, B., Pendry, J., and Tsai, D. (2006). Directed subwavelength imaging using a layered metal-dielectric system, *Phys. Rev. B*, **74**(11), p. 115116.
5. Huang, B., Wang, W., Bates, M., and Zhuang, X. (2008). Three-dimensional super-resolution imaging by stochastic optical reconstruction microscopy, *Science*, **319**(5864), pp. 810–813.
6. Feigenbaum, E., and Orenstein, M. (2007). Modeling of complementary (void) plasmon waveguiding, *J. Lightwave Technol.*, **25**(9), pp. 2547–2562.
7. Govyadinov, A., and Podolskiy, V. (2006). Metamaterial photonic funnels for subdiffraction light compression and propagation, *Phys. Rev. B*, **73**(15), p. 155108.
8. Chen, H.-T., Padilla, W. J., Zide, J. M. O., Gossard, A. C., Taylor, A. J., and Averitt, R. D. (2006). Active terahertz metamaterial devices, *Nature*, **444**(7119), pp. 597–600.
9. Kabashin, A. V., Evans, P., Pastkovsky, S., Hendren, W., Wurtz, G. A., Atkinson, R., Pollard, R., Podolskiy, V. A., and Zayats, A. V. (2009). Plasmonic nanorod metamaterials for biosensing, *Nat. Mater.*, **8**(11), pp. 867–871.

10. Abbe, E. (1873). Beiträge zur Theorie des Mikroskops und der mikroskopischen Wahrnehmung, *Archiv für Mikroskopische Anatomie*, **9**(1), pp. 413–418.
11. Lewis, A., Isaacson, M., Harootunian, A., and Muray, A. (1984). Development of a 500 Å spatial resolution light microscope: I. Light is efficiently transmitted through $\lambda/16$ diameter apertures, *Ultramicroscopy*, **13**(3), pp. 227–231.
12. Yildiz, A., Forkey, J. N., McKinney, S. A., Ha, T., Goldman, Y. E., and Selvin, P. R. (2003). Myosin V walks hand-over-hand: single fluorophore imaging with 1.5-nm localization, *Science*, **300**(5628), pp. 2061–2065.
13. Joannopoulos, J., Johnson, S., Winn, J., and Meade, R. (2011). *Photonic Crystals: Molding the Flow of Light* (Princeton University Press, Princeton, USA).
14. Bergman, D. (1978). The dielectric constant of a composite material: a problem in classical physics, *Phys. Rep.*, **43**(9), pp. 377–407.
15. Cai, W., Shalaev, V., and Paul, D. K. (2010). Optical metamaterials: fundamentals and applications, *Phys. Today*, **63**(9), p. 57.
16. Pendry, J. (2000). Negative refraction makes a perfect lens, *Phys. Rev. Lett.*, **85**(18), p. 3966.
17. Fang, N., Lee, H., Sun, C., and Zhang, X. (2005). Sub-diffraction-limited optical imaging with a silver superlens, *Science*, **308**(5721), pp. 534–537.
18. Ramakrishna, S., and Pendry, J. (2003). Imaging the near field, *J. Mod. Opt.*, **50**(9), pp. 1419–1430.
19. Smith, D., and Schurig, D. (2003). Electromagnetic wave propagation in media with indefinite permittivity and permeability tensors, *Phys. Rev. Lett.*, **90**(7), p. 077405.
20. Cortes, C. L., Newman, W., Molesky, S., and Jacob, Z. (2012). Quantum nanophotonics using hyperbolic metamaterials, *J. Opt.*, **14**(6), p. 063001.
21. Jacob, Z., Kim, J.-Y., Naik, G. V., Boltasseva, A., Narimanov, E. E., and Shalaev, V. M. (2010). Engineering photonic density of states using metamaterials, *Appl. Phys. B*, **100**(1), pp. 215–218.
22. Smith, D. R., Schurig, D., Mock, J. J., Kolinko, P., and Rye, P. (2004). Partial focusing of radiation by a slab of indefinite media, *Appl. Phys. Lett.* **84**(13), pp. 2244–2246.
23. Belov, P., and Hao, Y. (2006). Subwavelength imaging at optical frequencies using a transmission device formed by a periodic layered

metal-dielectric structure operating in the canalization regime, *Phys. Rev. B*, **73**(11), p. 113110.

24. Schilling, J. (2006). Uniaxial metallo-dielectric metamaterials with scalar positive permeability, *Phys. Rev. E*, **74**(4), p. 046618.
25. Jacob, Z., Alekseyev, L., and Narimanov, E. (2006). Optical hyperlens: far-field imaging beyond the diffraction limit, *Opt. Express*, **14**(18), pp. 8247–8256.
26. Salandrino, A., and Engheta, N. (2006). Far-field subdiffraction optical microscopy using metamaterial crystals: theory and simulations, *Phys. Rev. B*, **74**(7), p. 075103.
27. Smolyaninov, I. I., Hung, Y.-J., and Davis, C. C. (2007). Magnifying superlens in the visible frequency range, *Science*, **315**(5819), pp. 1699–1701.
28. Liu, Y., Bartal, G., Genov, D., and Zhang, X. (2007). Subwavelength discrete solitons in nonlinear metamaterials, *Phys. Rev. Lett.*, **99**(15), p. 153901.
29. Peleg, O., Segev, M., Bartal, G., Christodoulides, D., and Moiseyev, N. (2009). Nonlinear waves in subwavelength waveguide arrays: evanescent bands and the "phoenix soliton," *Phys. Rev. Lett.*, **102**(16), p. 163902.
30. Johnson, P., and Christy, R. (1972). Optical constants of the noble metals, *Phys. Rev. B*, **6**(12), p. 4370.
31. Morandotti, R., Eisenberg, H., Silberberg, Y., Sorel, M., and Aitchison, J. (2001). Self-focusing and defocusing in waveguide arrays, *Phys. Rev. Lett.*, **86**(15), p. 3296.
32. Pertsch, T., Zentgraf, T., Peschel, U., Bräuer, A., and Lederer, F. (2002). Anomalous refraction and diffraction in discrete optical systems, *Phys. Rev. Lett.*, **88**(9), p. 4.
33. Chebykin, A., Orlov, A., Vozianova, A., Maslovski, S., Kivshar, Y., and Belov, P. (2011). Nonlocal effective medium model for multilayered metal-dielectric metamaterials, *Phys. Rev. B*, **84**(11), p. 115438.
34. Elser, J., Podolskiy, V. A., Salakhutdinov, I., and Avrutsky, I. (2007). Nonlocal effects in effective-medium response of nanolayered metamaterials, *Appl. Phys. Lett.*, **90**(19), p. 191109.
35. Silveirinha, M. G. (2013). Effective medium response of metallic nanowire arrays with a Kerr-type dielectric host, *Phys. Rev. B*, **87**(16), p. 165127.
36. Nikolaev, V. V., Sokolovski, G. S., and Kaliteevski, M. A. (1999). Bragg reflectors for cylindrical waves, *Semiconductors*, **33**(2), p. 147.

37. Chew, W. (1995). *Waves and Fields in Inhomogenous Media* (IEEE Press, New York), p. 46249.

38. Born, M., andWolf, E. (1980). *Principles of Optics: Electromagnetic Theory of Propagation, Interference and Diffraction of Light* (Cambridge University Press archive).

Chapter 8

Nanoparticle-Assisted Stimulated Emission Depletion (STED) Super-Resolution Nanoscopy

Yonatan Sivan[a] **and Yannick Sonnefraud**[b]
[a]*Unit of Electro-Optics Engineering, Faculty of Engineering Sciences, Ben-Gurion University of the Negev, P.O. Box 653, Israel, 8410501*
[b]*Institut Ne'el, CNRS UPR 2940, 25 rue des Martyrs BP 166, 38042 Grenoble Cedex 9, France*
sivanyon@bgu.ac.il

8.1 Prologue: A Eulogy to a Friend

This chapter was written by me together with my colleague and friend Yannick Sonnefraud. Shortly after finishing the writing of the chapter, Yannick was diagnosed with a cancerous tumor in his brain. Tragically, the emergency operation that ensued was not successful, and Yannick passed away peacefully shortly after, on September 17, 2014, at age 33.

Yannick completed a PhD, working in Laboratoire de Spectrométrie Physique (LiPhy) at Institut Néel in Grenoble under the supervision of Prof. Serge Huant. Yannick then joined Imperial College

Plasmonics and Super-Resolution Imaging
Edited by Zhaowei Liu

ISBN 978-981-4669-91-7 (Hardcover), 978-1-315-20653-0 (eBook)
www.panstanford.com

London as a postdoctoral research associate in the plasmonics group of Prof. Stefan Maier. In 2010 he became a Leverhulme Research Fellow in the Centre for Plasmonics and Metamaterials. In early 2014 he returned to Institut Néel as head of research at the French National Center for Scientific Research (CNRS) 1st class, where he was developing a research program in the nano-optics group, focusing on quantum plasmonics and super-resolution microscopy.

Those who knew Yannick would describe him as nothing short of an all-around great person. He was incredibly enthusiastic about both work and play, and his gentle, easygoing nature and sense of humor made him easily approachable and lovable to his friends and colleagues. Cooking, traveling, board games, and salsa dancing were just some of his well-known endeavors, and he never hesitated to encourage his friends to join him.

At the personal level, Yannick introduced me to STED nanoscopy, a topic which gradually became central to both our careers. In that regard, he has had a long-lasting influence on my career path, more than any other high-ranking scientist I was fortunate to work with so far. I shall be indebted to him for that for many years to come. More so, I will miss him as a companion for a journey taken together, the excellent conversation partner, his systematic approach and vast knowledge, and the balance and calm he brought in to anything he did.

The following chapter is one of the last he wrote; it is dedicated to his memory.

Yonatan Sivan
November 18, 2014

8.2 Introduction

The limit that diffraction puts on imaging was first understood in the days of Ernst Abbe [1]. Since then, it is considered as one of the most fundamental problems in wave physics—the image of a point source cannot be sharper than about half a wavelength. This means that one cannot resolve two sources if they are separated by less than that distance. Breaking the diffraction limit has challenged scientists since then, and it was only in the 1980s that the diffraction limit was broken using *near*-field microscopes.

Figure 8.1 Dr. Yannick Sonnefraud, 1981–2014.

In the early 2000s, with the renewed interest in plasmonic systems (see, for example, Refs. [2, 3] for some recent reviews), several plasmonics-based approaches to beat the diffraction limit were proposed theoretically and tested experimentally. Specifically, these approaches include Pendry's superlens [4, 5, 6], the hyperlens [7, 8, 9] and other versions of meta(l)-lenses (see, for example, Ref. [10]), super-oscillations [11], plasmonic structured illumination (SIM) [12, 13, 14], etc. The overall goal of using plasmonic nanostructures for these super-resolution imaging techniques is to exploit their unique ability to support large wavevector components (the evanescent wave spectrum), that is, the components which encode information on the finest details of the image. In standard optical systems, the amplitude of these modes decays exponentially away from the sample, over a distance of the order of the wavelength. Thus, their absence from the far-field wave pattern is the origin of the limitation of achievable resolution. A detailed description of the state of the art of these approaches is the topic of this book and will not be discussed further in this chapter. An alternative approach for obtaining super-resolution is based on introducing ideas from information theory in order to recover the lost information encoded in the evanescent wave spectrum using prior knowledge on the image [15]. This approach lies within the realm of computational optics and is out of the scope of the current book.

At about the same time, in the late 1990s to early 2000s, several ways to attain diffraction-unlimited resolution (super-resolution) using a *far*-field microscope were developed in the context of *bioimaging*. By far, bioimaging, which encompasses most applications of microscopy to date, relies on fluorescence microscopy [16, 17], a technique in which a fluorescent label, usually a dye molecule, is attached to the object one wants to image, for example, a protein, virus, cell, DNA, etc. Far-field fluorescence microscopy is widely used in the life sciences due to a number of rather exclusive advantages such as noninvasive access to the interior of the (living) cells, the specific and sensitive detection of cellular constituents, and simple sample preparation. The super-resolution techniques in fluorescence microscopy (accordingly referred to as "fluorescence nanoscopy") are typically based on the ability to switch the fluorophore between a bright (emitting) and a dark (nonemitting) state with high spatial accuracy [18]. In the first implementation of this idea, invented and developed by Stefan Hell, the switching-off (depletion) mechanism used was stimulated emission depletion (or STED in short) [19, 20]. Later implementations frequently relied on a variation on that common theme, that is, using a variety of different physical mechanisms for switching off the fluorescence, including, for example, ground-state depletion [21], reversible saturable optical linear fluorescence transitions (RESOLFT) [22], and point accumulation for imaging in nanoscale topography (PAINT) [23], or on somewhat related concepts such as SIM [24, 25], stochastic methods such as PALM or STORM [26, 27], saturated excitation microscopy (SAX) [28], super-resolution optical fluctuation imaging (SOFI) [29], etc.

To date, most of these techniques are mature, and many have also been commercialized. However, STED nanoscopy probably remains the most prominent technique of them all as it is the fastest and the only one giving access to details buried deep under the surface of the biological sample (typically, up to a few tens of microns).

STED nanoscopy has already unraveled several key biological phenomena; see, for example, Refs. [30, 31] to name a few. The STED nanoscope, however, has still not become a widespread tool, partially because of the complicated setup required and the nanoscope's high cost. Another critical serious problem of these nanoscopes is the high intensities they require; this causes photobleaching, a process in which the fluorophores are damaged under the intense illumination, as well as damage to the biological tissue.

Over the past few years, tremendous progress has been achieved by improving the original illumination system; see, for example, [32, 33, 34] and Section 8.3 later. In this chapter, we describe in detail a *complementary* approach in which instead of modifying the illumination scheme, the fluorescent labels are modified. Indeed, there is ongoing work on the development of various novel STED fluorophores (see, for example, Refs. [35, 36]); however, we propose a different approach, specifically, to achieve high depletion intensities by exploiting the near-field enhancement occurring near small metallic nanoparticles (NPs). Thus, instead of using standard (i.e., stand-alone) fluorescent labels, we suggest the use of "hybrid" labels that comprise a fluorophore (or fluorophores) and a small NP; see Fig. 8.2. Accordingly, we refer to our scheme as NP-assisted STED nanoscopy, or, in short, NP-STED [37, 38, 39]. In that sense, our strategy differs from that of the plasmonic-based lensing schemes (see earlier)—we exploit the strong near-field enhancement near metal NPs rather than their ability to support high **k**-vectors (i.e., the short spatial extent of that field-enhanced regime). In this context, we note that plasmonic near-field enhancement of the process of stimulated emission was also explored in the context of nonlinear SIM near a multilayer planar geometry using focused plane-wave beams [40]. In that paper, it was shown theoretically that such an enhancement may allow for deep subdiffraction resolution with available laser power. However, due to the geometrical configuration, that approach will be valid only for imaging thin biological samples.

This chapter is dedicated to a detailed description of our approach, NP-STED. It is written in a modular, self-contained way, hopefully enabling the reader to be able to focus on the parts most relevant to him or her. With respect to previous journal publications [38, 39], we expanded the discussion on the choice of ideal NP-STED fluorescent labels and illumination, with emphasis on the complications arising in estimating the field enhancement with doughnut-shaped beams, as initially studied in Ref. [41]. Finally, we describe recent measurements of the STED resolution attained with metal nanoshell NP-STED labels showing a fourfold reduction of the intensity required for subdiffraction resolution [42]. This serves as an experimental proof-of-concept for NP-STED.

This chapter is organized as follows. In Section 8.3 we introduce the concept of STED nanoscopy both qualitatively and quantitatively, as introduced by Stefan Hell et al. In Section 8.4 we review the principles of the interaction of light with metal NPs. In Section 8.5 we combine these two approaches—we show how STED theory is modified in the presence of metal NPs, how the presence of a metal NP improves STED imaging performance, and what strategy is required to achieve optimal performance in NP-STED, including some specific possible realizations. Section 8.6 describes the first experimental proof-of-concept measurements of NP-STED. Section 8.7 provides a detailed description of the analytical and experimental methods in use in NP-STED, and finally, Section 8.8 provides a summary and outlook for future theoretical and experimental work, as well as estimating the future impact of NP-STED on fluorescence nanoscopy in general.

8.3 Principles of STED Nanoscopy

8.3.1 Qualitative Description

Let us now try to understand how STED nanoscopy works.

To date, STED was performed predominantly with dye molecules as the fluorescent emitters, and only much later with nitrogen vacancy (NV) centers in diamond [43, 44] and relatively recently with semiconductor nanocrystals (also known as quantum dots) [45, 46]. Thus, to keep the generality of the discussion, in what follows, we refer to the emitters as *fluorophores* and adopt a terminology most suitable to dye molecules as the fluorescent labels for the STED operation. Specifically, the fluorophores can be described as having two electronic states broadened by vibrational modes into level manifolds. These manifolds are manifested in the spectral domain as absorption and emission cross sections, typically shifted by no

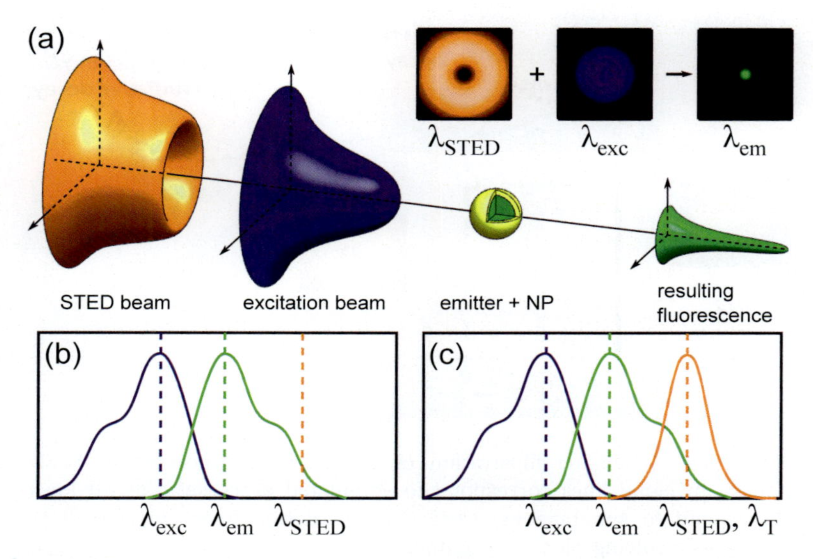

Figure 8.2 (a) Schematic illustration of a STED nanoscope. Insets show the transverse cross sections of the field distributions and signal. (b) Schematic illustration of the spectral configuration for standard STED measurements specifying the absorption (blue line) and emission (σ_{em}, green line) cross sections of the dye, as well as the spectral location of the depletion wavelength λ_{STED} (dashed orange line). (c) Same as panel (b) for NP-STED measurements with the local field enhancement associated with the plasmon resonance of the NP (orange line) centered around the depletion wavelength λ_{STED}. Adapted with permission from Ref. [39]. Copyright (2012) American Chemical Society.

less than a few tens of nanometers; see Fig. 8.2. The physics and the detailed implementation for other emitters may differ slightly.

The physical processes involved in STED nanoscopy are described in Fig. 8.3. The STED nanoscope is based on a scanning confocal fluorescence microscope, a configuration in which the biological system is illuminated by a beam strongly focused to a spot about half of the excitation wavelength λ_{exc}. The absorbed photon energy pumps electrons from the electronic ground-state S_0 to the excited-state manifold S_1. Then, they relax nonradiatively to the lower vibrational level of that manifold on a subpicosecond timescale $1/k_{vib}$, and eventually, they decay spontaneously back to the ground-state manifold S_0 by emitting photons (spontaneous

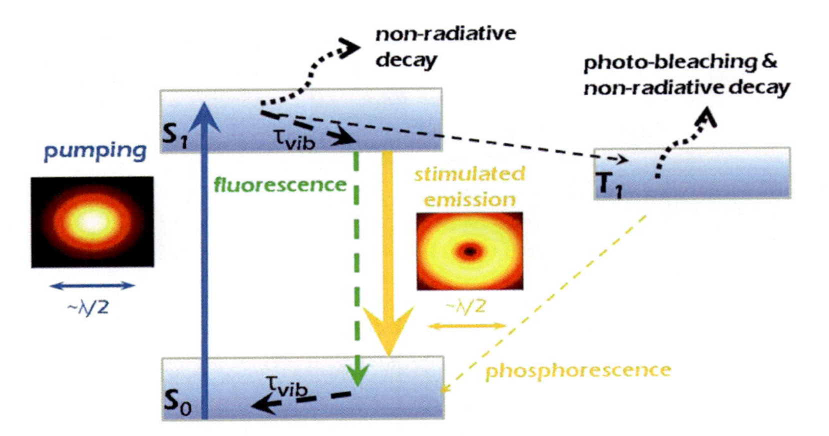

Figure 8.3 Schematic illustration of the relevant processes in STED nanoscopy. Solid lines correspond to stimulated processes, dashed lines correspond to spontaneous processes, and dotted lines correspond to processes involving a loss of brightness.

emission), on typical timescales of a few nanoseconds, $1/k_{S_1}$. Note that in fact, all fluorophores are nonideal (i.e., their quantum yield is lower than 100%, $q_i < 1$) so that the total spontaneous decay rate consists of contributions from both radiative and nonradiative decay channels.

The emitted photons are collected at each point of the scan, and an image is formed. From standard imaging theory (see, for example, Ref. [47]), it is easy to show that due to the translation invariance of the illumination scheme (see Section 8.7), the resulting image S_{conf} is given by a convolution of the fluorophore distribution $F(\mathbf{r})$ and the point-spread function (PSF) of the microscope Ψ_{conf}, where Ψ_{conf} can be approximated as an Airy disc [48] or simply a diffraction-limited Gaussian beam; see Section 8.7 for more details.

In STED nanoscopy, the sample is also illuminated by a second beam, usually a doughnut-shaped beam [32, 49], which is perfectly aligned with the Gaussian-shaped excitation beam. The depletion wavelength is tuned to the long-wavelength tail of the emission spectrum of the fluorophore; see Fig. 8.3 and Fig. 8.2b, thus forcing the fluorophores to decay to the ground-state manifold via stimulated emission. The emission depletion detuning minimizes re-excitation by the STED beam and allows a distinction to be made (usually via spectral filtering) between stimulated emission (at λ_{STED}) and spontaneous emission (around λ_{em}). Since the probability of that process is proportional to the depletion intensity, it is high at the doughnut crest and negligible at the very center of the doughnut where the depletion field vanishes. Doing so, only the fluorophores located at the very center of the doughnut, in a volume much smaller than the diffraction limit, are still free to emit light by spontaneous emission.

To understand the effect of stimulated emission on the image acquisition process, it is customary to look at the probability for spontaneous emission η_{sp}; see Fig. 8.4a and Eq. 8.10 below. Obviously, the partial removal of the population of the excited-state manifold causes the spontaneous emission probability to be low around the crest of the doughnut and close to 100% away from it, especially near the doughnut center; see Fig. 8.4a. Importantly, the central peak of the spontaneous emission probability becomes narrower in proportion to the number of depleting photons. This is crucial for the image formation because the spontaneous emission probability now weighs the confocal PSF; see Sections 8.3.2 and 8.7 below for more details. Accordingly, the depletion gives rise to a narrower (effective) STED PSF the more intense the doughnut is; see Fig. 8.4b. A detailed derivation of the resolution in STED nanoscopy can be found in Section 8.3.2.

Efficient depletion should ideally occur before a significant part of the population of the excited singlet state relaxes back (fluoresces) to the ground-state manifold by spontaneous emission. This dictates minimal delays between the excitation and depletion pulses [50]. More importantly, it requires that the depletion rate be substantial with respect to the rate of spontaneous emission. This will happen if the depletion intensity I_{STED} (or equivalently, depletion power P_{STED}) exceeds the saturation intensity

$$I_{\text{sat}} = \frac{hck_{S_1}(\lambda_{\text{STED}})}{\lambda_{\text{STED}}\sigma_{\text{em}}(\lambda_{\text{STED}})}, \tag{8.1}$$

which is the intensity level at which the rates of spontaneous and stimulated emission are equal. In Eq. 8.1, h is the Planck constant, c is the speed of light in vacuum, k_{S_1} is the total decay rate of the first excited singlet-level manifold of the fluorophore (including both

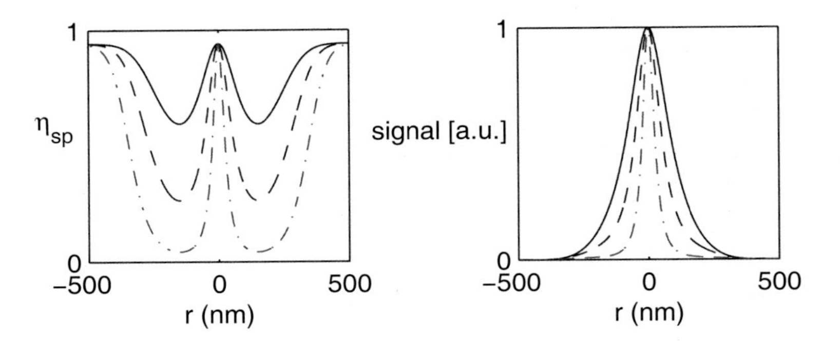

Figure 8.4 (a) Schematic illustration of the spontaneous emission probability (after the termination of the depletion pulse) (Eq. 8.10) in STED nanoscopy for various levels of depletion intensity. The width of the central peak is reduced as the depletion intensity increases. (b) The corresponding STED PSF (Eq. 8.14).

radiative and nonradiative decay channels), λ_{STED} is the depletion wavelength, and σ_{em} is the emission cross section of the fluorophore evaluated at the depletion wavelength.

The typical saturation level of dye molecules is of the order of several $\mathrm{MW/cm^2}$.[a] Thus, the required depletion intensities are quite high (around 0.1–1 $\mathrm{GW/cm^2}$) and are typically achieved with pulsed lasers. Such illumination intensities cause a reduction of brightness and photodamage to both the biological environment [51] and the fluorophores themselves. Such damage occurs in fluorescence microscopy with dye molecules in general[b]; its origin is population transfer to the triplet-state manifold T_1 an effect which is absent for Q-dots and NVs; see Fig. 8.3. These nonzero spin states are only weakly coupled to the singlet (zero-spin) states S_j via the weak spin–orbit coupling. Therefore, population transfer from the S_1 manifold to the triplet-state manifold T_1 (known as intersystem coupling) and the consequent decay back to S_0 (known as phosphorescence) occur on a timescales several orders of magnitude longer

[a]Note the relevant saturation intensities in STED nanoscopy are higher than typical because they are evaluated away from the peak of the emission cross section.

[b]Other fluorophores such as quantum dots or NVs in diamond are less susceptible to high intensities but come along with their own limitations on the applicability as STED labels.

than the fluorescence, that is, on the microsecond timescale or even longer. As a result, once a molecule decays into the triplet-state manifold, it may linger there for a long time without being able to return to the excitation-emission cycle. This process is thus called blinking and is manifested as a random, complete loss of brightness of a single fluorophore. On the macroscopic level, if the population of triplet-state manifold accumulates over time, the relative population of the singlet-state manifold decreases, so the overall brightness of the dye system decreases. This process is usually referred to as ground-state depletion [51].

A second related problem, of low photostability, is associated with the fact that the triplet-state manifold is the starting point for bleaching processes in which the molecule is modified chemically and irreversibly becomes optically inactive, thus limiting the signal brightness and the possible scan duration [32, 54, 55, 56]. The probability of this to occur is proportional to the number of illuminating photons.

While photodamage, ground-state depletion, and photobleaching are generic to fluorescence microscopy with dye molecules, we emphasize that the low brightness and photostability of the standard dye system are especially problematic in STED nanoscopy, since by depleting a substantial portion of the excited molecules, the overall signal is much weaker compared with the signal obtainable from a confocal microscope. In addition, the high intensities used in STED nanoscopy are substantially higher than those used in standard confocal microscopy or any other type fluorescence imaging.

Due to all the above, reducing the STED intensities and photodamage has been one of the major challenges in STED nanoscopy. Tremendous progress has been achieved by modifying the original illumination system, namely by using a lower excitation-depletion repetition rate that minimizes photobleaching [32], by replacing the original pulsed STED illumination with a continuous STED wave illumination [34], employing time gating in order to optimize the resolution and reduce the intensity requirements [57, 58, 59], as well as optimizing various other parameters [50, 59].

8.3.2 Quantitative Description

The resolution in STED nanoscopy was worked out in detail in Ref. [52] based on a solution of the rate equations for the population of the various singlet- (and triplet-) level manifolds. Specifically, it was shown that the improvement with respect to the (transverse) resolution of a confocal microscope is given by [38, 58]

$$\Gamma_{\text{res}} \cong \sqrt{1 + p\left(1 + k_{S_1} T_{\text{G}} - e^{-\alpha + k_{S_1} T_{\text{G}}}\right)}. \tag{8.2}$$

Here, we assumed that the STED intensity has a simple doughnut shape, $I_{\text{STED}}(r) = \frac{4P_{\text{STED}}}{\pi w^4} r^2 e^{-2r^2/w^2}$ where $r = |\mathbf{r}|$ and $w \approx \lambda/2NA$[a]; $P_{\text{STED}} = \int I_{\text{STED}}(r) d^2 r$ is the STED power and

$$p = \frac{P_{\text{STED}}}{P_{\text{sat}}}, \tag{8.3}$$

where $P_{\text{sat}} = \frac{\pi w^2}{4} I_{\text{sat}}$ is the saturation power; $\alpha \equiv k_{S_1} T_{\text{STED}}$ is the product of the STED pulse duration and the excited singlet-state decay rate. We also allow for time gating whereby only photons collected a prescribed time delay T_{G} after the onset of the STED pulse are taken into account in the image formation process.

The dependence of the resolution improvement (Eq. 8.2) on its various parameters is demonstrated in Fig. 8.5. The parameter α defines two relevant limits. In the first limit, for STED pulses which are short with respect to the excited singlet-level lifetime (i.e., when $\alpha \ll 1$) [20, 32], spontaneous emission during the STED pulse is negligible, so the spontaneous emission probability (see Fig. 8.4 and Eq. 8.10 below), and hence the resolution improvement (Eq. 8.2), depends on the depletion intensity I_{STED} but not on k_{S_1}. In this case, the resolution improvement, given by $\Gamma_{\text{res}} \cong \sqrt{1 + p\alpha}$, is optimal and resolution improvement by time gating is negligibly small [58]. However, on the opposite limit, for which the STED pulses become longer, spontaneous emission can compete with stimulated emission and cause the depletion to be less effective. Eventually, in the continuous-wave STED (CW-STED) limit [34] ($\alpha \gg 1$), the resolution improvement is given by $\Gamma_{\text{res}} \cong \sqrt{1 + p}$, that is, it depends on both I_{STED} and k_{S_1}.

The reduction of resolution improvement with growing k_{S_1} or T_{STED} can be understood in terms of the competition between

[a]See discussion in Section 8.7 regarding the value of the wavelength here.

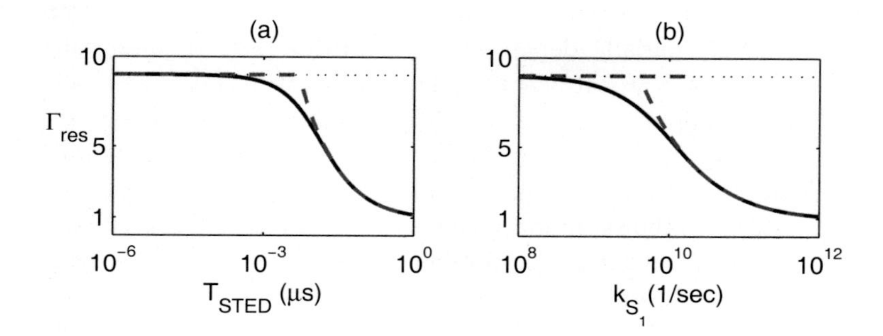

Figure 8.5 Improved resolution (Eq. 8.2) offered by STED nanoscopy at fixed STED pulse energy without (solid blue) and with (dotted black) time gating ($T_G = T_{STED}$) as a function of (a) STED pulse duration and (b) spontaneous decay rate. Also shown are the analytic approximations for the pulsed and CW regime (dashed red). The former is indistinguishable from the resolution improvement when time gating is employed. Adapted with permission from Ref. [39]. Copyright (2012) American Chemical Society.

spontaneous and stimulated emission. In particular, assuming a fixed STED pulse energy, when a short STED pulse (with respect to the excited singlet-state manifold decay rate k_{S_1}, that is, $\alpha \ll 1$) is used, the amount of spontaneous emission during the STED pulse is negligible. As a result, the contribution of light originating from the depleted regions (i.e., those regions corresponding to the outer wings of the excitation spot) to the fluorescence signal is small (and diminishing with increasing STED pulse duration or strength). In other words, the depletion pulse creates spatial variations of "active" fluorophores. This very effect is responsible for the resolution improvement. However, as the STED pulse duration increases, increasing numbers of photons are collected from these outer regions, thus giving rise to a wider collection spot and, hence, to poorer resolution. This is especially problematic if the timing between the excitation and depletion pulses is not ideal. In this latter case, the PSF attains a wide pedestal, that is, it no longer has a simple gaussian shape [50].

Nevertheless, time gating can be used to remove the pedestal and, in particular, to remove the dependence of Γ_{res} on k_{S_1} and thus to counteract the limitation a fast spontaneous decay puts on the resolution improvement [33, 50, 57, 58]. This can be interpreted in terms of spatially dependent decay rates [58]. A potentially simpler, more physical way to interpret it is in terms of spatially dependent population densities [38], as noted above. Indeed, for $T_G = T_{STED}$, that is, when any spontaneously emitted photons arriving before the end of the STED pulse are discarded, the dependence of the spontaneous emission probability (Eq. 8.10) and resolution improvement (Eq. 8.2) on the decay rate k_{S_1} is effectively removed and the optimal (i.e., k_{S_1}-free) resolution of the $\alpha \ll 1$ case is restored. For CW-STED, that is, when $\alpha \gg 1$ (so that necessarily $T_G < T_{STED}$), the dependence on k_{S_1} is only partially removed. In this case, the resolution improvement with optimal time gating is given by $\Gamma_{res} \cong \sqrt{1 + p\left(1 + k_{S_1} T_G\right)}$. Clearly, time gating comes at the cost of a weaker signal [33, 58]: it decays as $\sim e^{-k_{S_1} T_G}$. Accordingly, to date, only modest durations of time gating were employed [33, 50, 57]; these provided a clear yet modest improvement of resolution.

8.4 Light Interaction with Metal Nanoparticles

Most of the physics of light–metal NP interactions can be understood with the physics of damped, forced oscillators. The light acts as an oscillating driving force on the conduction electron gas of the NP, displacing it away from the atom cores of the crystal lattice. The electron cloud leaves positive charges behind, which, in turn, act on the gas as a restoring force, pulling it back to its rest position. Of course, the displacement of the electrons in the crystal is accompanied by a damping of the movement by Joule's effect (via various electron collision mechanisms). The combination of the light acting as driving force and crystal ion lattice acting as restoring force allows for the apparition of resonances in the oscillation of the electron cloud. These resonances are known as localized surface plasmon (LSP) resonances and give rise to several properties that can be used for our purpose and that we will review here. The spectral position of the resonances depends on several factors: the dielectric function of the medium surrounding the NP (which can be used in sensing applications [60]), the dielectric function of the metal used, and the shape of the NP.

An analytical description of the optical response of metal NPs of arbitrary shape is, in general, not available. Thus, their response is usually studied numerically. Nevertheless, some prototypical configurations are amenable to analytical solutions. Specifically, the optical response of metal NPs which are small compared to the illumination wavelength ($<$50 nm for visible light) can be studied under the quasi-static approximation. For example, in this approximation, an external electric field $\mathbf{E_0}$ incident on a *spherical* NP of radius a induces an electric dipole moment $\mathbf{p} = \epsilon_0 \epsilon_\mathrm{d} \alpha \mathbf{E_0}$, where the polarizability α is [61]

$$\alpha = 4\pi a^3 \frac{\epsilon(\omega) - \epsilon_\mathrm{d}}{\epsilon(\omega) + 2\epsilon_\mathrm{d}}, \tag{8.4}$$

with $\epsilon(\omega)$ and ϵ_d being the dielectric functions of the metal and the dielectric environment, respectively. It follows that in this situation, the field is homogeneous over the whole particle. In addition, the resonance condition—also called the Fröhlich condition—only depends on the dielectric functions of the surrounding medium and on the metal used. If $\mathrm{Im}[\epsilon_\mathrm{m}]$ is small or varies slowly in the vicinity of the resonance, the LSP resonance occurs when $\mathrm{Re}[\epsilon(\omega)] = -2\epsilon_\mathrm{d}$. For example, LSP resonance in a few-nanometer Ag spheres occurs at $\lambda \approx 350$ nm, while for Au spheres it is at longer wavelengths, $\lambda \approx 550$ nm and for aluminum spheres it is in the UV range [62].

When the particle considered becomes too large for the quasi-static approximation to be valid, one needs to solve the complete electrodynamic problem. For spherical NPs an analytical solution exists [63]. In this case, the position of the (dipolar) resonance does not simply follow the Fröhlich condition anymore: it shifts to the red (lower energies) as the particle size is increased. This can be understood intuitively with the analogy with a forced oscillator made earlier: when the particle becomes larger, the electron cloud, negatively charged, is effectively at a larger distance from the positive ion cores left behind during the motion. The restoring force is thus smaller, leading to the red-shift of the resonance. The larger distance between the charges also means that retardation starts playing a role: the phase of the incident field (and eventually, the amplitude as well) varies across the volume of the NP. This corresponds to a broadening of the resonance because of increased loss in the radiative channel [64]. This radiation damping also leads to a reduction of the near-field enhancement. In addition, the spectral response can become more complex with the appearance of higher-order modes (multipoles).

The quasi-static solutions, as well as Mie's analytical solution for a spherical metal NP can be extended to elongated ellipsoids or (multi-)layered core–shell structures [61]. These geometries provide additional means to control the spectral position of the LSP resonance. For instance, the ellipsoidal geometry gives rise to multiple resonances, with those associated with the longer dimensions shifted to longer wavelengths. Similarly, a hollow metal sphere, or nanoshell, tends to exhibit its dipolar plasmon resonance at a wavelength much larger than that of a plain sphere of identical outer diameter—changing the inner diameter of the nanoshell allows fine-tuning of the LSP resonance's position [65, 66]. One way to interpret this red shift is to view the dielectric core within the metal shell as means to dilute the electron density and thus to lower the plasma frequency and, hence, the restoring force and the frequency of the LSP resonance. A more detailed description of the behavior of LSP resonances in metallic NPs can be found in Refs. [67, 68].

LSP resonance increases the strength with which the NPs interact with incident radiation. This is manifested both in the far field and in the near field. Indeed, in the far field, the scattering and absorption cross sections are enhanced at resonance, reaching values up to a factor more than 1 order of magnitude larger than the physical cross section [69]. In the near field, another consequence of the resonance is that the NP captures and concentrates the incident electromagnetic energy, leading to enhancements of the electromagnetic field in the immediate vicinity of the NP. Typical enhancements range from 3 to 10 for single NPs, but can reach several orders of magnitude for groups of NPs separated by small gaps of a couple of tens of nanometers or less. This property is used to enhance spectroscopic methods relying on the magnitude of the field to boost the signal collected, such as surface-enhanced Raman scattering (SERS) [70–75] or surface-enhanced infrared absorption

(SEIRA) [76–78]. This property can also be used to produce local heating that has found applications in thermal imaging [79] and cancer cure [80]. For these reasons, often metal NPs are called optical nanoantennas: they do act similarly to their microwave counterparts and capture electromagnetic radiation of wavelengths much larger than their dimension and focusing it in the near field.

Plasmonic near-field enhancement is also relevant when nano-emitters, for example, fluorophores, are used. Its effect is twofold. First, the capacity of a nanoemitter to absorb electromagnetic energy is proportional to $|\mu\mathbf{E}|^2$, with μ its intrinsic transition dipole moment and **E** the electric field at its position. Hence, a nanoemitter placed close to a metal NP can benefit from the large cross section of the NP, seeing its own absorption cross section increase. Second, it is known since the seminal work of E. M. Purcell in 1946 that the decay rates of emitters depend on their environment [81]. This effect is usually attributed to the local density of states available at the emitters' position. In particular, in the presence of a nanoantenna, the local density of states increases such that the energy of an excited emitter can decay into photons that propagate in the far field but also into surface plasmon waves on the NP surface. From a plasmon, the energy can be lost in heat by the Joule effect (leading to an enhancement of the *nonradiative* decay rate) or be converted into a photon (in which case it contributes to increasing the *radiative* decay rate). The ratio of radiative decay rate with the metal NP, and without it, is usually called the Purcell factor. The ratio between the radiative decay rate of the emitter and the total decay rate is called quantum yield, q_i. The quantum yield of very efficient emitters can usually not be improved by the use of a nanoantenna, whereas the emission of low-quantum-yield species can benefit greatly from the combination with metal NPs. To date, the highest emission enhancement reported demonstrated an increase of emission by 3 orders of magnitude, thanks to a combination of the increase of the absorption cross section of the dye used and enhancement of a quantum yield which is low for the bare dye [82]. In practical applications, observable modifications of the fluorescent properties of nanoemitters are attained using relatively large NPs, for example, nanorods or nanoshells of several tens of nanometers or more [65,

83, 84]. A more detailed description of this effect and the balance between radiative and nonradiative processes can be found in Ref. [69].

The analogy between metal NPs and antennas has other aspects: The NPs can be designed to present an asymmetric scattering pattern, scattering the light in specific directions, thanks to interferences between different elements or antenna modes as done with rake-type TV antennas found on the roofs of houses [85, 86]. Nanoemitters placed appropriately in the vicinity of these nanoantennas are then forced, by coupling to the NPs, to emit in these specific modes, and see their emission become directional [87, 88].

In any instance, in the present chapter the property of metal NPs that will be used most is the ability to enhance an electromagnetic field locally. The change in the emission properties of nanoemitters will be shown to be detrimental for resolution improvement purposes, but may be exploited to improve the photostability of the emitter.

8.5 Principles of NP-STED Nanoscopy

8.5.1 Qualitative Discussion

The essence of NP-STED is ultimately quite simple—to combine the two methods described so far, namely to use metal NPs in STED nanoscopy; see Fig. 8.2. Our goal is to improve the resolution or reduce the intensity requirements and, when possible, to improve the image brightness and the photostability of the fluorophores.

The use of metal NPs in STED nanoscopy has several effects. First, the presence of the metal particle leads to a local enhancement of the depletion intensity experienced by the fluorophores, denoted by Γ_I. In that regard, the depletion power which is focused on the fluorophores is used in a more efficient manner, namely it focuses light on the fluorophores rather than on unlabeled tissue, thus effectively increasing the power of the STED beam and reducing photodamage to the tissue. Simultaneously, the penetration to the tissue may be deeper due to the higher intensities. Thus, specifically, the STED intensity now becomes $\Gamma_I(\mathbf{r}', \mathbf{r}; \lambda_{\text{STED}}) I_{\text{STED}}(\mathbf{r}' - \mathbf{r})$ so that it now varies with both the scan coordinate $\mathbf{r}$ and the sample space coordinate $\mathbf{r}'$, that is, the STED intensity will no longer be translation invariant. At the same time, the decay rate of the excited singlet level is enhanced as well due to the Purcell effect (see Section 8.4), so that $k_{S_1} \rightarrow \Gamma_k(\mathbf{r}', \lambda_{\text{em}}) k_{S_1}$, that is, the decay rate depends now on the fluorophore position with respect to the metal NP. These two enhancement effects correspond to an effective rescaling of the normalized STED power (p in Eqs. 8.2–8.3) by $\Gamma_p \equiv \Gamma_I / \Gamma_k$, as well as a rescaling of α to $\alpha\gamma k$; see Eqs. 8.5 and 8.12 as well as [38].

To boost the relative STED intensity, it is clear that one needs to choose conditions where the field enhancement is substantial and ideally dominates over the decay rate enhancement. The way to do that is to tune the plasmon resonance wavelength to the depletion wavelength λ_{STED} (see Fig. 8.2c), that is, to the long-wavelength tail of the emission line. The depletion will then be correspondingly more efficient for a given incident intensity and, hence, the resolution will be better compared to standard STED. Alternatively, the plasmonic enhancement can allow the use of weaker and, hence, cheaper depletion sources. With lower input intensities, there will be less photons incident upon the regions not in the immediate vicinity of the illuminated hybrid fluorescent label, thus lowering photodamage to the biological environment.

In practice, the plasmon resonance is sufficiently broad (typically $\sim$50–100 nm) such that some residual enhancement of the singlet-state decay rate is very likely. When the decay rate enhancement is dominated by the absorption in the metal (i.e., by nonradiative decay), it is accompanied by a reduction of the quantum yield [3], a highly undesired effect. Nevertheless, this residual Purcell enhancement may have *useful* effects. Indeed, by shortening the lifetime of the excited singlet-level manifold, the Purcell effect may lead to a reduction of the absolute number of photons that manage to decay to the triplet-state manifold within the excited-state lifetime, thus limiting photobleaching [89]; this effect was already demonstrated experimentally in various different configurations; see, for example, Refs. [90, 91, 92] and references therein. In addition, since Purcell enhancement also applies to triplet (i.e., phosphorescent) emission, a significant enhancement of the nonradiative triplet decay rate can lead to an overall lowering of the triplet-state population. In fact, this may happen quite naturally in NP-STED since the triplet wavelength λ_T is typically spectrally red-shifted with respect to the fluorescence wavelength λ_{em} by about 100–200 nm [53]. Since λ_{STED} is also red-shifted with respect to the fluorescence wavelength λ_{em} (see Figs. 8.2 and 8.3), with a proper design, it can be matched to the triplet wavelength λ_T. In that sense, a plasmon resonance at the wavelength of triplet emission may allow to transform the detrimental absorption in the metal into an *advantage* [93], namely as a means to mediate photobleaching. Indeed, spectral schemes where the plasmon resonance overlaps the triplet emission λ_T were used before in the context of confocal microscopy and organic semiconductor lasing [92, 94, 95]. Then, even rather modest triplet decay rate enhancements have been shown experimentally to lead to significantly weaker photobleaching[a]. Such an improved photostability not only can allow brighter images and longer scans but would also allow using higher intensities (hence better resolution) [32] and better contrast [52] compared to standard STED nanoscopy. A further advantage may be that the restriction to perform CW-STED only with a low-triplet-yield dye could be relaxed. Importantly, all the above can be achieved without the need to employ low repetition rates, as is customary in standard STED nanoscopy [32] or, more generally, in fluorescence microscopy [96]. The relative importance of the singlet and triplet decay rate enhancements and their effect on the overall photobleaching rate depends of many variables and may be easily controlled via the emission depletion wavelength detuning.

8.5.2 Quantitative Discussion

We would now like to study quantitatively how the performance improvement offered by NP-STED depends on the STED intensity enhancement $\Gamma_I(\mathbf{r}, \mathbf{r}', \lambda_{STED})$ and decay rate enhancement $\Gamma_k(\mathbf{r}', \lambda_{em})$ as well as on the duration of the STED pulse and time gate. Such estimate requires the calculation of the fields, populations, and

[a]The observed reduction of photobleaching may have also been related to the higher mobility of the triplet states and not only to the absolute value of the triplet emission rate; see Ref. [93] for an elaborate discussion.

decay rates for each specific geometry, material composition, and spectral configuration of the hybrid metal NP emitter fluorescent labels; see Section 8.7 for more details. Nevertheless, it is possible to describe any such hybrid label in a single plot and in a simple manner using a simple approximation—that the field and decay rate enhancements used to estimate the resolution represent the *averaged* values experienced by the fluorophore(s) in the given label. Specifically, for a single fluorophore, the decay rate enhancement should be viewed as the spectral average rate over the emission line[a], whereas the decay rate of a spatial distribution of many fluorophores should be averaged also in space; we denote this enhancement as $\bar{\Gamma}_k$. Similarly, the intensity enhancement should be viewed as the average over the position and scan coordinate; we denote this enhancement as $\bar{\Gamma}_I$. When this average intensity enhancement is weakly dependent on the illumination pattern, see [41] and Section 8.5.4 for a detailed discussion, it can even be extracted from the quasi-static solution for plane-wave illumination. A prerequisite for performing such averaging is that the nonuniformity of the field enhancement will not distort the PSF of the system such that its width becomes ambiguous. Indeed, it was shown in Ref. [42] that the PSF maintains a smooth, localized shape, even in the presence of substantial enhancement variations.

Once the spectral and spatial averaging are performed, Eq. 8.2 provides a measure of the system's resolution (i.e., the reduction of the PSF width), although the illumination pattern is no longer strictly translation-invariant. Following Eq. 8.2, the resolution improvement offered by NP-STED compared with standard confocal microscopy now reads

$$\Gamma_{\text{res}}^{\text{NP-STED}} \cong \sqrt{1 + \bar{\Gamma}_p p \left(1 + \bar{\Gamma}_k k_{S_1} T_{\text{G}} - e^{-\bar{\Gamma}_k\left(\alpha + k_{S_1} T_{\text{G}}\right)}\right)}, \tag{8.5}$$

where $\bar{\Gamma}_p \equiv \bar{\Gamma}_I / \bar{\Gamma}_k$. The resolution improvement offered by NP-STED compared with standard STED is, thus, $\Gamma_{\text{res}}^{\text{NP-STED}} / \Gamma_{\text{res}}^{\text{STED}}$. For pulsed STED, it is easy to show that when field enhancement dominates the decay rate enhancement (i.e., for $\zeta \gg 1$, see Eq. 8.12

[a]Typical values given in the literature are indeed usually measured by integrating over the whole emission line.

below), one gets that

$$\Gamma_{\text{res}}^{\text{NP-STED}} / \Gamma_{\text{res}}^{\text{STED}} \to \sqrt{1+\bar{\Gamma}_p \alpha p} / \sqrt{1+\alpha p} \to \sqrt{\bar{\Gamma}_p}. \qquad (8.6)$$

In practice, the size of the NP limits the maximal achievable resolution. Accordingly, if the resolution calculated from Eq. 8.2 is smaller than the NP size, as would typically occur for sufficiently high field enhancements (see, for example, the upper parts of all the subplots in Fig. 8.6 below), one should exploit the (residual) field enhancement in order to lower the incident STED intensity. For brevity, we refer below only to resolution improvement and only plots of the resolution improvement are shown here.

As two illustrative examples, we now look at the two extreme regimes of STED pulse duration; see Fig. 8.6. The STED resolution serving as reference is set to be $\Gamma_{\text{res}}^{\text{STED}} \approx 6$ for the pulsed configuration and $\Gamma_{\text{res}}^{\text{STED}} \approx 3-4$ for the CW configurations. Such resolutions are provided by commercially available STED nanoscopes.

For pulsed STED operation ($T_{\text{STED}} = 250$ ps), Fig. 8.6a shows that when the decay rate enhancement is small, the field enhancement provides a clear improvement of resolution and almost no sensitivity to the decay rate enhancement, manifested by the small slope of the constant resolution improvement contour maps. Thus, in this regime, there is improvement of resolution (or equivalently, a reduction of required intensity) for any level of field enhancement even if it is smaller than the decay rate enhancement, that is, even when $\bar{\Gamma}_p < 1$. The reason is that for short STED pulses, the total amount of spontaneous emission is negligible, even if enhanced. However, once the decay rate enhancement becomes sufficiently high, it clearly limits the resolution improvement. In this case, the slope of the constant resolution improvement contours grows from 0 to 1. Similarly, the contour slope is 1 also in the CW-STED regime (see Fig. 8.6b), in which case improved resolution can be attained only when the field enhancement dominates the decay rate enhancement (i.e., only for $\bar{\Gamma}_p > 1$).

The improved performance in time-gated NP-STED is demonstrated in Fig. 8.6c,d. For pulsed STED, time gating removes the dependence on the decay rate and enables to exploit the intensity enhancement in its full extent for resolution improvement. This is

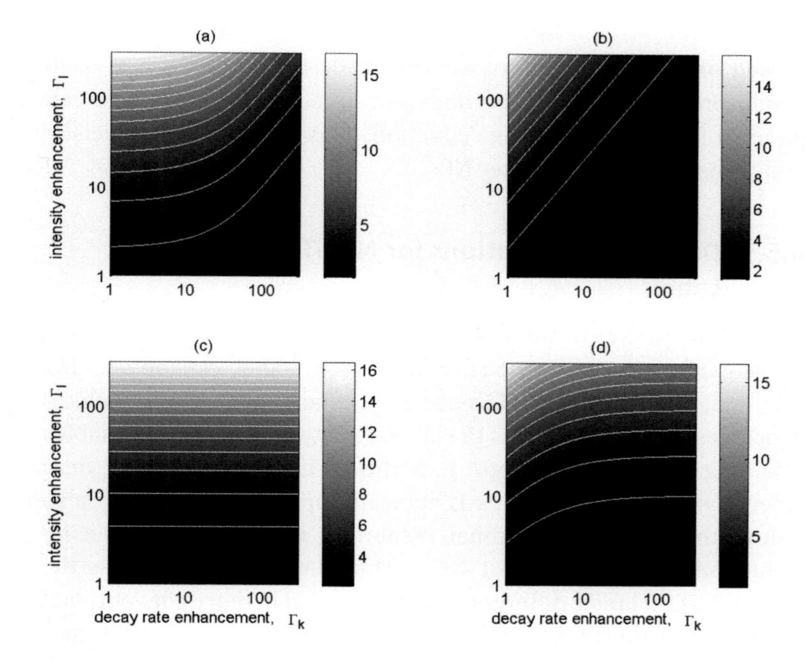

Figure 8.6 Gray-colored maps overlayed with contours of resolution improvement provided by NP-STED compared to standard STED (see Eq. 8.6) for (a) pulsed operation ($T_{\text{STED}} = 250$ ps and $p \approx 700$), (b) CW-STED ($p \approx 10$), (c) time-gated pulsed STED ($T_{\text{G}} = T_{\text{STED}} = 250$ ps and $p \approx 700$) and (d) time-gated CW-STED ($T_{\text{G}} = 2$ ns and $p \approx 10$). In all cases, $1/k_{S_1} = 5$ ns. The axes represent the field and decay rate enhancements, respectively in *logarithmic scale*. Adapted with permission from Ref. [39]. Copyright (2012) American Chemical Society.

only partially correct for the CW case; see Fig. 8.6d. The reduction of signal level may be minimized for slower scanning, potentially at the cost of further photobleaching.

According to the discussion above, we can identify two scenarios for the use of metal NPs in STED nanoscopy for resolution improvement/intensity reduction, depending on the value of $\bar{\Gamma}_p$. NPs for which $\bar{\Gamma}_p \approx 1$ or lower should be used in pulsed STED. In this case, the resolution improvement originates purely from near-field enhancement, whereas enhancement of the decay rate enhancement leads to a reduction of the signal level and photobleaching rate. In the second scenario, NPs for which $\bar{\Gamma}_p > 1$ can be used either in pulsed or CW-STED. Otherwise, especially in the absence of field enhancement (i.e., when Γ_I does not exceed unity), the NPs can be used for photobleaching reduction. This approach is especially appealing for few-nanometer NPs.

8.5.3 Design Considerations for NP-STED Fluorescent Labels

The ideal NP-STED fluorescent label is a metal NP as small as possible. Such an NP-STED fluorescent label can enable the minimization of the interference of the label with the biological processes; it also maximizes the label density which, in turn, enables attaining optimal resolution.[a] In addition, field enhancement near a few-nanometer metal NPs is typically optimal [3, 64], see also below, thus enabling maximal resolution improvement/intensity reduction; see Sections 8.5.1–8.5.2. These considerations make the ideal NP-STED labels different from those used in more conventional fluorescence enhancement techniques that rely on much larger metal NPs (e.g., in the range of 50–200 nm) [3]; see also Section 8.4. The reason for that is that NP-STED relies on *local, near-field* enhancement in small volumes, whereas fluorescence enhancement typically requires enhancement over a large volume (high average fields; see also Ref. [97]), hence requiring larger particles whose scattering cross sections are detectable with standard optical microscopy.

However, hybrid fluorescent labels which are too small are undesired as well, because in those, the fluorophores will be too close to the metal surfaces, so their energy will be absorbed in the metal (a process usually called quenching). Indeed, the Purcell effect for fluorophores at nanometer proximity to a metal surface gives rise to increasing decay rate enhancement and decreasing quantum yield [3] as the fluorophore gets closer to the metal surface. These effects hamper the super-resolution performance (i.e., $\Gamma_p \equiv \Gamma_I/\Gamma_k$ drops) and quench the measurable signal. Conversely, as the NP-STED label gets larger, the fluorophores

[a] Indeed, the size of the fluorescent label is the ultimate resolution limit.

gets further away from the metal surface and the decay rate enhancement is suppressed. Although larger metal NPs typically show also decreasing field enhancement; see, for example, Ref. [64], the decay rate enhancement drops faster, giving rise to higher value of Γ_l and an optimal size of about a few tens of nanometers.

This makes the ideal NP-STED labels of a comparable or somewhat larger size with respect to other NPs which are currently used as fluorescent or contrast labels in microscopy, such as functionalized quantum dots [98], Cornell (C)-dots [99] and nanodiamond particles with NV color centers [43, 100, 101].

Another critical consideration in the design of NP-STED labels is their spectral response. We recall that the plasmon resonance of the NPs should be tuned to the long-wavelength tail of the emission line; see Fig. 8.3. The original STED nanoscopes, based on pulsed STED configuration, worked with specific dye molecules that emit in the range of 650–700 nm and, hence, are depleted in the 750–800 nm range. Since gold is usually the metal of choice,[a] the spheres of which resonate at around 550 nm, one has to resort to somewhat nontrivial geometries. For example, one could use thin metal nanoshells, surrounding a dielectric core [65, 66], with a radii aspect ratio of about 4:5; the fluorophores can then be either inside the NP [38, 39] or around it (with a proper dielectric spacer layer). Such shells provide good and uniform field enhancement; however, they are difficult to fabricate, especially if few-nanometer shells are desired. An alternative is gold nanorods, with a similar aspect ratio; again, they would have to come along with a dielectric spacer layer separating the dye from the metal surface. Such nanorods provide a rather nonuniform field enhancement but have the potential advantage to enable a simultaneous enhancement of both depletion and excitation wavelengths [37]; they may also be more inert in terms of transferring heat to the biological environment and oxidizing. More recent STED nanoscopes operate with a CW depletion source and can work with a much wider range of dye molecules, including with depletion wavelengths as

---

[a]This is because of its chemical stability: contrary to the others, it does not oxidize, it is nontoxic to the biological environment, and it has favorable surface chemistry—oxide-free, clean Au surfaces can be easily modified by thiols.

short as 550–600 nm [102] which enable one to perform STED with green and yellow fluorescent proteins (GFPs/YFPs), which are naturally produced in cells. This wavelength regime is favorable in terms of fabrication of NP-STED labels since it enables the use of simple gold spheres which resonate at that spectral range. The field enhancement provided by these particles is, however, typically lower.

8.5.4 Ideal NP-STED Illumination

Once the shape of the NP-STED fluorescent label is optimized, one may ask, What is the optimal beam shape to use for NP-STED? Indeed, there is a variety of doughnut beams, including topological or nontopological doughnuts; see, for example, [49, 103, 104, 105, 106]. We would like to choose the beam for which the plasmonic field enhancement is maximized. Clearly, the relevance of this question goes beyond the context of NP-STED, especially since (1) most field enhancement studies are performed under typical illumination patterns such as a plane wave or focused field and since (2) plasmonics-assisted techniques employing complex illumination patterns, such as STED or SIM, attain growing popularity.

In Ref. [41], we studied the dependence of the plasmonic field enhancement on the illumination pattern and polarization for NPs of *spherical* shapes, for example, metal spheres or (multi-)nanoshells. We showed that for such geometries, the (spatial average of the) field enhancement is typically weakly dependent on the illumination pattern and polarization. For particles of more complicated shapes, the dependence of the field enhancement on illumination structure is still weak as long as the particles are small enough. This happens because for such geometries, the scattered wave is dominated by the electric dipole contribution, and the scattering of all other multipoles is substantially weaker.

However, a doughnut-shaped beam with a *true* zero at its center (e.g., a circularly polarized Laguerre–Gaussian beam [49] or an azimuthally polarized beam), such as used in the context of NP-STED, *does not* have an electric dipole component. Thus, the enhancement will be typically dominated by the next multipole in the hierarchy, for example, the magnetic dipole (for the azimuthally polarized beam) or the electric quadrupole (for the circularly polarized Laguerre–Gaussian beam). Nevertheless, these beams acquire an electric dipolar contribution *in the coordinate frame centered on the NP* if the NP is not centered at the center of the illumination beam. Thus, as the STED beam is scanned across the NP, the enhancement level changes from the one typical to the electric dipolar moment to the magnetic dipole / electric quadrupole moment and back. It is difficult to quantify the relative magnitudes of the enhancements associated with each of these multipoles. However, the differences can be quite small such that the (local as well as average) field enhancement is roughly uniform as the beam is scanned across the NP. In these cases, the field enhancement can be estimated from the solution for a plane-wave illumination [41]. If the NP is sufficiently small and has a simple shape, the solution of the quasi-static problem may be applicable. This allows for an easy estimate of the field enhancement, making the intensive computations of the interactions of the doughnut beam with the scatter unnecessary [38]. However, these detailed computations will be necessary for large spherical particles [42], for dense labeling in which case interactions between adjacent metal NPs may not be negligible anymore, and, obviously, for particles of more complicated shapes. At least in the former case, exact calculations show that the PSF remains smooth, even in the presence of substantial variations of the enhancement as a function of the scan coordinate [42]. This justifies the use of the PSF width as a measure for the resolution of the scheme, despite the formal lack of translation invariance.

Despite the relative weak dependence of the enhancement on the illumination pattern, we emphasize that the ideal NP-STED beam should be a beam that provides super-resolution also in the axial direction; in that sense, this is the same as in standard STED. Indeed, improving the axial resolution is of great importance, especially since it is initially poorer compared with the lateral resolution even for confocal microscopy. To achieve axial super-resolution in the context of STED nanoscopy, it is clear that a rapid variation of the STED intensity along the axial direction is required. Standard doughnut beams (such as circularly polarized Laguerre–Gaussian beams [49]) do not introduce any improvement of axial resolution, because such doughnut beams have zero intensity everywhere along their axial direction. However, there are different illumination approaches, such as a dark-sphere beam [103], the double-helix PSF [107] or 4π microscopy [108, 109] can, in principle, enable axial super-resolution. To date, only the latter and some variations of it were demonstrated experimentally in the context of STED [110–112].

8.5.5 Example: Metal Nanoshells

As a specific example, we study a configuration of metal nanoshells [66], (i.e., dielectric core, metal shell NPs) with fluorophores placed at the core center [89, 113–115]; see Fig. 8.7a. Metal nanoshells have several merits. First, there is a wealth of experience in fabricating and using them in biological and medical applications where they have shown biocompatibility and chemical stability [65, 83, 92, 116, 117]. Second, by varying the thicknesses of the core and shell one can tune λ_{PR} across the visible and near-IR spectrum [65, 115]. This facilitates matching the plasmon resonance to the STED fluorophore. Third, metal nanoshells offer substantial field enhancements in the core (see Fig. 8.7b), while keeping the fluorophore at a sufficient distance from the metal. As a result, the decay rate enhancement is minimized (Fig. 8.7c) and quantum yield quenching is not too strong (Fig. 8.7d). Fourth, they have already been shown to result in increased photostability [92, 94] which originates from a combination of the enhanced excited-state decay rate discussed above and chemical isolation of the encapsulated dye from oxygen.

As a first demonstration, Fig. 8.8a compares the signal obtained from a confocal STED and NP-STED imaging schemes at a given depletion intensity; see Section 8.7 for the details of the calculation. We study a 26 nm diameter particle with a silica core ($r_{\mathrm{core}} = 10$ nm, $\epsilon_{\mathrm{d}} = 2.25$) and a gold shell ($r_{\mathrm{shell}} = 13$ nm, dielectric data ϵ_{m} taken from Refs. [118, 119]). For simplicity, we assume a single emitter at the center of the NP. While STED clearly provides a narrower signal and, hence, better resolution than the confocal scheme, the improvement provided by NP-STED is far more significant. For a more systematic comparison, Fig. 8.8b shows the signal width (full-width at half-maximum, FWHM) of NP-STED, $d_{\mathrm{NP\text{-}STED}}$, as a function of the signal width of (standard) STED, d_{STED}. The improved

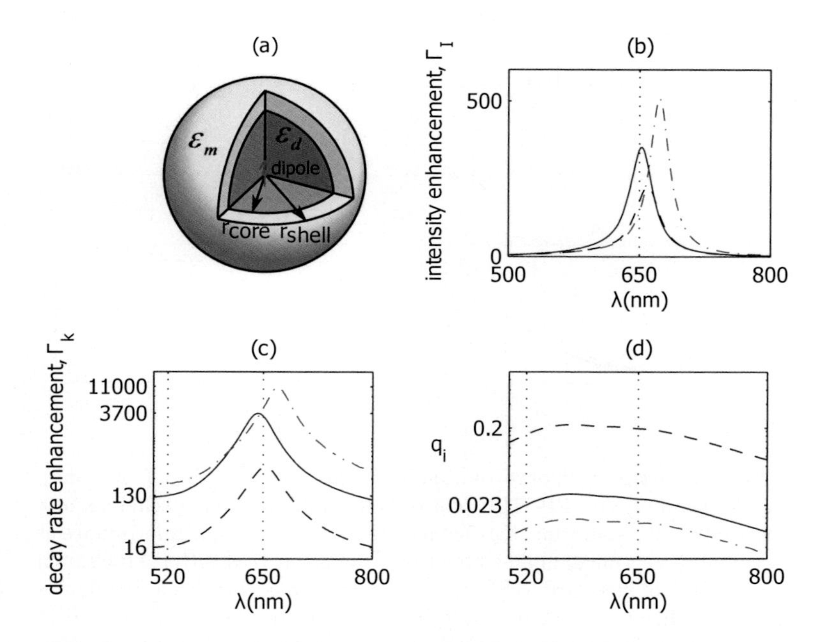

Figure 8.7 (a) Geometry of the metal nanoshell. (b) Electric field enhancement Γ_I, (c) total decay rate enhancement Γ_k, and (d) quantum yield q_i as a function of wavelength at the center of a 52 nm diameter (black dashed line), 26 nm diameter (blue line), and 20 nm diameter (red dash-dotted line) metal nanoshell. Dashed lines denote the enhancement levels at the chosen emission ($\lambda = 520$ nm) and depletion ($\lambda \approx 650$ nm) wavelengths. Adapted with permission from Ref. [39]. Copyright (2012) American Chemical Society.

resolution provided by NP-STED is quite striking—up to 7 times better compared to STED (see Fig. 8.8d), thus providing resolution which may even exceed the NP size.

Since in such cases the NP size limits the achieved resolution, one should exploit the (residual) field enhancement for lowering the STED intensities. This is shown in Fig. 8.8c, where the resolution provided by both schemes is plotted as a function of the peak normalized depletion intensity $\zeta^{(m)} = \max(\zeta)$ (see Eq. 8.12 below). It can be seen that *for any given resolution*, NP-STED requires about 80 times lower intensity compared with STED nanoscopy (see Fig. 8.8e).

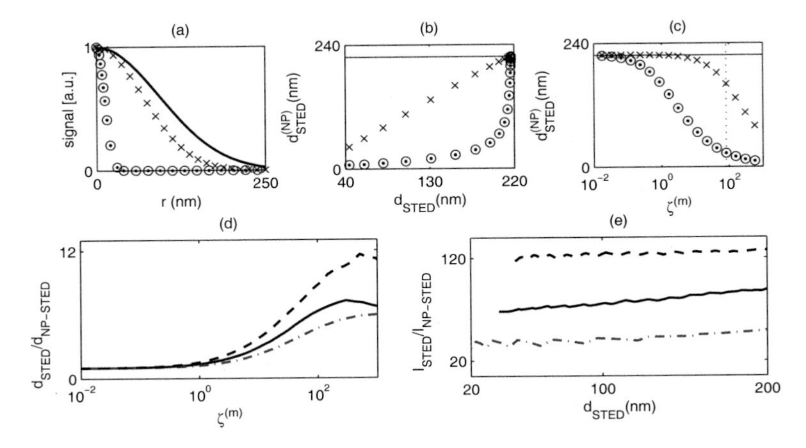

Figure 8.8 Comparison of resolution obtained using a confocal (solid line), STED (x symbol) and NP-STED (core–shell symbol) imaging schemes in a pulsed STED configuration. (a) Signal collected from a single fluorophore at the center of a 26 nm diameter silica/gold nanoshell. (b) FWHM of the signal as a function of standard STED resolution. (c) Same as (b) as a function of peak-normalized incident intensity $\zeta^{(m)} = \max(\zeta)$; see Eq. 8.12. The vertical dashed line corresponds to the parameters used in (a). (d) Resolution improvement $d_{\text{STED}}/d_{\text{NP-STED}}$ and (e) intensity reduction $(\Gamma_{\text{res}}^{\text{NP-STED}}/\Gamma_{\text{res}}^{\text{STED}})^2 \cong \Gamma_{\text{p}}^{\text{NP-STED}}/\Gamma_{\text{p}}^{\text{STED}} = I_{\text{STED}}/I_{\text{NP-STED}}$ obtained with nanoshells of a 52 nm diameter (black dashed line), 26 nm diameter (blue line), and 20 nm diameter (red dash-dotted line). Adapted from Ref. [38].

As for the signal intensity, we note that for this geometry, the fluorescence quantum yield is reduced by about 28 times at $\lambda_{\text{em}} = 520$ nm (see Fig. 8.7d). However, the up to 80-fold intensity reduction should be sufficient to compensate and even overcompensate for that loss by allowing faster repetitions rates, longer scans, or stronger excitation levels than in standard STED nanoscopy. Moreover, since the triplet state decay rate enhancement is quite high (~3700 times; see Fig. 8.7c), then for phosphorescence yields of $q_i \approx 10^{-4} - 10^{-1}$ [53], we can have $\Gamma_k(\lambda_{\text{STED}} \approx \lambda_{\text{T}})q_i \gg 1$, in which case, the triplet population decay rate is enhanced by a factor $\sim\Gamma_k(\lambda_{\text{STED}} \approx \lambda_{\text{T}})q_i$ [93]. The photobleaching rate is then further reduced by about the same factor [93], thus providing even further improved photostability. Unfortunately, the triplet emission quantum yield is frequently lower, depending heavily on the molecule and its environment. In this case, $\Gamma_k q_i$ is not necessarily appreciable so that the triplet lifetime will not be shortened.

We also examine the dependence of the improved performance on nanoshell size. As shown in Fig. 8.8d,e, larger nanoshells provide better improvement in imaging performance (e.g., 12-fold improvement of resolution or a 120-fold reduction of required intensities for a 52 nm diameter nanoshell); this is a result of the lower decay rate enhancement ($\Gamma_k(\lambda_{\text{em}})$ is $\sim$16 compared with $\sim$130 for the 26 nm nanoshell; see Fig. 8.7d). This very effect leads, however, also to a much weaker potential photostabilization ($\Gamma_k(\lambda_{\text{STED}} \approx \lambda_{\text{T}})$ $\sim$350 compared with $\sim$3700 for the 26 nm nanoshell). Smaller nanoshells show the opposite trend.

8.6 Experimental Results

As a first experimental implementation of the principles of NP-STED, core–shell structures have been used in Ref. [42]. In that work, silica cores of radius $r_{\text{core}} = 60$ nm containing the fluorescent emitters (ATTO 647N) are covered with a 20 nm thick shell of gold [66, 120]. Figure 8.9a presents the optical properties of the system, and a transmission electron microscope image of five core shells is shown as inset in Fig. 8.9b. The red curve corresponds to the emission line of ATTO 647N, and the gray curves are dark-field spectra [121] of individual nanoshells, indicating the position of their dipolar plasmon resonance. To match the conditions for optimal NP-STED operation, as discussed in the previous sections, the thickness of the gold has been tuned so that the plasmon resonance mainly enhances the field at the level of the fluorescent emitters, at the wavelength of the depletion beam (780 nm in the STED nanoscope used) [39]. To prepare the samples, bare cores and corresponding core shells are diluted and drop-cast on a glass coverslip and the solvent is left to evaporate. The side of the coverslip with the particles is immersed in index-matching oil and placed in contact with a glass slide. Images of both NP samples are taken with an inverted STED nanoscope, through the coverslip. The resolution obtained in confocal (no depletion beam used) and in STED mode are compared, with the percentage improvement for different depletion powers shown in

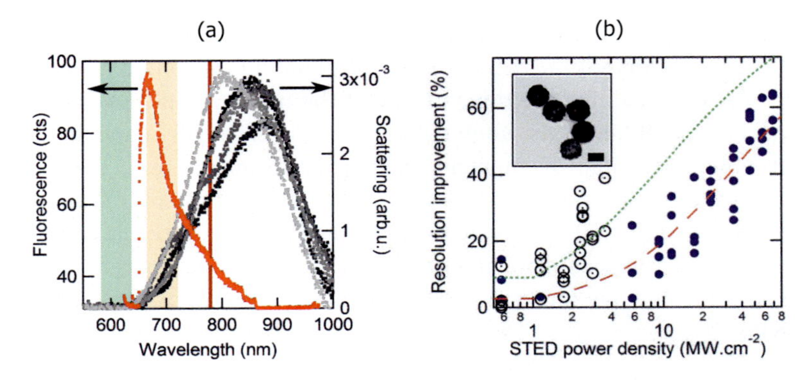

Figure 8.9 Experimental demonstration of the principles of NP-STED. (a) Spectral properties of the core–shell NPs used; they consist in a 120 nm diameter silica core doped with the fluorescent dye ATTO 647N, covered with a 20 nm thick gold shell. Red curve: emission of the dye, measured when illuminated at 634 nm. Gray curves: dark-field scattering spectra of five individual NPs. The areas shaded in green, orange, and dark red, respectively, indicate the spectral ranges of the pump beam, the detection filter, and the depletion laser. (b) Resolution improvement obtained with bare silica cores (blue disks) and core–shell NPs (black dotted open disks). Inset: TEM image of five core shells; scale bar 100 nm. Adapted with permission from Ref. [42]. Copyright (2014) American Chemical Society.

Fig. 8.9b. In the case of the bare cores, the optimal improvement, corresponding to a resolution of approx. 120 nm matching the size of the core, is reached for STED power densities above 60 MW.cm^{-2}. The optimal improvement is not reached for the core shells, as the fluorescence levels they emit drop for power densities of the order of 3 MW.cm^{-2}. However, for lower power densities, the increase of resolution with increasing power density occurs at about four times lower intensity than for the bare cores, which corresponds to the intensity enhancement provided by the NP and leads to a reduction of the power needed to reach super-resolution. This proof-of-concept experiment validates the principles of NP-STED but also highlights several issues that will need to be addressed in order to make it a viable technique:

Particle size. To be used for the labeling of biological samples, the NPs should be as small as possible. It is challenging to grow the very thin (nanometer range) gold shells required to tune the plasmon resonance to the correct spectral position; thus, future NP-STED labels might need to move from this core–shell geometry to metal NPs coated with the dye (cf. Section 8.5.3 and Refs. [37, 66]).

Metal luminescence. When illuminated with a pulsed laser, the gold shell of these NP-STED labels luminesces because of two-photon absorption [42, 122, 123]. This luminescence has a very broad spectrum which covers the wavelength range of the fluorescence of the dye used in that study. It is detrimental to the resolution, as it is produced by the STED depletion beam. Fortunately, the luminescence also decays much faster than the fluorescence of the dye: in Ref. [42], time-gating the signal allowed for the rejection of most of this parasitic signal. However, this comes at the cost of a reduced detected intensity; see Section 8.3.2. This problem could be mitigated, or even solved altogether, by increasing the duration of the depletion pulse or even by moving to CW depletion mode, which involves lower intensities [34].

Damage to the metal and bioenvironment. With a pulsed depletion beam, the high peak intensity/power can lead to damage to the gold NPs, which, in that study, prevented the observation of optimal resolution with the core shells. Again, a modification of the depletion beam parameters could improve this situation, as well as faster scanning speeds to leave time to the system to dissipate some heat, as already demonstrated in Refs. [124, 125].

8.7 Methods

8.7.1 Numerical Calculations

In general, the STED image is acquired by scanning the illumination and depletion beams across the sample and counting the emitted photons for every point of the scan. The total signal arriving at the detector at a given point in the scan $\mathbf{r}$ (the center of the illumination and depletion beams) is given by the product of the excitation probability $\eta_{\text{exc}}(I_{\text{exc}}(\mathbf{r}', \mathbf{r}); \lambda_{\text{exc}})$, the spontaneous emission probability $\eta_{\text{sp}}\,(I_{\text{STED}}(\mathbf{r}', \mathbf{r}); \lambda_{\text{STED}})$ and the detection PSF

$\Psi_{\mathrm{det}}(\mathbf{r}' - \mathbf{r}; \lambda_{\mathrm{em}})$, such that the complete signal is given by

$$S_{\mathrm{STED}}(\mathbf{r}) = \int \eta_{\mathrm{exc}}(\mathbf{r}', \mathbf{r}; \cdot)\eta_{\mathrm{sp}}\left(\mathbf{r}', \mathbf{r}; \cdot\right) \Psi_{\mathrm{det}}(\mathbf{r}' - \mathbf{r}; \cdot)d^3\mathbf{r}', \tag{8.7}$$

where $\mathbf{r}'$ is the coordinate in the object space. The excitation probability is given by

$$\eta_{\mathrm{exc}}(\mathbf{r}', \mathbf{r}; \lambda_{\mathrm{exc}}) = F(\mathbf{r}')N_1(I_{\mathrm{exc}}(\mathbf{r}', \mathbf{r}); \lambda_{\mathrm{exc}}), \tag{8.8}$$

where F is the distribution function of fluorophores in the object space and N_1 is the number density of the electrons in the excited-state manifold S_1. For excitation well below saturation, as is typical in fluorescence microscopy, the latter is proportional to the excitation PSF of the microscope. The excitation PSF as well as the detection PSF can be estimated as Gaussian-like functions. For simplicity, they both can be lumped into a single entity, defined as the confocal PSF, namely

$$\Psi_{\mathrm{conf}}(\mathbf{r}', \mathbf{r}; \lambda_{\mathrm{exc}}, \lambda_{\mathrm{em}}) = \Psi_{\mathrm{exc}}(\mathbf{r}', \mathbf{r}; \lambda_{\mathrm{exc}})\Psi_{\mathrm{det}}(\mathbf{r}' - \mathbf{r}; \lambda_{\mathrm{em}}). \tag{8.9}$$

Generally, the PSF of a typical confocal microscope has a lateral resolution of $\sim\lambda/2NA$ and longitudinal (axial) resolution of $\sim 1.4n\lambda/\mathrm{NA}^2$, where λ has a value roughly between the excitation and emission wavelengths and NA is the numerical aperture of the microscope [48].

The spontaneous emission probability can be calculated from the rate equations governing the population of the energy levels of the fluorophore. In the general case, it was shown that it is given by [52]

$$\eta_{\mathrm{sp}}(\mathbf{r}', \mathbf{r}; \lambda_{\mathrm{STED}}) = q_i \frac{1 + \gamma(\mathbf{r}', \mathbf{r}; \lambda_{\mathrm{STED}})e^{-k_{S_1}(\mathbf{r}')T_{\mathrm{STED}}(1+\gamma(\mathbf{r}', \mathbf{r}; \lambda_{\mathrm{STED}}))}}{1 + \gamma(\mathbf{r}', \mathbf{r}; \lambda_{\mathrm{STED}})}. \tag{8.10}$$

Here, q_i is the quantum yield of the fluorophore, and the depletion efficiency factor is given by

$$\gamma(\mathbf{r}', \mathbf{r}; \lambda_{\mathrm{STED}}) = \frac{\zeta(\mathbf{r}', \mathbf{r}; \lambda_{\mathrm{STED}})k_{\mathrm{vib}}}{\zeta(\mathbf{r}', \mathbf{r}; \lambda_{\mathrm{STED}})k_{S_1} + k_{\mathrm{vib}}}, \tag{8.11}$$

where

$$\zeta(\mathbf{r}', \mathbf{r}; \lambda_{\mathrm{STED}}) = \frac{I_{\mathrm{STED}}(\mathbf{r}', \mathbf{r}; \lambda_{\mathrm{STED}})}{I_{\mathrm{sat}}}, \tag{8.12}$$

is the STED intensity normalized by the saturation intensity (see Eq. 8.1). Equation 8.10 takes into account stimulated emission (via

ζ) as well as the undesired effect of re-excitation from the S_0 manifold to the S_1 manifold. Additional parameters used above are the total decay rate of the singlet level k_{S_1}, the vibrational decay rate k_{vib} and the STED pulse duration T_{STED}. For vanishing STED intensity, we recover the emission probability of the confocal microscope, $\eta_{\text{sp}} = q_i$.

Equation 8.10 was derived under two conditions, namely that the vibrational decay rate within the excited singlet-level manifold is much shorter than either the STED pulse duration and the singlet-state decay rate, that is, $k_{\text{vib}} \gg 1/T_{\text{STED}}, k_{S_1}$, and that the fluorophores all decay to their ground state between consequent illumination pulses. These conditions are valid in basically all STED configurations to date and for any value of STED intensity.

In the absence of any scattering (i.e., in the absence of NPs), the excitation and depletion fields are translation invariant (i.e., all functions of $\mathbf{r}' - \mathbf{r}$ rather than on $\mathbf{r}'$ and $\mathbf{r}$ separately).[a] In this case, the signal is given by a convolution between the fluorophore distribution F and the product of the emission probability and the excitation and detection PSF, namely

$$S_{\text{STED}}(\mathbf{r}) = \int F(\mathbf{r}')\Psi_{\text{conf}}(\mathbf{r}' - \mathbf{r}; \cdot)\eta_{\text{sp}}\left(\mathbf{r}' - \mathbf{r}; \cdot\right) d^3\mathbf{r}', \qquad (8.13)$$

where, for brevity, the dependence on the wavelengths is not written explicitly anymore. It is convenient to define an (effective) STED PSF, Ψ_{STED}, as the product of the confocal PSF (Eq. 8.9) and the emission probability (Eq. 8.10), namely

$$\Psi_{\text{STED}}(\mathbf{r}' - \mathbf{r}; \cdot) = \Psi_{\text{conf}}(\mathbf{r}' - \mathbf{r}; \cdot)\eta_{\text{sp}}\left(I_{\text{STED}}(\mathbf{r}' - \mathbf{r}); \cdot\right). \qquad (8.14)$$

Note that the high spontaneous emission probability on the outer parts of the beam is of little consequence as the excitation in these regions is very weak.

[a] In this context, we should note that the approach described above ignores scattering from the biological medium. In practice, however, scattering proves to be the limiting factor on the possible depth into the tissue at which STED still works due to the distortion of the doughnut-shaped depletion beam.

Equations 8.13–8.14 allows us to avoid any integral calculations and simply look at the FWHM of the STED PSF at the focal plane, d_{STED}, as a measure for resolution.

However, in the presence of a scattering element, namely a metal NP (or NPs), the field distributions, and hence excitation and depletion profiles, become spatially dependent. In this case, Eq. 8.7 becomes

$$S_{\text{NP-STED}}(\mathbf{r}) = \int F(\mathbf{r}')\Psi_{\text{conf}}(\mathbf{r}', \mathbf{r}; \cdot)\eta_{\text{sp}}(\mathbf{r}', \mathbf{r}; \cdot)d^3\mathbf{r}'. \quad (8.15)$$

Thus, the image is no longer a simple convolution of the fluorophore distribution and the PSF of the microscope. Instead, the image should be obtained by explicit evaluation of the integral (Eq. 8.15) at varying relative alignments of the incident doughnut-shaped depletion beam and the NP. Then, each point of the scan constitutes a pixel of the final image. This process mimics the raster scanning used in a (NP-)STED measurement.

Nevertheless, we emphasize that for sufficiently small NPs, the overall doughnut-field distribution is unaltered, and only a local-field enhancement is induced in the immediate vicinity of the NP; see Fig. 8.10. Accordingly, the probability of spontaneous emission maintains its shape (even if the depth of the valleys (see Fig. 8.4) is varying slightly with $\mathbf{r}$). This allows us to still regard the width of

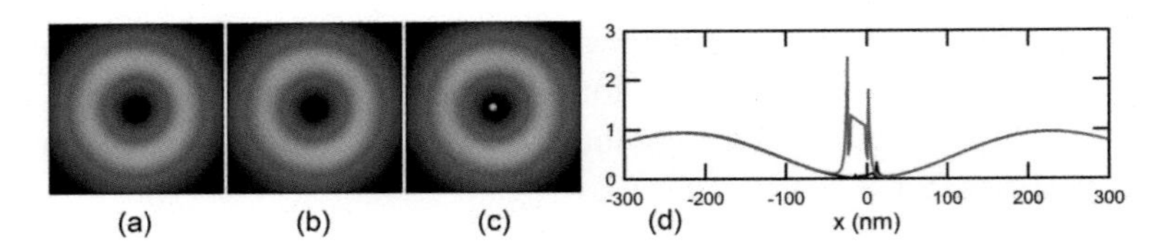

Figure 8.10 (a) Intensity distribution in the focal plane of an incident Laguerre–Gaussian beam propagating in free space. (b) Intensity distribution in the focal plane of a Laguerre–Gaussian beam incident on a small metal nanoshell at the beam center; very little scattering is observed. (c) Same as (b) for a metal nanoshell shifted 10 nm away from the beam center. Significant local-field enhancement is observed, yet the overall doughnut-shaped field pattern is not modified. Scale bar length is 200 nm. (d) Cross sections along the center for (a), (b), and (c) with black, blue, and red colors, respectively. In accordance with the local character of the scattering, the different lines are nearly indistinguishable except near the NP. Adapted with permission from Ref. [39]. Copyright (2012) American Chemical Society.

the obtained signal, $d_{\text{NP-STED}}$, as a measure of the resolution in NP-STED, as in standard STED, despite the formal lack of translation invariance. We have verified that in the absence of metal NPs, this procedure reproduces the well-known confocal PSF and the resolution scaling of STED nanoscopy [52].

In the specific calculations presented in Section 8.5.5, the field distributions were calculated in the frequency domain using rigorous theory for vectorial beam focusing [126] using standard commercial software. These calculations can also be performed in the time domain or using rigorous Mie theory calculations (see, for example, Ref. [41]) for spherical particles; the latter approach is the most efficient computationally. The illumination system is assumed to consist of an NA $= 1.2$ objective lens focusing into water with optimal aperture filling by the incident beam. The depletion pulse is assumed to be 70 ps long and to have the spatial doughnut profile of a circularly polarized Laguerre–Gaussian beam [49]. Decay rate and quantum yield calculations were performed in the time domain using standard commercial software for calculating the field distribution in the presence of a point dipole source [3]. The fluorophore was chosen to be a dye molecule having a Lorentzian-shaped absorption and emission cross sections with a 50 nm spectral width and magnitude of 10^{-16}cm^2. The intrinsic fluorophore lifetime is assumed to be 5 ns, and the intrinsic fluorescence quantum yield to have the rather high value of $q_i = 0.65$. These values correspond to those of ATTO647N, a dye which is popular in STED nanoscopy. The emission depletion detuning is chosen to be 130 nm, as in standard STED.

For simplicity, we assume that the system consists of a single metal nanoshell with a single emitter at its center, that is, $F(r) = \delta(r)$. The set of calculations of a scan along the radial direction thus provides a cross section of the obtained signal. The resolution of the scheme is taken to be the width of signal. In more realistic cases, the core will be doped with more than one emitter. In this case, one can simply assume a uniform distribution $F(r) =$ const. within the volume containing the emitters (typically smaller than the overall core size). The calculation would be performed in the same way, namely using Eq. 8.7. Since in the case of a uniform doping of the core the average decay rate enhancement increases, the performance improvement would be somewhat lower than with a single emitter. However, the signal will be naturally brighter compared to the case of a single emitter. This was confirmed in the measurements done in Ref. [42].

8.7.2 Experimental STED Nanoscopy System

The STED nanoscope employed in Ref. [42] is based on a single 80 MHz Ti:Sapphire source tuned to 780 nm [127, 128]. Figure 8.11 shows a schematic of the system. The beam exiting the laser source is split via a Glan–Taylor polarizer and half-wave plate to adjust the ratio of intensity going into the two beam paths. To provide excitation light a portion of the pulses is coupled into a

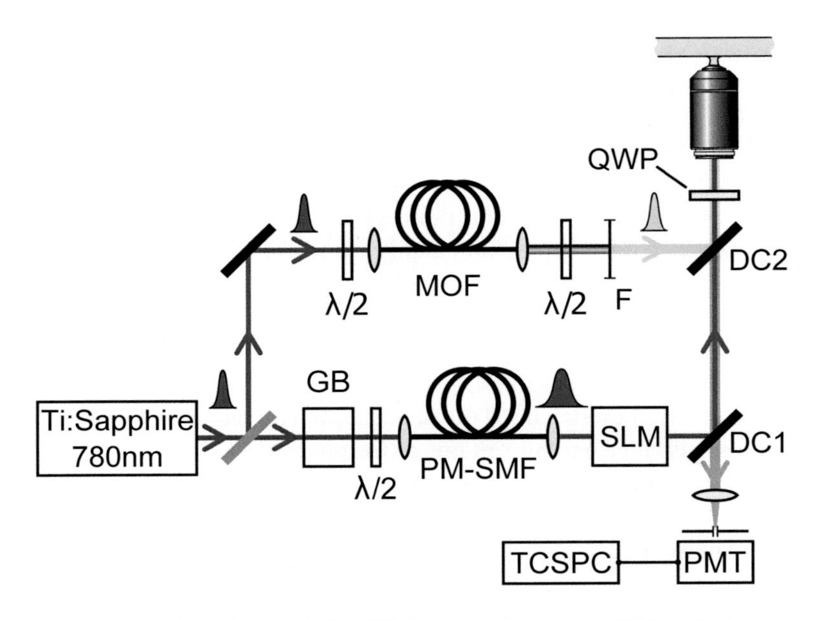

Figure 8.11 Description of the STED nanoscopy setup. MOF: microstructured optical fiber; QWP: quarter-wave plate; GB: 1 m long SF57 glass block to stretch the pulse; PM-SMF: polarization-maintaining single-mode fiber; SLM: spatial light modulator; DC1 and DC2: dichroic mirrors; PMT: photomultiplier tube; TCSPC: time-correlated single-photon counting unit. Adapted with permission from Ref. [42]. Copyright (2014) American Chemical Society.

microstructured optical fiber (MOF) in which a supercontinuum is formed spanning from ~500 nm to above 1 μm. Band-pass filters (F) are then used to select an appropriate waveband from the supercontinuum. A 54 nm waveband centered at 609 nm was used for excitation. For depletion, the remaining pulses from the Ti:sapphire are stretched from ~100 fs to ~300 ps using a combination of 1 m of SF57 glass (GB) and 100 m of polarization-maintaining single-mode fiber (PM-SMF). The depletion beam is then modulated with a liquid crystal spatial light modulator (SLM) in order to impart the helical phase profile required to generate the doughnut for STED. The SLM is also used to correct for aberrations present in the depletion beam due to optical misalignments/imperfections. The excitation and depletion beams are combined using dichroic mirrors (DC1 and DC2), circularly polarized via an achromatic quarter-wave plate (QWP) and focused into the sample with an 100×, 1.4 NA Leica objective lens. The dichroic mirrors also serve to separate scattered excitation and depletion light from fluorescence excited in the sample. Residual scattered photons are filtered from the fluorescence signal via further band-pass filters. Fluorescence is coupled into a 50 μm-core fiber which acts as a confocal pinhole and is detected via a hybrid photomultiplier tube (PMT) connected to time-correlated single-photon counting electronics. Sample scanning is performed via a three-axis piezo stage.

8.8 Summary and Outlook

We have reviewed in detail the theoretical aspects of NP-STED—the use of hybrid fluorescent labels consisting of a fluorescent emitter and a metal NP in STED nanoscopy—and described the first experimental demonstration of this concept. We showed that this technique has great promise in allowing for a significant improvement in all aspects of the imaging performance of STED nanoscopy, namely allowing low STED power, high resolution, or improved photostability. This variety allows one to design an NP to suit the imaging requirements and object sizes.

The proposed technique relies on challenging yet demonstrated NP fabrication technology for fluorescence microscopy. In addition, it is complementary to the improvements made to STED nanoscopes; thus, it is compatible with any STED mode of operation such as CW [34] and time-gated [33, 58] STED modes.

Importantly, as the scheme relies on local, *near-field* enhancement rather than on far-field properties such as the scattering cross section, one can use very small NPs and thus enjoy the large field enhancement they provide, minimize the interference of the NP with the biological processes, and not limit the resolution improvement by using exceedingly large labels.

One of the most appealing merits of NP-STED, in comparison to other plasmonics-based applications, is that it is not sensitive to absorption in the sense that the resolution improvement can be made *independent* of the absorption. However, absorption may indirectly affect the performance via a reduction of brightness and heating (hence damaging) the NPs.

A proof-of-principle experiment has confirmed the NP-STED concepts using core–shell nanostructures. With these labels, a fourfold reduction of the depletion power required to reach super-resolution has been reached. The experiment also revealed several issues that will need to be addressed to perform efficient NP-STED: the fabrication of labels with sizes compatible with the needs of biological systems will have to be developed, and ways to circumvent both luminescence and heating of the metallic NPs will be necessary. All these problems have, however, suitable solutions.

Due to the need to avoid placing the fluorophores too close to the metal surface (in order to avoid substantial decay rate enhancements), the NP sizes in NP-STED are not expected to be much smaller than about 20 nm, that is, only slightly larger than large quantum dots [45, 46]. Taking into account the limit this puts on label density, we conclude that NP-STED nanoscopy is expected to reduce intensities and photobleaching for 3D imaging at resolution levels of 20–50 nm; nevertheless, this resolution exceeds that provided by most commercial STED systems. Moreover, although the NP size limits the type of biological problems that can be studied with NP-STED, the technique can still be used for the imaging of many systems such as immobile structures, processes occurring at cell membranes, actin and neuron chains, vesicles, etc. Moreover, NPs of such sizes are also ideal for live-cell imaging via endocytosis [129]. The associated reduction of the intensity requirement will allow reduction of the cost of pulsed STED systems and improvement of the performance of cheaper STED systems [130]. In particular, it may prove to be useful for the recently developed parallel STED schemes using several doughnuts at the same time. This approach increases the speed of the measurements but also requires more laser power to generate the depletion beams—a drawback that could be alleviated by NP-STED [131, 132]. Finally, NP-STED may give access imaging at depths which were so far inaccessible in in vivo experiments.

Further implementations of the concepts of plasmonics-assisted STED may be dedicated to CW-STED systems. In that regard, it would be intriguing to study the optical response of the metal itself to the high field intensities. This response was shown to be very complex and strong (see Refs. [124, 125]). Another direction would be to study super-resolution imaging near a patterned metal surface. Such nanostructures may allow for substantial (average) field enhancement, directional emission (hence increased collection efficiencies), and easy spectral control. More importantly, with regard to the challenges encountered so far, the field enhancement occurring near patterned metal surfaces may allow using standard (i.e., metal-free) fluorescent labels (hence no impact on the biology) and much better heat management and mechanical stability. It remains to be seen how the doughnut beam would interact with such patterned surfaces and what the overall gain in imaging performance parameters in this approach would be. Further implementations of the concepts of plasmonics-assisted STED may be found in enhancing two-photon excitation STED microscopy [37, 133], multicolor STED imaging, and CARS-STED [134, 135], as well as potentially in other super-resolution techniques such as nonlinear SIM or SAX. Ultimately, NP-STED opens the way to perform correlated optical electron microscopy on the same sample, currently a major challenge in the field of bioimaging.

References

1. Abbe, E. (1873). Beiträge zur Theorie des Mikroskops und der mikroskopischen Wahrnehmung, *Arch. Mikrosk. Anat.*, **9**, pp. 413–418.
2. Barnes, W. (2011). Metallic metamaterials and plasmonics, *Philos. Transact. A: Math. Phys. Eng. Sci.*, **369**, pp. 3431–3433.
3. Giannini, V., Fernández-Domí nguez, A., Heck, S., and Maier, S. (2011). Plasmonic nanoantennas: fundamentals and their use in controlling the radiative properties of nanoemitters, *Chem. Rev.*, **111**, pp. 888–3912.
4. Fang, N., Lee, H., Sun, C., and Zhang, X. (2005). Sub-diffraction-limited optical imaging with a silver superlens, *Science*, **308**, pp. 534–537.
5. Pendry, J. (2000). Negative index makes a perfect lens, *Phys. Rev. Lett.*, **85**, p. 3966.
6. Smith, D., Padilla, W., Vier, D., Nemat-Nasser, S., and Schultz, S. (2000). Composite medium with simultaneously negative permeability and permittivity, *Phys. Rev. Lett.*, **84**, p. 4184.
7. Jacob, Z., and Narimanov, E. (2006). Optical hyperlens: far-field imaging beyond the diffraction limit, *Opt. Express*, **14**, pp. 8247–8256.
8. Liu, Z., Lee, H., Xiong, Y., Sun, C., and Zhang, X. (2007). Far-field optical hyperlens magnifying sub-diffraction-limited objects, *Science*, **315**, p. 1686.
9. Salandrino, A., and Engheta, N. (2006). Far-field subdiffraction optical microscopy using metamaterial crystals: theory and simulations, *Phys. Rev. B*, **74**, p. 075103.
10. Lemoult, F., Fink, M., and Lerosey, G. (2011). Revisiting the wire medium: an ideal resonant metalens, *Waves Random Complex Media*, **21**, pp. 591–613.
11. Rogers, E., and Zheludev, N. (2013). Optical super-oscillations: sub-wavelength light focusing and super-resolution imaging, *J. Opt.*, **15**, p. 094008.
12. Fernández-Domínguez, A., Liu, Z., and Pendry, J. (2015). Coherent four-fold super-resolution imaging with composite photonicplasmonic structured illumination, *ACS Photonics*, **2**(3), pp. 341–348.
13. Wei, F., and Liu, Z. (2010). Plasmonic structured-illumination microscopy, *Nano Lett.*, **10**, pp. 2531–2536.
14. Wei, F., Lu, D., Shen, H., Ponsetto, J., Huang, E., and Liu, Z. (2014). Wide field super-resolution surface imaging through plasmonic structured illumination microscopy, *Nano Lett.*, **14**, pp. 4634–4639.

15. Shechtman, Y., Eldar, Y., Szameit, A., and Segev, M. (2011). Sparsity based sub-wavelength imaging with partially incoherent light via quadratic compressed sensing, *Opt. Express*, **19**, pp. 14807–14822.

16. Diaspro, A. (2002). *Confocal and Two-Photon Microscopy* (Wiley-Liss).

17. Lakowicz, J. (2006). *Principles of Fluorescence Spectroscopy*, 3rd ed. (Springer, NY, USA).

18. Hell, S. (2007). Far-field optical nanoscopy, *Science*, **316**, pp. 1153–1158.

19. Hell, S., and Wichmann, J. (1994). Breaking the diffraction resolution limit by stimulated emission: stimulated-emission-depletion fluorescence microscopy, *Opt. Lett.*, **19**, pp. 780–782.

20. Klar, T., Jakobs, S., Dyba, M., Egner, A., and Hell, S. (2000). Fluorescence microscopy with diffraction resolution barrier broken by stimulated emission, *Proc. Natl. Acad. Sci. U S A*, **97**, pp. 8206–8210.

21. Fölling, J., Bossi, M., Bock, H., Medda, R., Wurm, C., Hein, B., nd C. Eggeling, S. J., and Hell, S. (2008). Fluorescence nanoscopy by ground-state depletion and single-molecule return, *Nat. Methods*, **5**, pp. 943–945.

22. Hofmann, M., Eggeling, C., Jakobs, S., and Hell, S. (2005). Nonlinear structured illumination microscopy by surface plasmon enhanced stimulated emission depletion, *Proc. Natl. Acad. Sci. U S A*, **102**, pp. 17565–17569.

23. Sharonov, A., and Hochstrasser, R. (2006).Wide-field subdiffraction imaging by accumulated binding of diffusing probes, *Proc. Natl. Acad. Sci. U S A*, **103**, pp. 18911–18916.

24. Gustafsson, M. (2000). Surpassing the lateral resolution limit by a factor of two using structured illumination microscopy, *J. Microsc.*, **198**, pp. 82–87.

25. Gustafsson, M. (2005). Nonlinear structured-illumination microscopy: wide-field fluorescence imaging with theoretically unlimited resolution, *Proc. Natl. Acad. Sci. U S A*, **102**, pp. 13081–13086.

26. Bates, M., Huang, B., and Zuang, X. (2008). Super-resolution microscopy by nano-scale localization of photo-switchable fluorescent probes, *Curr. Opin. Chem. Bio.*, **12**, pp. 505–514.

27. Moerner,W. (2007).New directions in single-molecule imaging and analysis, *Proc. Natl. Acad. Sci. U S A*, **104**.

28. Fujita, K., Kobayashi, M., Kawano, S., Yamanaka, M., and Kawata, S. (2007). High-resolution confocal microscopy by saturated excitation of fluorescence, *Phys. Rev. Lett.*, **99**, p. 228105.

29. Dertinger, T., Colyer, R., Iyer, G., Weiss, S., and Enderlein, J. (2009). Fast, background-free, 3D super-resolution optical fluctuation imaging (SOFI), *Proc. Natl. Acad. Sci. U S A*, **106**, pp. 22287–22292.
30. Eggeling, C., Ringemann, C., Medda, R., Schwarzmann, G., Sandhoff, K., Polyakova, S., Belov, V., Hein, B., von Middendorff, C., Schönle, A., and Hell, S. (2009). Direct observation of the nanoscale dynamics of membrane lipids in a living cell, *Nature*, **457**, pp. 1159–1162.
31. Willig, K., Rizzoli, S., Westphal, V., Jahn, R., and Hell, S. (2006). STED microscopy reveals that synaptotagmin remains clustered after synaptic vesicle exocytosis, *Nature*, **440**, pp. 935–939.
32. Donnert, G., Keller, J., Medda, R., Andrei, M., Rizzoli, S., Luhrman, R., Jahn, R., Eggeling, C., and Hell, S. (2006). Macromolecular scale resolution in biological fluorescence microscopy, *Proc. Natl. Acad. Sci. U S A*, **103**, pp. 11440–11445.
33. Vicidomini, G., Moneron, G., Han, K., Westphal, V., TA, H., Reuss, M., Engelhardt, J., Eggeling, C., and Hell, S. (2011). Sharper low-power STED nanoscopy by time-gating, *Nat. Methods*, **8**, pp. 571–573.
34. Willig, K., Harke, B., Medda, R., and Hell, S. (2007). STED microscopy with continuous wave beams, *Nat. Methods*, **4**, pp. 915–918.
35. Lukinavicius, G., Umezawa, K., Olivier, N., Honigmann, A., Yang, G., Plass, T., Mueller, V., Reymond, L., Jr., I. C., Luo, Z., Schultz, C., Lemke, E., Heppenstall, P., Eggeling, C., Manley, S., and Johnsson, K. (2013). A near-infrared fluorophore for live-cell superresolution microscopy of cellular proteins, *Nat. Chem.*, **132**, p. 5.
36. Rittweger, E., Rankin, B., Westphal, V., and Hell, S. (2007). Fluorescence depletion mechanisms in super-resolving STED microscopy, *Chem. Phys. Lett.*, **442**, pp. 483–487.
37. Balzarotti, F., and Stefani, F. (2012). Plasmonics meets far-field optical nanoscopy, *ACS Nano*, **6**, p. 4580.
38. Sivan, Y. (2012). Performance improvement in nanoparticle-assisted stimulated emission depletion nanoscopy, *Appl. Phys. Lett.*, **101**, p. 021111.
39. Sivan, Y., Sonnefraud, Y., Kéna-Cohen, S., Pendry, J., and Maier, S. (2012). Nanoparticle-assisted stimulated-emission-depletion nanoscopy, *ACS Nano*, **6**, pp. 5291–5296.
40. Zhang, H., Zhao, M., and Peng, L. (2011). Nonlinear structured illumination microscopy by surface plasmon enhanced stimulated emission depletion, *Opt. Express*, **19**, pp. 24783–24794.

41. Foreman, M., Sivan, Y., Maier, S., and Török, P. (2012). Independence of plasmonic near-field enhancements to illumination beam profile, *Phys. Rev. B*, **86**, p. 155441.
42. Sonnefraud, Y., Sinclair, H., Sivan, Y., Foreman, M., Dunsby, C., Neil, M., French, P., and Maier, S. (2014). Experimental proof of concept of nanoparticle assisted STED, *Nano Lett.*, **14**, pp. 4449–4453.
43. Han, K., Willig, K., Rittweger, E., Jelezko, F., Eggeling, C., and Hell, S. (2009). Three-dimensional stimulated emission depletion microscopy of nitrogen-vacancy centers in diamond using continuous-wave light, *Nano Lett.*, **9**, p. 3323.
44. Rittweger, E., Irvine, K. H. S., Eggeling, C., and Hell, S. (2009). STED microscopy reveals crystal colour centres with nanometric resolution, *Nat. Photonics*, **3**, p. 144.
45. Irvine, S., Staudt, T., Rittweger, E., Engelhardt, J., and Hell, S. (2008). Direct light-driven modulation of luminescence from Mn-doped ZnSe quantum dots, *Angew. Chem., Int. Ed. Engl.*, **47**, pp. 2685–2688.
46. Lesoine, M., Bhattacharjee, U., Guo, Y., Vela, J., Petrich, J., and Smith, E. (2013). Subdiffraction, luminescence-depletion imaging of isolated, giant, CdSe/CdS nanocrystal quantum dots, *J. Phys. Chem. C*, **117**, pp. 3662–3667.
47. Goodman, J. (2006). *Introduction to Fourier Optics*, 3rd ed. (Roberts, CO, USA).
48. Novotny, L., and Hecht, B. (2006). *Principles of Nano-Optics* (Cambridge University Press).
49. Török, P., and Munro, P. (2004). The use of Gauss–Laguerre vector beams in STED microscopy, *Opt. Express*, **12**, pp. 3605–3617.
50. Galiani, S., Harke, B., Vicidomini, G., Lignani, G., Benfenati, F., Diaspro, A., and Bianchini, P. (2012). Strategies to maximize the performance of a STED microscope, *Opt. Express*, **20**, pp. 7362–7374.
51. Hopt, A., and Neher, E. (2001). Highly nonlinear photodamage in two-photon fluorescence microscopy, *Biophys. J.*, **80**, pp. 2029–2036.
52. Leutenegger, M., Eggeling, C., and Hell, S. (2010). Analytical description of STED microscopy performance, *Opt. Express*, **18**, pp. 26417–26429.
53. Turro, N. (1978). *Modern Molecular Photochemistry*, (Benjamin/Cummings, Menlo Park, CA).
54. Bout, D., and Deschenes, L. (2002). Single molecule photobleaching: increasing photon yield and survival time through suppression of two-step photolysis, *Chem. Phys. Lett.*, **365**, pp. 387–395.

55. Eggeling, C., Volkmer, A., and Seidel, C. (2005). Molecular photobleaching kinetics of Rhodamine 6G by one- and two-photon induced confocal fluorescence microscopy, *ChemPhysChem*, **6**, pp. 791–804.
56. Eggeling, C.,Widengren, J., Rigler, R., and Seidel, C. (1998). Photobleaching of fluorescent dyes under conditions used for single-molecule detection: evidence of two-step photolysis, *Anal. Chem.*, **70**, pp. 2651–2659.
57. Auksorius, E., Boruah, B., Dunsby, C., Lanigan, P., Kennedy, G., Neil, M., and French, P. (2008a). Stimulated emission depletion microscopy with a supercontinuum source and fluorescence lifetime imaging, *Opt. Lett.*, **33**, p. 113.
58. Moffitt, J., Osseforth, C., and Michaelis, J. (2011). Time-gating improves the spatial resolution of STED microscopy, *Opt. Express*, **19**, pp. 4242–4254.
59. Vicidomini, G., Moneron, G., Eggeling, C., Rittweger, E., and Hell, S. (2012). STED with wavelengths closer to the emission maximum, *Opt. Express*, **20**, pp. 5225–5236.
60. Homola, J. (2008). Surface plasmon resonance sensors for detection of chemical and biological species, *Chem. Rev.*, **108**(2), pp. 462–493.
61. Bohren, C., and Huffman, D. (1983). *Absorption and Scattering of Light by Small Particles* (Wiley).
62. Hylton, N., Li, X., Giannini, V., Lee, K.-H., Ekins-Daukes, N., Loo, J., Vercruysse, D., Dorpe, P. V., Sodabanlu, H., Sugiyamaand, M., and Maier, S. (2013). Loss mitigation in plasmonic solar cells: aluminium nanoparticles for broadband photocurrent enhancements in GaAs photodiodes, *Sci. Rep.*, **3**, p. 2874.
63. Mie, G. (1908). Beiträge zur optik trüber medien, speziell kolloidaler metallösungen, *Annalen der Physik (Leipzig)*, **330**, p. 377.
64. Meier, M., and Wokaun, A. (1983). Enhanced fields on large metal particles: dynamic depolarization, *Opt. Lett.*, **8**, pp. 581–583.
65. Halas, N. (2002). The optical properties of nanoshells, *Opt. Photon. News*, **August**, pp. 26–30.
66. Oldenburg, S. J., Averitt, R. D., Westcott, S. L., and Halas, N. J. (1998). Nanoengineering of optical resonances, *Chem. Phys. Lett.*, **288**(2–4), pp. 243–247.
67. Prodan, E., Radloff, C., Halas, N. J., and Nordlander, P. (2003). A hybridization model for the plasmon response of complex nanostructures, *Science*, **302**, 5644, pp. 419–422.

68. Sonnefraud, Y., Leen Koh, A., McComb, D., and Maier, S. (2012). Nanoplasmonics: engineering and observation of localized plasmon modes, *Laser Photonics Rev.*, **6**, p. 277.

69. Giannini, V., Fernández-Domí nguez, A. I., Sonnefraud, Y., Roschuk, T., Fernández-Garcí a, R., and Maier, S. A. (2010). Controlling nanoscale light with designed nanoplasmonics, *Small*, **6**, p. 2498.

70. Adato, R., Yanik, A. A., Amsden, J. J., Kaplan, D. L., Omenetto, F. G., Hong, M. K., Erramilsun., and Altug, H. (2009). Ultra-sensitive vibrational spectroscopy of protein monolayers with plasmonic nanoantenna arrays, *Proc. Natl. Acad. Sci.*, **106**, 46, pp. 19227–19232.

71. Kneipp, K., Kneipp, H., Itzkan, I., Dasari, R., and Feld, M. (2002). Surface-enhanced Raman scattering and biophysics, *J. Phys.: Condens. Matter*, **14**, p. R597.

72. Le, F., Brandl, D.W., Urzhumov, Y. A.,Wang, H., Kundu, J., Halas, N. J., Aizpurua, J., and Nordlander, P. (2008). Metallic nanoparticle arrays: a common substrate for both surface-enhanced Raman scattering and surface-enhanced infrared absorption, *ACS Nano*, **2**, 4, pp. 707–718.

73. Ru, E. L., and Etchegoin, P. (2009). *Principles of Surface-Enhanced Raman Spectroscopy and Related Plasmonic Effects* (Elsevier, Amsterdam).

74. Willets, K., and Duyne, R. V. (2007). Localized surface plasmon resonance spectroscopy and sensing, *Annu. Rev. Phys. Chem.*, **58**, pp. 267–297.

75. Zhu, W., Banaee, M., Wang, D., Chu, Y., and Crozier, K. (2011). Lithographically fabricated optical antennas with gaps well below 10 nm, *Small*, **7**, pp. 1761–1766.

76. Kundu, J., Le, F., Nordlander, P., and Halas, N. J. (2008). Surface enhanced infrared absorption (seira) spectroscopy on nanoshell aggregate substrates, *Chem. Phys. Lett.*, **452**(1–3), pp. 115–119.

77. Neubrech, F., Pucci, A., Cornelius, T. W., Karim, S., Garcí a-Etxarri, A., and Aizpurua, J. (2008). Resonant plasmonic and vibrational coupling in a tailored nanoantenna for infrared detection, *Phys. Rev. Lett.*, **101**, 15, p. 157403.

78. Osawa, M. (2001). Surface-enhanced infrared absorption in near-field optics and surface plasmon polaritons, *Top. Appl. Phys.*, **81**, p. 163.

79. Boyer, D., Tamarat, P., Maali, A., Lounis, B., and Orrit, M. (2002). Photothermal imaging of nanometer-sized metal particles among scatterers, *Science*, **297**, p. 1160.

80. Lal, S., Clare, S. E., and Halas, N. J. (2008). Nanoshell-enabled photothermal cancer therapy: impending clinical impact, *Acc. Chem. Res.*, **41**, p. 1842.
81. Purcell, E. M. (1946). Spontaneous emission probabilities at radio frequencies, *Phys. Rev.*, **69**, p. 681.
82. Kinkhabwala, A., Yu, Z., Fan, S., Avlasevich, Y., Mullen, K., and Moerner, W. E. (2009). Large single-molecule fluorescence enhancements produced by a bowtie nanoantenna, *Nat. Photonics*, **3**, p. 654.
83. Bardhan, R., Grady, N., Cole, J., Joshi, A., and Halas, N. (2009). Fluorescence enhancement by Au nanostructures: nanoshells and nanorods, *ACS Nano*, **3**, pp. 744–752.
84. Bardhan, R., Grady, N., and Halas, N. (2008). Nanoscale control of nearinfrared fluorescence enhancement using Au nanoshells, *Small*, **4**, pp. 1716–1722.
85. Kosako, T., Kadoya, Y., and Hofmann, H. F. (2010). Directional control of light by a nano-optical Yagi–Uda antenna, *Nat. Photonics*, **4**, p. 312.
86. Vercruysse, D., Sonnefraud, Y., Verellen, N., Fuchs, F. B., Di Martino, G., Lagae, L., Moshchalkov, V. V., Maier, S. A., and Van Dorpe, P. (2013). Unidirectional side scattering of light by a single-element nanoantenna, *Nano Lett.*, **13**, 8, pp. 3843–3849.
87. Curto, A. G., Volpe, G., Taminiau, T. H., Kreuzer, M. P., Quidant, R., and van Hulst, N. F. (2010). Unidirectional emission of a quantum dot coupled to a nanoantenna, *Science*, **329**, 5994, pp. 930–933.
88. Taminiau, T., Stefani, F., Segerink, F., and van Hulst, N. (2008). Optical antennas direct single-molecule emission, *Nat. Photonics*, **2**, p. 234.
89. Enderlein, J. (2002b). Theoretical study of single molecule fluorescence in a metallic nanocavity, *Appl. Phys. Lett.*, **80**, pp. 315–317.
90. Cang, H., Liu, Y., Wang, Y., Yin, X., and Zhang, X. (2013). Giant suppression of photobleaching for single molecule detection via the Purcell effect, *Nano Lett.*, **13**, pp. 5949–5953.
91. Pellegrotti, J., Acuna, G., Puchkova, A., Holzmeister, P., Gietl, A., Lalkens, B., Stefani, F., and Tinnefeld, P. (2014). Controlled reduction of photobleaching in DNA origami–gold nanoparticle hybrids, *Nano Lett.*, **14**, pp. 2831–2836.
92. Zaiba, S., Lerouge, F., Gabudean, A.-M., Focsan, M., Lermé, J., Gallavardin, T., Maury, O., Andraud, C., Parola, S., and Baldeck, P. (2011). Transparent plasmonic nanocontainers protect organic fluorophores against photobleaching, *Nano Lett.*, **11**, pp. 2043–2046.

93. Kéna-Cohen, S., Wiener, A., Sivan, Y., Stavrinou, P., Bradley, D., Horsefield, A., and Maier, S. (2011). Plasmonic sinks for the selective removal of longlived states, *ACS Nano*, **5**, pp. 9958–9965.
94. Hale, G., Jackson, J., Shmakova, O., Lee, T., and Halas, N. (2001). Enhancing the active lifetime of luminescent semiconducting polymers via doping with metal nanoshells, *Appl. Phys. Lett.*, **78**, pp. 1502–1504.
95. Wang, W., Xiong, T., and Cui, H. (2008). Fluorescence and electrochemiluminescence of luminol-reduced gold nanoparticles: photostability and platform effect, *Langmuir*, **24**, pp. 2826–2833.
96. Donnert, G., Eggeling, C., and Hell, S. (2007). Major signal increase in fluorescence microscopy through dark-state relaxation, *Nat. Methods*, **4**, pp. 81–86.
97. Sivan, Y., Xiao, S., Chettiar, U., Kildishev, A., and Shalaev, V. (2009). Frequencydomain simulations of a negative-index material with embedded gain, *Opt. Express*, **17**, p. 24060.
98. Bruchez, M., Moronne, M., Gin, P., Weiss, S., and Alivisatos, A. (1998). Semiconductor nanocrystals as fluorescent biological labels, *Science*, **281**, p. 2013.
99. Burns, A., Ow, H., and Wiesner, U. (2006). Fluorescent core-shell silica nanoparticles: towards "lab on a particle" architectures for nanobiotechnology, *Chem. Soc. Rev.*, **35**, pp. 1028–1042.
100. Faklaris, O., Joshi, V., Irinopoulou, T., Tauc, P., Sennour, M., Girard, H., Gesset, C., Arnault, J.-C., Thorel, A., Boudou, J.-P., Curmi, P., and Treussart, F. (2009). Photoluminescent diamond nanoparticles for cell labeling: study of the uptake mechanism in mammalian cells, *ACS Nano*, **3**, p. 3955.
101. Fu, C., Lee, H., Chen, K., Lim, T., Wu, H., Lin, P., Wei, P., Tsao, P., Chang, H., and Fann, W. (2007). Characterization and application of single fluorescent nanodiamonds as cellular biomarkers, *Proc. Natl. Acad. Sci. U S A*, **104**, p. 727.
102. Honigmann, A., Eggeling, C., Schultze, M., and Lepert, A. (2012). Superresolution STED microscopy advances with yellow CW OPSL, *Laser World Focus*.
103. Bokor, N., and Davidson, N. (2007). Tight parabolic dark spot with high numerical aperture focusing with a circular pi phase plate, *Opt. Commun.*, **270**, p. 145.
104. Bokor, N., Iketaki, Y., Watanabe, T., and Fujii, M. (2005). Investigation of polarization effects for high-numerical-aperture first-order Laguerre-

Gaussian beams by 2D scanning with a single fluorescent microbead, *Opt. Express*, **13**, p. 10440.

105. Hao, X., Kuang, C., Wang, T., and Liu, X. (2010). Effects of polarization on the de-excitation dark focal spot in sted microscopy, *J. Opt.*, **12**, p. 115707.

106. Iketaki, Y., Watanabe, T., Bokor, N., and Fujii, M. (2007). Investigation of the center intensity of first- and second-order Laguerre-Gaussian beams with linear and circular polarization, *Opt. Lett.*, **32**, p. 2357.

107. Pavani, S., Thompson, M., Biteen, J., Lord, S., Liu, N., Twieg, R., Piestun, R., and Moerner, W. (2009). Three-dimensional single-molecule fluorescence imaging beyond the diffraction limit using a double-helix point spread function, *Proc. Natl. Acad. Sci. U S A*, **106**, pp. 2995–2999.

108. Hell, S., Lindek, S., Cremer, C., and Stelzer, E. (1994). Measurement of the 4pi-confocal point spread function proves 75 nm resolution, *Appl. Phys. Lett.*, **93**, pp. 1335–1338.

109. Hell, S., and Stelzer, E. (1992). Fundamental improvement of resolution with a 4pi-confocal fluorescence microscope using two-photon excitation, *Opt. Commun.*, **93**, pp. 277–282.

110. Dyba, M., and Hell, S. (2002). Focal spots of $\lambda/23$ open up far-field fluorescence microscopy at 33nm axial resolution, *Phys. Rev. Lett.*, **88**, p. 163901.

111. Harke, B., Ullal, C., Keller, J., and Hell, S. (2008). Three-dimensional nanoscopy of colloidal crystals, *Nano Lett.*, **8**, pp. 1309–1313.

112. Schmidt, R., Wurm, C., Jakobs, S., Engelhardt, J., Egner, A., and Hell, S. (2008). Spherical nanosized focal spot unravels the interior of cells, *Nat. Methods*, **5**, pp. 539–544.

113. Enderlein, J. (2002a). Spectral properties of a fluorescing molecule within a spherical metallic nanocavity, *Phys. Chem. Chem. Phys.*, **4**, pp. 2780–2786.

114. Gordon, J., and Ziolkowsky, R. (2007). The design and simulated performance of a coated nano-particle laser, *Opt. Express*, **15**, pp. 2622–2653.

115. Miao, X., Brener, I., and Luk, T. (2010). Nanocomposite plasmonic fluorescence emitters with core/shell configurations, *J. Opt. Soc. Am. B*, **27**, pp. 1561–1570.

116. Li, W., Zhang, J., Zhoub, Y., and Zhang, P. (2011). Highly enhanced fluorescence of fluorophores inside a metallic nanocavity, *Chem. Commun.*, **47**, pp. 5834–5836.

117. Sun, Y., and Xia, Y. (2003). Gold and silver nanoparticles: a class of chromophores with colors tunable in the range from 400 to 750 nm, *Analyst*, **128**, pp. 686–691.

118. Nehl, C., Grady, N., Goodrich, G., Tam, F., Halas, N., and Hafner, J. (2004). Scattering spectra of single gold nanoshells, *Nano Lett.*, **4**, pp. 2355–2359.

119. Palik, E. (1998). *Handbook of Optical Constants of Solids* (Academic Press).

120. Ow, H., Larson, D. R., Srivastava, M., Baird, B. A., Webb, W. W., and Wiesner, U. (2005). Bright and stable core-shell fluorescent silica nanoparticles, *Nano Lett.*, **5**, 1, p. 113, pMID: 15792423.

121. Lei, D. Y., Fernández-Domí nguez, A. I., Sonnefraud, Y., Appavoo, K., Haglund, R. F., Pendry, J. B., and Maier, S. A. (2012). Revealing plasmonic gap modes in particle-on-film systems using dark-field spectroscopy, *ACS Nano*, **6**(2), pp. 1380–1386.

122. Beversluis, M. R., Bouhelier, A., and Novotny, L. (2003). Continuum generation from single gold nanostructures through near-field mediated intraband transitions, *Phys. Rev. B*, **68**, p. 115433.

123. Park, J., Estrada, A., Sharp, K., Sang, K., Schwartz, J. A., Smith, D. K., Coleman, C., Payne, J. D., Korgel, B. A., Dunn, A. K., and Tunnell, J. W. (2008). Two-photon-induced photoluminescence imaging of tumors using near-infrared excited gold nanoshells, *Opt. Express*, **16**, 3, pp. 1590–1599.

124. Chu, S.-W., Su, T.-Y., Oketani, R., Huang, Y.-T., Wu, H.-Y., Yonemaru, Y., Yamanaka, M., Lee, H., Zhuo, G.-Y., Lee, M.-Y., Kawata, S., and Fujita, K. (2014a). Measurement of a saturated emission of optical radiation from gold nanoparticles: application to an ultrahigh resolution microscope, *Phys. Rev. Lett.*, **112**, p. 017402.

125. Chu, S.-W., Wu, H.-Y., Huang, Y.-T., Su, T.-Y., Lee, H., Yonemaru, Y., Yamanaka, M., Oketani, R., Kawata, S., and Fujita, K. (2014b). Saturation and reverse saturation of scattering in a single plasmonic nanoparticle, *ACS Photonics*, **1**, pp. 32–37.

126. Richards, B., and Wolf, E. (1959). Electromagnetic diffraction in optical systems. II. Structure of the image field in an aplanatic system, *Proc. R. Soc. London Ser. A*, **253**, pp. 358–379.

127. Auksorius, E., Boruah, B., Dunsby, C., Lanigan, P., Kennedy, G., Neil, M., and French, P. (2008b). Stimulated emission depletion microscopy with a supercontinuum source and fluorescence lifetime imaging, *Opt. Lett.*, **33**, 2, pp. 113–115.

128. Lenz, M., Sinclair, H., Savell, A., Clegg, J., Brown, A., Davis, D., Dunsby, C., Neil, M., and French, P. (2014). 3D stimulated emission depletion microscopy with programmable aberration correction, *J. Biophotonics*, **7**, pp. 29–36.

129. Albanese, A., Tang, P., and Chan, W. (2012). The effect of nanoparticle size, shape, and surface chemistry on biological systems, *Annu. Rev. Biomed. Eng.*, **4**, pp. 1–16.

130. Schrof, S., Staudt, T., Rittweger, E., Wittenmayer, N., Dresbach, T., Engelhardt, J., and Hell, S. (2011). STED nanoscopy with mass-produced laser diodes, *Opt. Express*, **19**, pp. 8066–8072.

131. Bingen, P., Reuss, M., Engelhardt, J., and Hell, S. (2011). Parallelized STED fluorescence nanoscopy, *Opt. Express*, **19**, p. 23716.

132. Chmyrov, A., Keller, J., Grotjohann, T., Ratz, M., d'Este, E., Jakobs, S., Eggeling, C., and Hell, S. (2013). Nanoscopy with more than 100,000 doughnuts, *Nat. Methods*, **10**, p. 737.

133. Moneron, G., and Hell, S. (2009). Two-photon excitation STED microscopy, *Opt. Express*, **17**, pp. 14567–14573.

134. Beeker, W., Gross, P., Lee, C., Cleff, C., Offerhaus, H., Fallnich, C., Herek, J., and Boller, K. (2006). A route to sub-diffraction-limited CARS microscopy, *Opt. Express*, **17**, p. 22632.

135. Cleff, C., Gross, P., Fallnich, C., Offerhaus, H., Herek, J., Kruse, K., Beeker, W., Lee, C., and Boller, K. (2013). Stimulated emission pumping enabling sub-diffraction-limited spatial resolution in coherent anti-Stokes Raman scattering microscopy, *Phys. Rev. A*, **87**, p. 033830.

Chapter 9

Lab-on-Antennas: Plasmonic Antennas for Single-Molecule Spectroscopy

Yongmin Liu[a] and Hu Cang[b]

[a] *Deparment of Mechanical and Industrial Engineering and Department of Electrical and Computer Engineering, Northeastern University, 360 Huntington Avenue, Boston, MA 02115, USA*

[b] *Waitt Advanced Biophotonics Center, Salk Institute for Biological Studies, 10010 North Torrey Pines Road, La Jolla, CA 92037, USA*

y.liu@northeastern.edu, hucang@salk.edu

Single-molecule spectroscopy (SMS) has transformed the field of optical spectroscopy, providing an indispensable approach for extremely sensitive imaging and detection. However, SMS still encounters substantial challenges that prevent its extensive applications: The concentration of fluorescence molecules in a typical SMS is usually lower than physiological relevant concentrations, and the resolution of SMS is not sufficient to catch the transient conformation of a molecule. We review here a few recent breakthroughs in plasmonic antennas that can help SMS overcome these concentration and resolution barriers. Specifically, the nanofocusing power of plasmonic antennas can reduce the focusing volume of SMS and increase the working concentration of fluorescence molecules from the nano- to the millimolar level, and the Purcell effect can enhance spontaneous emission (fluorescence) rate of

Plasmonics and Super-Resolution Imaging
Edited by Zhaowei Liu

ISBN 978-981-4669-91-7 (Hardcover), 978-1-315-20653-0 (eBook)
www.panstanford.com

single fluorescence molecules and suppress photobleaching. With the rapid progress in numerical simulations and nanofabrication, high-performance antennas could be developed and implemented at a large scale, rendering this "lab on antennas" approach feasible.

9.1 Introduction

The invention of single-molecule spectroscopy (SMS) [5, 6] has made it possible to follow biochemical reactions of single enzyme molecules in real time [7–9], which has led to the success of sequencing a single DNA molecule with long reads (10–15 kb) [10, 11] and detecting base modification (methylation) for epigenetics [12]. These achievements have transformed many areas of biological research, ranging from basic science [9] to translational studies [13].

The success of state-of-the-art SMS is the result of enormous efforts over the past two decades to improve the sensitivity of single-molecule detection. For examples, superior fluorescence probes [14] have been developed to yield more photons, oxygen scavengers [15, 16] to suppress photobleaching and enhance the life span of single fluorescence molecules, and new detection schemes such as confocal fluorescence microscopy [17], total internal reflection microscopy (TIRF) [18], zero-mode waveguides (ZMWs) [11], and whispering gallery mode (WGM) resonators [19, 20] to reduce background fluorescence.

However, two key barriers still remain, hampering the expanded use of SMS:

- **Concentration barrier**. To detect a single fluorescence molecule requires, on average, no more than one molecule within the detection volume at a time. Since, the majority of current SMS experiments are carried out with diffraction-limited optics whose detection volume is on the order of 0.1–1 cubic micrometers, the concentration of the fluorescence molecules must be kept at the pico- to the nanomolar level, which is out of the range of physiologically relevant concentrations of a high micromolar-to-millimolar range.

- **Resolution barrier**. The temporal resolution of current SMS is below 1 ms, limited by the photon flux of single fluorescence molecules, whereas the dynamics of most protein molecules is on a nanoseconds-to-submilliseconds timescale, too fast for SMS to capture. While advanced algorithms have been developed to enhance the resolution of existing SMS [21–23], these algorithms are still constrained by the information that can be obtained from the experimental data, which is ultimately determined by the number of photons that one can harvest from a single molecule. Therefore, to improve the resolution of SMS requires new techniques.

Plasmonic antennas have emerged as a powerful solution to overcome these barriers. The plasmonic antennas (Fig. 9.1) are the optical counterparts of radio antennas, which can enhance the photon coupling efficiency in and out of an emitter [4, 24–33]. The principle of plasmonic antennas was proposed in the 1900s, but because nanometer-scale precision is required to build such a device, it took a long time to realize the extreme power of the antennas. Van Duyne [34], Fleischman [35], and Creighton [36] discovered the antenna effect accidently in the 1970s, when a piece of roughed metal film was shown to enhance the Raman signal of molecules adsorbed on its surface. It was termed "surface-enhanced Raman spectroscopy" (SERS). Subsequent studies by Moskovits [37], Creighton [38], and Stockman [39–41] revealed that the metal film could form various nanometer-scale structures on its surface, functioning as plasmonic antennas and giving rise to giant enhancement of spectroscopy signals. It is only recently—thanks to progress in nanotechnology—that we are able to design and fabricate plasmonic antennas to manipulate light at the nanometer scale [42]. For instance, a plasmonic antenna can focus the omnidirectional fluorescence of a single quantum dot into a single-directional beam [32], accelerate the rate of fluorescence [26, 43], and enhance fluorescence of a single molecule [3, 4]. In this chapter, we will review the unique properties of plasmonic antennas that can be utilized for high-resolution SMS.

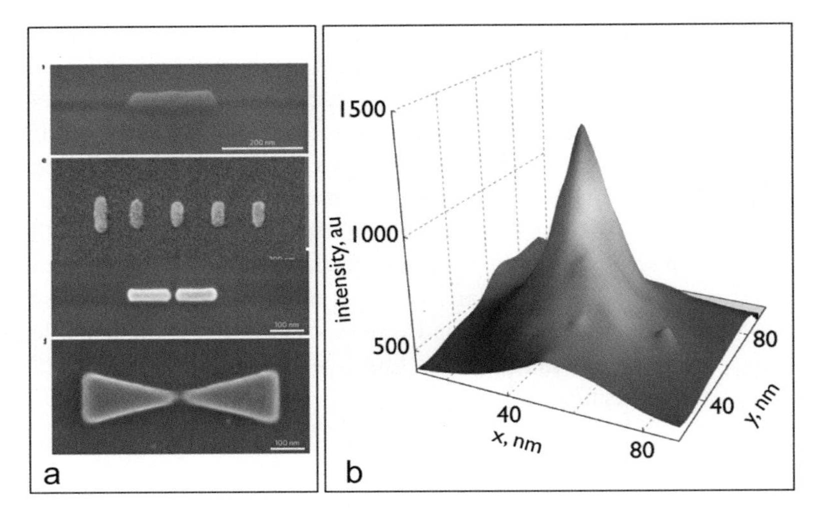

Figure 9.1 Plasmonic antennas. (a) A variety of antennas have been fabricated recently. From top to bottom, they are monopole, Yagi-Uda, bar-dipole, and bowtie-dipole antennas. (b) An antenna can focus light down to a spot well below the diffraction limit of the light. Shown here is the intensity profile of the electromagnetic field within the center of a dipole antenna, whose width is 24 nm.

9.1.1 Nanofocusing

Plasmonic antennas can confine light to a nanometer deep-subwavelength scale. The reduction of focus volume will allow an increase of the concentration of fluorophores for single-molecule analysis to millimolar, physiological relevant ligand/substrate concentrations. For example, a common antenna, bowtie antenna, consists of two metallic triangles, facing tip to tip, that are separated by a small gap (Fig. 9.1). The two triangles form a nanoresonator and confine light within the gap. The focus volume is no longer limited by the wavelength but determined by the size of the gap. The gap can be less than 10 nm in size, which corresponds to working at a concentration as high as 1.7 mM and still with less than one molecule per volume. This antenna can enable SMS at physiological relevant ligand concentrations of most protein enzymes.

Purcell enhancement can suppress photobleaching for high-resolution SMS. Plasmonic antennas enhance the rate of fluorescence (spontaneous emission) of fluorophores. This is the so-called Purcell effect, after E. M. Purcell [44]. The temporal resolution of a SMS is governed by the brightness of a fluorescence molecule. A key factor that limits the brightness of a fluorophore is photobleaching. While it is tempting to increase the excitation power of the laser in an SMS experiment for more signals, stronger excitation leads to stronger photodamage to the fluorophore. The bleaching reduces the time window of observation. In addition, the total number of photons that a fluorophore can emit before it bleaches is a constant. Therefore, how to effectively suppress bleaching is the key to increase the photon output from a fluorophore. Conventional strategies to improve the time and space resolution focus on optimizing molecular structures for brighter fluorophores. However, probing the conformation dynamics of single-protein molecules requires pushing the detection limits even further, demanding fluorescence molecules that are even brighter. Since, photobleaching is a chemical reaction activated by the energy a fluorophore receives from light, the enhanced fluorescence rate of a fluorophore reduces the time that the fluorophore dwells at an excited state, and hence reduces the probability of the molecule entering a photobleaching reaction. Over the years, plasmonic structures have been shown to reduce photo-oxidation of semiconducting polymer [45] and fluorescence dyes [46, 47], remove long-lived triplet states of fluorophores [48], suppress quantum dot blinking [49], and even manipulate selection rules [50]. We have recently demonstrated that a gold dimer antenna can suppress photobleaching and enhance the photon flux by up to 3 orders of magnitude at the single-molecule level [1]. The striking improvement in photon output from a single fluorophore could increase the time resolution of single-molecule fluorescence spectroscopy to less than 0.01 ms.

This lab-on-antenna approach is advantageous over other methods to enhance the resolution of SMS. For example, ZMWs are used in single-molecule DNA sequencing [10, 11] and studying of enzyme dynamics [51]. ZMWs are subwavelength pinholes whose cutoff wavelength is larger than the wavelength of the light, such that there is no propagating mode existing in the waveguide. The excitation light becomes evanescent and is confined within tens of nanometers from the bottom of the waveguides. A 50 nm diameter circular pinhole can confine light to a volume of 30 zL and allow SMS to work at concentrations as high as 75 μM. However, a ZMW doesn't provide any enhancement effect; rather, the metal cladding layer of the waveguide quenches the fluorescence. Instead, plasmonic antennas can strongly enhance the fluorescence rate, suppress photobleaching and increase the resolution of SMS. Furthermore, ZMWs smaller than 50 nm are hard to fabricate, which limits the highest fluorophore concentration to 75 μM. On the other hand, plasmonic antennas, which have different structures in comparison to ZMWs, can achieve 10 nm level confinement and allows mM physiological relevant ligand concentrations, representing up to 3 orders of magnitude improvement over ZMWs.

The lab-on-antenna approach is feasible right now because the progress in nanotechnology allows fabrication of devices with single-nanometer precision that is critical to determine the strength of the antennas. Top-down methods, including electron beam and focused ion beam lithography, can achieve ~10 nm resolution. Bottom-up approaches, including self-assembly and template-assisted self-assembly, could achieve resolution down to the molecule level. In addition, new hybrid approaches that combine both the top-down and bottom-up fabrications have been developed to fabricate devices with nanoscale (1–100 nm) resolutions on a large scale for high-throughput applications.

9.2 Reducing the Focus Volume

While it has long been predicted that plasmonic structures can confine light to the subwavelength scale [24, 27, 42, 52], experimental demonstrations have been difficult. The resolution of an optical microscope is limited to about half the wavelength, and thus not enough to measure the electromagnetic field inside. Although the current state-of-the-art techniques, including near-field scanning optical microscopy (NSOM) [53], electron energy-loss spectroscopy (EELS), and femtosecond microscopy, can achieve a resolution of tens of nanometers, they introduce a nonnegligible amount of perturbation to the surface and therefore complicate the data interpretation. As a result, after more than 30 years of the discovery of the surface enhancement effect, many fundamental aspects of the hotspots from the plasmonic focusing, such as how big these spots are and how the field is confined, remain unknown to us. We have utilized the Brownian motion of fluorescent molecules to scan the surface of plasmonic hotspots and measured the distribution of the electromagnetic field inside single hotspots for the first time. The angstrom-sized molecules can readily reach places inaccessible to NSOM, introduce minimum perturbations, and transfer the local field information [5, 6] directly to the far field by its fluorescence. The 2.1 nm resolution local field mapping inside the hotspots formed on the surface of aluminum thin-film and silver nanoparticle clusters reveals that light is confined with an exponential profile to a region as small as 15 nm [3].

9.2.1 Probing the Focus Volume with Single-Molecule Super-Resolution Imaging

Figure 9.2 illustrates the single-molecule technique that enables us to probe inside a plasmonic hotspot. A sample is submerged in a solution of freely diffusing fluorescent dyes (Fig. 9.2a). Since the diffusion of the dyes (a 1 nm diameter sphere diffuses through a 250 nm wide spot in about less than 0.1 ms in water at room temperature) is much faster than the image acquisition time (typically 50–100 ms), the fluorescence from the rapidly diffusing dyes contributes to a homogeneous background. When a dye molecule is transiently adsorbed onto the surface of a sample, it stops diffusing and appears as a bright spot, with the intensity of the spot reporting the local field strength. By fitting the image of the spot to a 2D Gaussian function, the molecule can then be localized with nanometer accuracy [54, 55]. After the dye molecule is bleached (typically within a few hundreds of milliseconds), the fluorescent spot disappears, and the spot is ready for next adsorption event. In this way, the Brownian motion of the dye molecules not only helps isolate a single emitter within the observation volume to ensure the necessary condition for single-molecule super-resolution imaging without the need for the more complex serial photoswitching of fluorescent dyes, it also assists in efficient sampling of the surface.

By employing millions of dye molecules to scan the surface of a sample concurrently, we are able to map the field profile within a single hotspot (Fig. 9.2b) for the first time, in just a few minutes. Every color sphere in the figure represents one single-molecule event, with the x and y coordinates corresponding to the location of the molecule's centroid and the z coordinate corresponding to the peak intensity of the molecule obtained from the Gaussian fitting. The hotspot is formed on the surface of a thin $\sim$15 nm aluminum film on a quartz slide fabricated by direct electron beam evaporation. Although we only use a diffraction-limited optical microscope to record the data, the reconstructed mapping offers a resolution well beyond the diffraction limit. A Gaussian kernel method was used to render the image of a hotspot. The hotspot is 15.4 nm and 15.2 nm wide in x and y directions, respectively. Most of the events observed locate within a region of about 15 nm wide, highlighting a strong near-field optical confinement.

The height of the peak corresponds to the maximum enhancement factor of the hotspot. By comparing this result with the same fluorescence dyes immobilized on the surface of a quartz slide under the same conditions, we determined that at the center of the hotspot, the enhancement is about 36 times. The modest enhancement reflects the high ohmic loss of aluminum at the visible wavelength [56]. Generally, a metal film effectively quenches the fluorescence from molecules adsorbed on the immediate surface [57]. The aluminum, however, oxidizes quickly in air and develops a dense oxidized layer on the surface with a thickness between 1 and 5 nm serving as a dielectric buffer to reduce the quenching efficiency.

We have also investigated hotspots formed on the surface of silver nanoparticle aggregates. A drop of 40 nm silver nanoparticles solution is deposited on the surface of a glass. Upon drying, the nanoparticles aggregate. Recent experiments [58] have shown that the hotspots coincide with the nanometer-sized gaps between the particles. A rendered image of a hotspot is shown in Fig. 9.3a. The enhancement decays exponentially in both spatial directions (x and y), highlighted in Fig. 9.3b. The strong confinement is also reflected in the distribution of the single-molecule events (Fig. 9.3c), with the majority of the observed events located within a region of

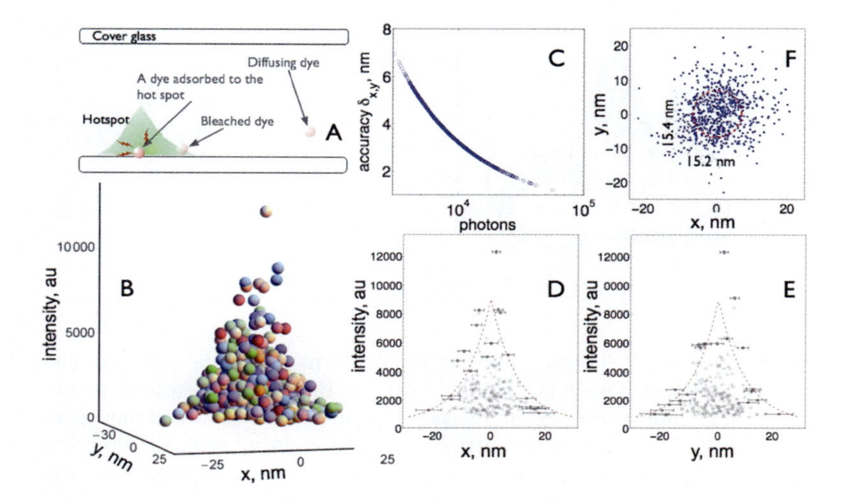

Figure 9.2 (A) We utilize the Brownian motion of fluorescence dye molecules that stochastically adsorb to the surface to probe the electromagnetic field in an antenna. By controlling the concentration of the dyes, the adsorption rate can be adjusted to ensure that within a diffraction-limited spot, only one molecule emits photons at a time; therefore, by using the adsorption locations as the *x* and *y* coordinates, and the fluorescence intensity as the *z* coordinate, we obtain a 3D scatter plot of the fluorescence enhancement profile of the hotspot (B), with each sphere representing one single-molecule event [3].

13.2 nm $\times$ 20.3 nm, about 30 times smaller than the wavelength of the excitation light of 644 nm.

9.2.2 Plasmonic Antennas for High-Concentration SMS

Kinkhabwala et al. have demonstrated utilizing the nanofocusing power of plasmonic antennas to overcome the concentration barrier [2]. They used top-down electron beam lithography to fabricate gold bowtie antenna arrays on a tin-oxide-coated glass substrate. Each bowtie is about 70 nm long, 20 nm thick, with a gap about 20 nm. The bowtie concentrated the electromagnetic field within the gap, which is of nanometer size, significantly smaller than a diffraction-limited focus volume. They have shown that the reduction of the focus volume allows fluctuation correlation spectroscopy (FCS) at a

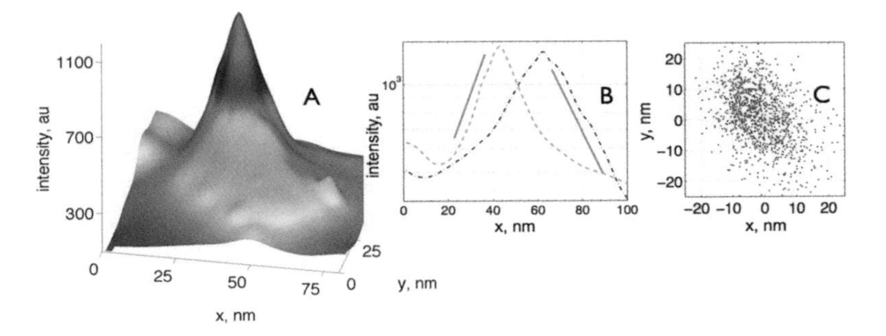

Figure 9.3 (a) A hotspot formed on silver nanoparticle clusters. The maximum enhancement factor at the center of the peak corresponds to 136 times of the fluorescence from the same dye molecules adsorbed on a glass surface. (b) The hotspot exhibits an exponential decay profile. Two cross sections of the hotspot along x (green) and y (blue) directions through the peak are plotted on a logarithmic scale, with the solid red lines as eye guides for the exponential profile. (c) The widths of the hotspot, estimated from the distribution of the single-molecule events on a scatter plot, are 13.2 nm and 20.3 nm along the two axes, respectively [3].

high concentration of fluorescence molecules up to the micromolar range (Fig. 9.4a,b).

Acuna et al. have recently demonstrated another beautiful application of plasmonic antennas for single-molecule Förster resonance transfer (SM-FRET) spectroscopy [4]. They used a self-assembly method to generate plasmonic antennas. Specifically, gold nanoparticles are attached to DNA molecules and form origami structures. This DNA origami also contains docking sites for plasmonic antennas. Carefully designing the DNA sequence allows single fluorescence dye molecules to be near or within two gold nanoparticles, which are equivalent as a dimer antenna. They have observed up to 117-fold enhancement of the fluorescence intensity of the fluorescence molecules positioned in a 23 nm gap between 100 nm gold nanoparticles. The small focus volume has allowed direct visualization of the binding and unbinding of short DNA strands, as well as the conformational dynamics of a DNA Holliday junction at up to micromolar concentrations (Fig. 9.4c,d).

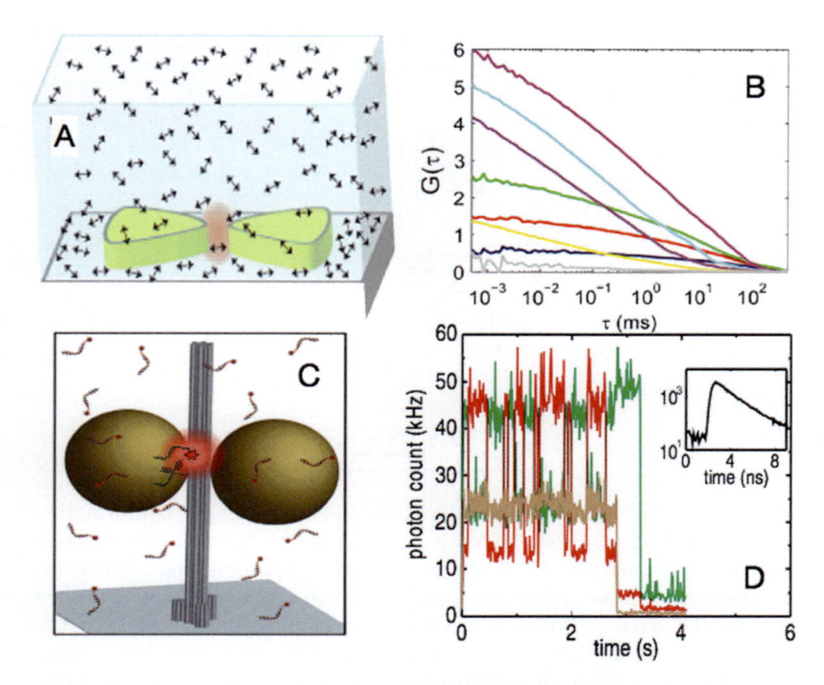

Figure 9.4 (a) A bowtie antenna (yellow) fabricated on a glass substrate. The excitation light is strongly confined within the gap of the bowtie (red), which is of subwavelength size. This hotspot can be utilized for fluctuation correlation spectroscopy (b) at high concentration. Reprinted from Ref. [2], Copyright (2012), with permission from Elsevier. (c) An antenna consists of two gold nanoparticles fabricated through a self-assembly approach. (d) The strong enhancement of fluorescence allows observation of conformation dynamics of the Holliday junction with FRET spectroscopy at high concentrations of the dye molecules. Adapted from Ref. [4]. Reprinted with permission from AAAS.

9.3 Suppressing Photobleaching

Fluorescence spectroscopy is one of the most used methods for chemical detections [59]. One of the biggest challenges in a fluorescence experiment is the photobleaching of fluorophores. An excited fluorophore is chemically reactive. It is likely to undergo a chemical reaction leading to its permanent damage and the loss of its fluorescence capability. As a result, the total number of photons

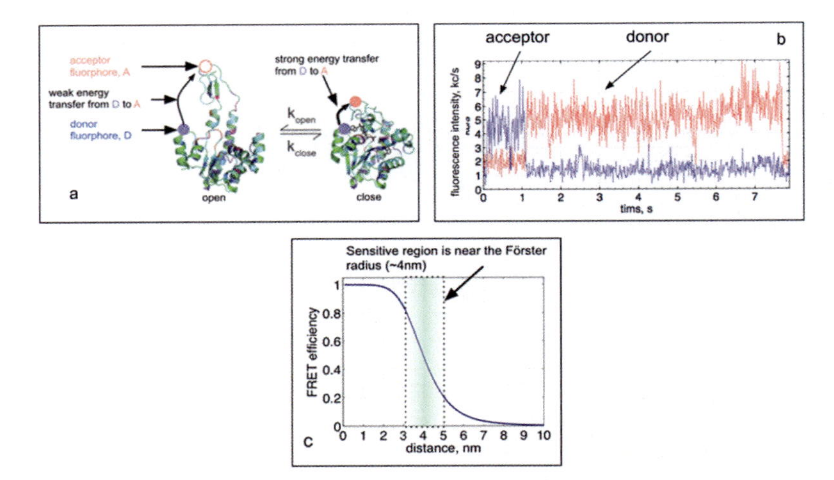

Figure 9.5 SM-FRET is a molecular ruler that measures the conformation dynamics of biomolecules. (a) As an example, adenylate kinase (AK) protein has two conformations, open and closed forms. Upon binding a substrate, AK changes from the open to the closed conformation, which can be probed by SM-FRET: a pair of fluorophores, including a donor fluorophore, D, and an acceptor fluorophore, A, are tagged at two residues of AK. The FRET signal measures the distance changes between the two residues of AK. (b) Time-resolved SM-FRET can record the conformation changes of the AK molecule. The red curve represents the signal from the donor channel, and the blue curve represents the acceptor channel.

that can be extracted out of a single fluorescence molecule before it is bleached has an upper limit [60]. This is the photobleaching limit, which determines the maximum signal-to-noise ratio (SNR) that a fluorescence-based detection method can reach and hence the temporal resolution of SMS.

9.3.1 Photobleaching Limits the Resolution of SMS

Figure 9.5 illustrates an example of the resolution barrier in measuring the conformation changes of a protein, adenylate kinase (AK) [21], by SM-FRET. The FRET signal provides a molecular ruler with which to measure the distance between two amino acid residues in AK. As shown in Fig. 9.5b, the temporal resolution is determined by the photon flux of each single-molecule event. In a typical SMS experiment, a fluorophore emits 1–10,000 photons per second. To achieve an SNR of 3 requires a minimum of $3 \times 3 = 9$ photons for a measurement. Therefore, at a photon rate of 1 per millisecond, one needs to wait 9 ms to collect enough photons. While higher laser excitation power could increase the brightness, it also accelerates the photobleaching of the fluorophore and reduces the length of the observation time; therefore, improving temporal resolution requires increasing the photon flux and suppressing photobleaching.

Since reactions of excited fluorescence molecules with oxygen radicals are the main cause of photobleaching [61], current efforts on suppressing the photobleaching have been largely focusing on one of two strategies, either by making the local environment where the fluorophore resides more chemically inner or by tweaking the structure of a fluorophore for better photostability. Enormous progress has been made along these two routes, such as the discovery of oxygen scavenger systems [15, 62] and the invention of quantum dots [63]. However, as the next-generation biotechnology is pushing the detection limit to the single-molecule level [10, 64], the current techniques cannot meet the demand.

9.3.2 The Plasmonic Purcell Effect

Plasmonics provides an entirely different and novel strategy to overcome the photobleaching limit. According to cavity quantum electrodynamics [65], the fluorescence—spontaneous emission—of a fluorophore is a process stimulated by the local vacuum fluctuations [44, 66], which depends on the local density of optical states (LDOS). In 1946, Purcell proposed that LDOS can be engineered with a resonator [44], leading to suppression or enhancement of spontaneous emission [43, 66, 67].

Figure 9.6 provides a classical physics interpretation of the Purcell effect. A fluorophore can be modeled as an oscillating electrical dipole. When the dipole is placed inside a cavity, it induces mirror dipoles. The interaction between the dipole and its mirror image modulates the dipole radiation pattern. Tuning the cavity resonance mode can make the dipole and its mirror image in or out of phase, thus enhancing or suppressing the dipole emission.

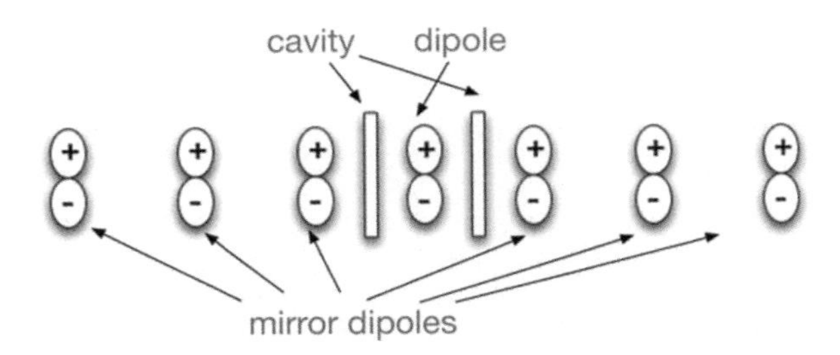

Figure 9.6 A classic interpretation of the Purcell effect. Consider a fluorophore as a dipole. When the dipole is placed in a cavity, mirror dipoles will be induced. The interaction between the dipole and its mirror images strongly modulates the dipole radiation. Hence by tuning the cavity, one can enhance or suppress the dipole emission.

Therefore, the spontaneous emission of a fluorophore in a cavity depends on the density of the allowed modes determined by the cavity. Purcell, and later Kleppner, derived an elegant formula stating that when on resonance, the spontaneous emission rate of a single fluorophore will be enhanced by

$$\beta \propto \frac{Q}{V/\lambda^3}$$

where β is the Purcell factor, Q the quality factor of the cavity, V the mode volume, and λ the wavelength of the emitted photons.

Since the mode volume of a plasmonic devices is of sub-wavelength size, due to the strong light confinement, the Purcell factor, β, can be extremely large. For example, from previous section, we measured that the size of a plasmonic hotspot is as small as 20 nm in width, 1/30th of the wavelength of light, the Purcell factor could be as large as 27,000. Such a giant Purcell effect offers an unprecedented opportunity to tune the spectroscopic properties of fluorophore that is impossible with other methods. For example, Kinkhabwala et al. have observed an up to 180 times faster spontaneous emission rate assisted by a bowtie antenna [26]. Hale et al. have observed reduction of photo-oxidation of a semiconducting polymer when doped with silica core-gold nanoshells [45], and Muthu et al. have observed decreasing photobleaching of fluoresce dyes on the surface of a thin silver film [68]. Recently, these observations have been further expanded on to tune the spectroscopic properties of dye molecules to selectively remove the long-lived triplet state [48]. Novel core–shell type of plasmonic nanocomposite structures have also been developed as nanocontainers to protect fluorescence dyes [47]. We have demonstrated a proof-of-concept implementation that turns an ordinary fluorescence molecule into a "super" fluorophore, emitting 1000-fold more photons than its photobleaching limit allows. Our calculations show that this strategy can, in principle, lead to even bigger improvement and will soon make the photobleaching no longer a limit in SMS [1].

9.3.3 The Kinetic Model of Photobleaching Suppression

To help better understand the photobleaching limit and the dynamic process of fluorescence, we have developed a simple kinetic model based on the Enderlein [56, 69, 70] model to understand how plasmonic devices can suppress photobleaching effectively. A simplified two-level Jablonski diagram [59] is shown in Fig. 9.7 and Fig. 9.8. A ground-state fluorophore, denoted as g, could be excited to an excited state, e, at a rate of $k_{\mathrm{A}}g$. The excited fluorophore, e, then relaxes back to the ground state, g, either through a radiative channel k_{R} by emitting a photon or through a nonradiative channel, k_{NR}, where the energy is dissipated as heat. The excited fluorophore, e, could also undergo a chemical reaction and lose its fluorescence capability permanently at a photobleaching rate with a rate constant of Γ. Thus, we can use a set of kinetic equations to model the excitation, the relaxation, and the bleaching process as follows:

$$\begin{aligned} \frac{d}{dt}t &= -(k_{\mathrm{R}} + k_{\mathrm{NR}} + \Gamma)e + k_{\mathrm{A}}g \\ \frac{d}{dt}t &= -k_{\mathrm{A}}g + (k_{\mathrm{R}} + k_{\mathrm{NR}})e \end{aligned} \tag{9.1}$$

From Eq. 9.1 we can calculate the average *life span* of a fluorophore, the time that a fluorophore continuously emits photons before being bleached:

$$\int te(t)dt \Big/ \int e(t)dt = (k_{\mathrm{R}} + k_{\mathrm{NR}} + k_{\mathrm{A}} + \Gamma)/\Gamma k_{\mathrm{A}} \tag{9.2}$$

We use the term "life span" here to distinguish it from another term, "lifetime," which is the time it takes for an excited fluorophore to

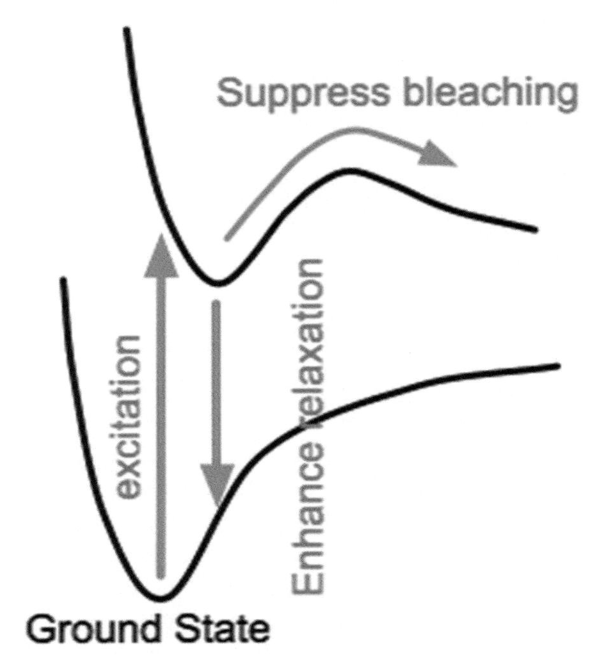

Figure 9.7 A kinetic model of photobleaching. The black curves represent the potential energy surface of a molecule. After adsorbing a photon, the molecule enters an excited state from a ground state. The excited molecule will most probably relax back to the ground state, but with a small probability, it suffers permanent damages as a result of a photochemical reaction—bleaching.

relax back to its ground state, defined as $1/(k_{NR} + k_R)$. For most fluorophores, the relaxation process k_{NR}, and k_R are on the order of 1 ns^{-1}, which is much bigger than the bleaching rate constant Γ and the excitation rate constant k_A, both of which are around 1 ms^{-1}; thus the average life span is $\tau \approx k_R/\Gamma Q k_A$, where Q is the quantum yield of the fluorophore defined as $Q = k_R/(k_R + k_{NR})$.

Since the observed fluorescence (spontaneous emission) intensity is $k_R e$, integrating it over time, we arrive at a simple result: the total number of photons that a fluorophore can emit before it

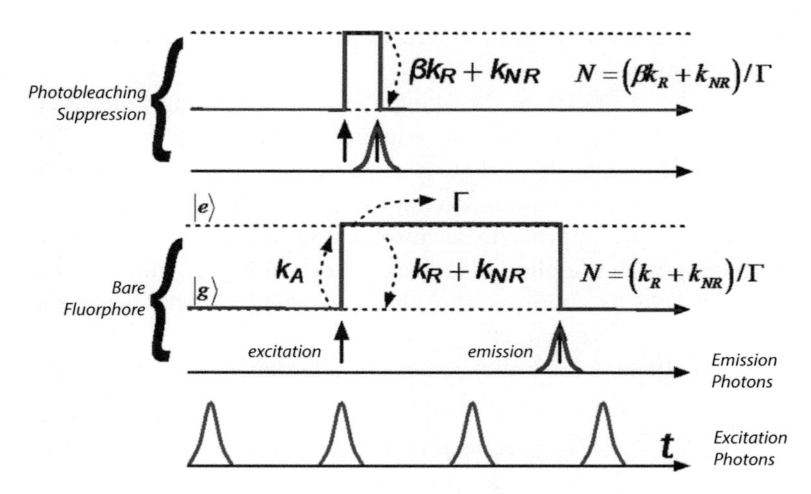

Figure 9.8 This figure explains how Purcell effect could suppress photobleaching. The bleaching involves excited fluorescence molecules. Therefore, the total time span of a molecule at the excited state is a constant $1/\Gamma$, where Γ is the bleaching rate constant. The Purcell effect enhances the relaxation rate and hence reduces duration of a molecule at the excited state after each excitation. As a result, the total number of excitation-relaxation cycles, N, increases. The total number of emitted photons is proportional to N.

bleaches, the *photobleaching limit*, is

$$\int k_{\mathrm{R}} e(t) dt = k_{\mathrm{R}} / \Gamma \tag{9.3}$$

The photobleaching limit is a useful quantity to characterize the performance of a fluorophore since it doesn't depend on the excitation intensity. Exciting a fluorophore harder by increasing the excitation powers I, although that can make the fluorophore emit more photons in a unit time. On the other hand, since the excitation rate constant k_{A} is linearly proportional to the excitation intensity I (i.e., $k_{\mathrm{A}} \propto I$), the stronger excitation make the fluorophore bleach faster as well. (The life span is $\tau \approx k_{\mathrm{R}}/\Gamma Q k_{\mathrm{A}} \propto 1/I$.) Therefore, the total number of photons emitted over the entire life span of the fluorophore remains to be the same as k_{R}/Γ, which is independent of the excitation intensity, I. In a demanding single-molecule-level experiment even a single photon is too dear to lose [22, 71]; the photobleaching limit hence determines the maximum signal to the SNR ultimately.

Decades of efforts on breaking the photobleaching limit k_R/Γ have been largely focusing on reducing the bleaching rate Γ. The progress although significant, is evolutionary and insufficient to meet the current demand. Solving the problem of photobleaching requires a new approach. Because the spontaneous emission is competing with the photobleaching, shown in Fig. 9.7 and Eq. 9.3, it is possible to suppress the photobleaching by enhancing the spontaneous emission rate.

Let us denote the enhancement of spontaneous emission rate constant, as $k_R\beta$. Thus, the *life span* of a fluorophore becomes

$$\tau = k_R(\beta + 1/Q - 1)/Q\Gamma k_A \quad (9.4)$$

Compared to the bare fluorophore under the same excitation intensity with identical k_A, we see a $(\beta + 1/Q - 1)$ times increase of the life span. To calculate the new total number of photons that can be collected, we need to correct the loss of photons as a by-product of the Purcell effect. Only a fraction of the spontaneous emission, here denoted as the radiation efficiency factor η, can radiate to the far field. Therefore, the average number of photons that a single fluorophore can emit before photobleaching is given by

$$\beta\eta k_R/\Gamma \quad (9.5)$$

Two important conclusions can be drawn from Eq. 9.5. First, for a system with a positive Purcell factor β, we should see a $\beta\eta$ times improvement beyond the photobleaching limit k_R/Γ. Second, the improvement is independent of the nonradiative decay, k_{NR}, of the fluorophore and thus independent of the quantum yield of the fluorophore, Q.

We can also calculate the new quantum yield of the photobleaching as

$$\Phi = \Gamma/(\beta k_R + k_{NR} + \Gamma) \approx Q\Gamma/k_R(\beta Q - Q + 1) \quad (9.6)$$

Because a commonly used fluorescent molecule usually exhibits high quantum yield, $Q \approx 1$, and both Γ and k_A of the molecule are much smaller than the relaxation constants k_f and k_{iNR}, with Γ and k_A around ms^{-1} and k_f and k_{iNR} around ns^{-1} [59, 61, 72], we can simplify the results by neglecting these factors. Hence, comparing Eqs. 9.4–9.6 with Eqs. 9.2–9.3, we can see that the life span and the total number of photons emitted can increase by β and $\beta\eta$ times, respectively, and the bleaching quantum yield reduces by about β times.

For weak fluorescent dyes, whose fluorescence quantum yield $Q<<1$, the improvement of the total number of photons is the same as $\beta\eta$ times. However, the improvement of the life span τ and the bleaching quantum yield Φ is about $Q\beta$ times. This is because the enhanced LDOS only affects the spontaneous emission of a molecule, which is Q fraction of the total relaxation process of the molecule. Because plasmonic nanostructures can have a high Purcell factor β >1000, suppression of photobleaching should be observed even for weak dyes whose fluorescence quantum yield $Q$is larger than $1/\beta \approx$ 0.001.

9.3.4 Numerical Simulation of Plasmonic Antennas for Photobleaching Suppression

We perform numerical simulations using a commercial finite-element solver (COMSOL Multiphysics) on a simple plasmonic antenna consisting of two gold nanoparticles, as depicted in the Fig. 9.9. The diameter of the gold spheres is 100 nm, and the

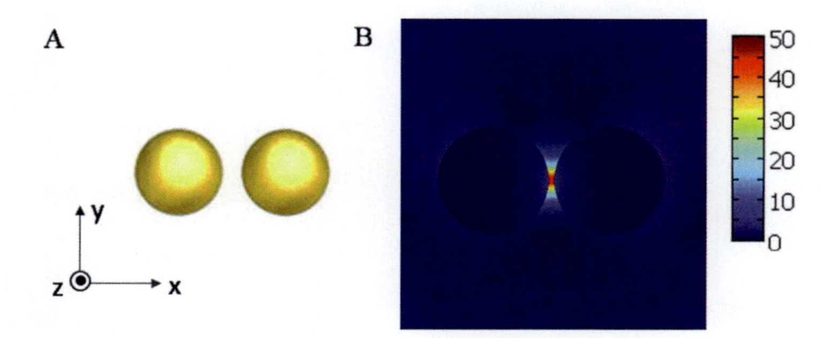

Figure 9.9 (a) Schematic of a gold dimer antenna consisting of two gold nanoparticles. (b) Electric field enhancement when the plane wave is incident along the y axis and polarized along the x axis. The dashed circles outline the gold nanoparticles.

gap between the spheres is 5 nm. At a 660 nm wavelength, the permittivity of gold is $\varepsilon_m = -13.6 + 1.035i$ [73]. First, a plane-wave excitation is applied with the propagation along the y axis and polarization along the x axis. Under this condition, the fundamental dipole modes of the two gold spheres are excited. The electric field within the tiny gap is dramatically enhanced up to 48.7 (Fig. 9.9b), corresponding to an intensity enhancement about 2500.

We then model a fluorophore as a dipole at different polar positions. When the dipole radiation matches the mode profile of the dipole antenna, that is, orients along the x axis, its oscillation will be enhanced significantly. The radiative power is obtained by integrating the Poynting vector over a surface that encloses the dipole and the gold nanoparticles. The nonradiative power due to the metal ohm loss is calculated by the volume integration of $j \times E$ for the two gold spheres, where j and E denote the current density and electric field, respectively. Compared to the dipole radiation in free space without the optical antenna, the maximum enhancement factors of radiative emission and nonradiative emission are 2800 and 1400, respectively, when the fluorophore is centered at the gap region (Fig. 9.10b–d). More importantly, the radiative emission

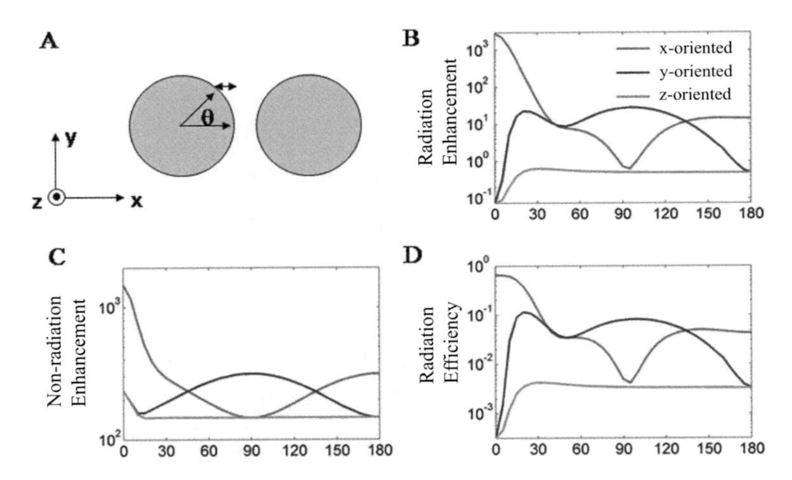

Figure 9.10 (a) Schematic of the simulation process. A dipole source with different orientations is placed 2.5 nm away from the nanoparticles at different polar angles. (b) Radiation enhancement, (c) nonradiation enhancement, and (d) radiation efficiency η for three dipole orientations.

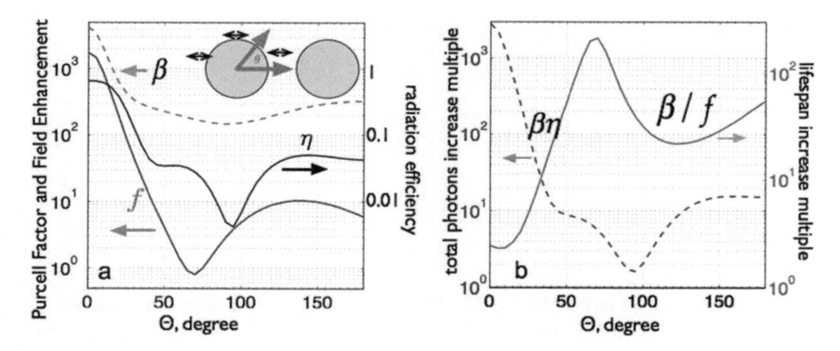

Figure 9.11 (a) A finite difference simulation for a plasmonic structure consisting of two 100 nm gold spheres with a gap of 5 nm (schematically shown in the inset). The local excitation field enhancement factor $f = |E_{\mathrm{LOC}}|^2/|E_0|^2$ (blue curve), Purcell factor β (red dashed curve), and radiation efficiency η (black curve) of a molecule are plotted against the location of the molecule adsorbed 2.5 nm above the surface of one sphere, represented by the polar angle θ. The increase of the life span β/f (green curve) and the increase of the total number of photons $\beta\eta$ (blue curve) are plotted in (b). A molecule located at the center of the gap, corresponding to $\theta = 0$ exhibits a 3 orders of magnitude increase of the total photons. Adapted with permission from Ref. [1]. Copyright (2013) American Chemical Society.

efficiency is still as large as 65.7%, indicating the antenna performs very efficiently.

Substituting these calculations results into the kinetics equations, we can estimate the enhancement of the photon output of single fluorescence molecules located in the plasmonic dipole antennas. When a fluorophore's radiation pattern matches the antenna's mode perfectly, the fluorophore experiences the strongest Purcell effect of $\beta \approx 3200$ at the center of the gap and the antenna exhibits the highest radiation efficiency as well of $\eta \approx 65.7\%$ (Fig. 9.11). Substituting these numbers into Eq. 9.5, we obtain a $\beta\eta \approx 2112$ times improvement beyond the photobleaching limit (Fig. 9.11).

The life span, however, exhibits a more complex behavior, as it is a result of two competing factors: an increased local field factor $f = E_{\mathrm{LOC}}^2/E_0^2$, making the fluorophore bleach faster, and a positive Purcell factor β, making the fluorophore bleach slower. As shown in Fig. 9.9, the antenna also concentrates the excitation field at the

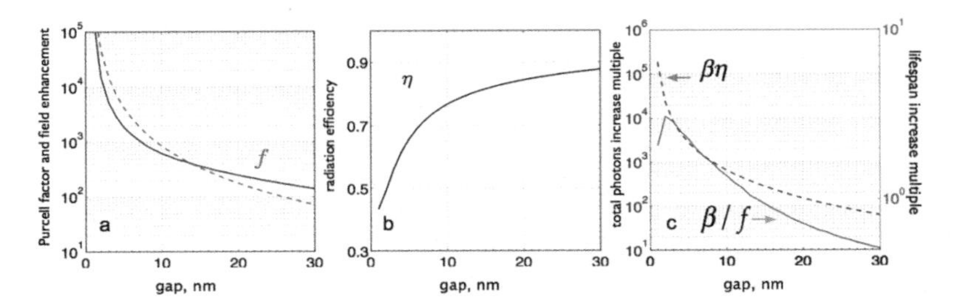

Figure 9.12 The Purcell factor β and the local field enhancement factor f (a), the radiation efficiency of the antenna, η (b), and the photobleaching suppression factor $\beta\eta$ and life span β/f (c) measured at the center of the gap of a dimer antenna plotted as a function of the gap.

center of the gap, with a local field enhancement factor of $f \approx 2500$ (Fig. 9.11). Thus, the change of life span of a fluorophore, assuming $Q = 1$, β/f is about 1.28 times (Fig. 9.11).

As the fluorophore moves away from the center of the gap, the antenna operates less efficiently; hence, we see a quick decrease of the Purcell factor β (red dashed line, Fig. 9.11) as well as the local field f (blue line, Fig. 9.11). Since the field enhancement factor f decays faster than the Purcell factor β (Fig. 9.11), the change of life span β/f appears actually increasing (green curve, Fig. 9.11). However, the radiation efficiency η drops quickly to below 0.1 as well. As a result, the strongest photobleaching suppression, that is, maximum $\beta\eta$, occurs only at the center of the antenna (blue dashed curve, Fig. 9.11b). For comparison, we also simulate dipoles oriented along the y axis and z axis, which are plotted in blue and green lines, respectively in Fig. 9.10b–d. One can see that the radiation enhancement is fairly low, especially for dipoles orientated in the z direction, which suggests that plasmonic antenna is hardly excited.

Figure 9.12 shows that the gap critically determines the efficiency of the plasmonic dimer. As the gap increases, both the Purcell factor F_{p} and the local field enhancement factor f decrease (Fig. 9.12a). Even the radiation efficiency of the dimer, η, increases (Fig. 9.12b); the incensement is much smaller than the drop of the Purcell factors. Hence, the suppression of photobleaching decreases (Fig. 9.12c) for dimers with a larger gap.

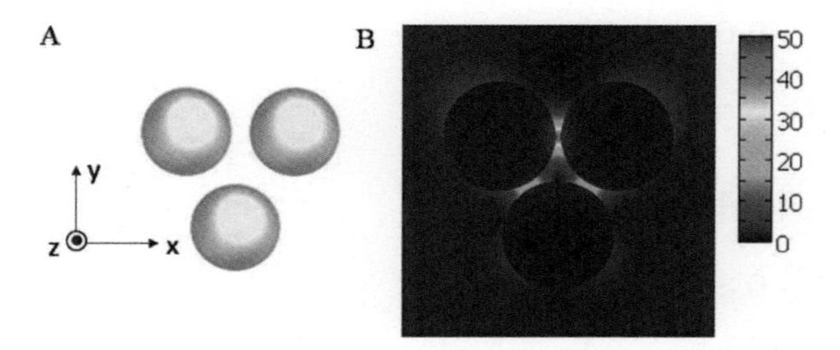

Figure 9.13 (a) Schematic of a trimer consisting of three gold nanoparticles. (b) Electric field enhancement when a plane wave is incident along the *y* axis and polarized along the *x* axis.

In addition to the dimer, we have simulated the timer consisting of three gold nanoparticles, as shown in Fig. 9.13. The field enhancement is about 47 when the gaps between particles are 5 nm, comparable with that of the dimer. Following the same procedure, we have calculated the maximum radiative emission enhancement, nonradiative emission enhancement, and radiative emission efficiency, which are 2408, 1645, and 59.5%, respectively. This result further confirms our physical approach to suppress photobleaching via the Purcell effect.

9.3.5 Experimental Observation

We have recently demonstrated the feasibility of our proposal experimentally by using single-molecule imaging. First, we can confirm that the photon output from a single fluorescence molecule is independent of the excitation power when not in the nonlinear excitation regime (Fig. 9.14).

The proposed antenna structure requires fabricating gaps between two nanospheres beyond our current fabrication capability on the basis of top-down techniques. Thus, we use a bottom-up approach instead [58]. Briefly, a drop of 100 nm gold nanoparticles is deposited on the surface of a quartz slide. Upon drying, the nanoparticles aggregate and tiny gaps arise between the nanoparticles. We then mount the quartz slide on a total internal reflection fluorescence (TIRF) [74] microscope and flow in a Chromeo 642 dye solution. The dye molecules adsorb onto the surface of the

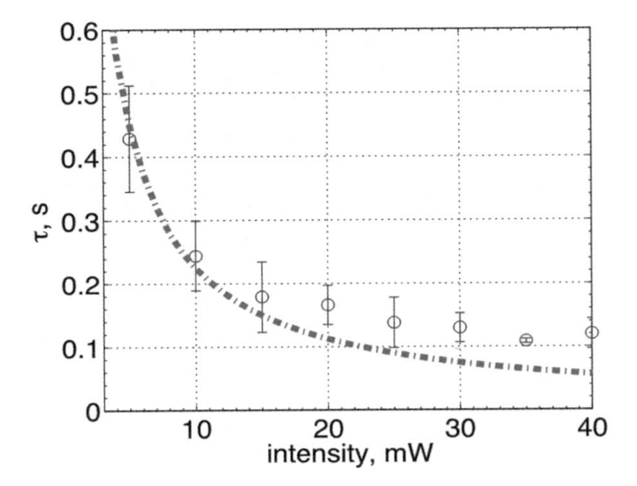

Figure 9.14 A control experiment measures the life span τ of the Chromeo 642 dye on glass slide as a function of the excitation power I. An inverse relation (red dashed-dotted curve), $1/I$, is evident in the figure. Adapted with permission from Ref. [1]. Copyright (2013) American Chemical Society.

nanoparticles. After a laser beam starts to excite the sample, the antennas appear as bright spots (Fig. 9.15a). The stronger-than-background fluorescence at an antenna is a combined effect of stronger adsorption of dyes onto the gold nanoparticles, the Purcell effect enhancing the emission rate, and the enhancement of the local field at the antenna by a factor of $f = E_{\text{LOC}}^2/E_0^2$. The fluorescence from each antenna decays right after the start of the excitation due to the photobleaching. However, compared to the dyes being excited at the same excitation power but on a quartz surface (Fig. 9.1b), the bleaching is much slower (Fig. 9.15b). The average life span of dyes from 24 spots is 2.1 s (Fig. 9.15c), which is a 5 times improvement. Since the local field is f times stronger than the excitation field, the suppression of photobleaching is actually much stronger than 5 times.

Because the f factor is hard to calibrate with our current setup, to better measure the suppression power of the photobleaching, we use a single-molecule method to measure the total number of photons that a single fluorophore can emit at an antenna.

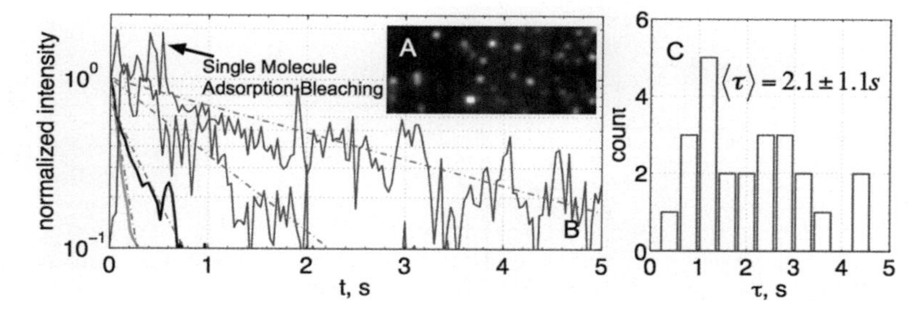

Figure 9.15 (a) One frame from a video measuring the life span of fluorescent molecules is shown here. The bright spots are fluorescence from the molecules on the nanoantennas. (b) The blue curves are decays of fluorescence from two bright spot due to the photobleaching. The red dashed-dotted lines are fit to a single exponential decay, yielding the decay constants of 0.97 ± 0.09 s and 2.76 ± 0.07 s, respectively. The transient spikes on the curve correspond to adsorption-bleaching events of single molecules during the experiments. The power of the excitation laser is 10 mW. As a comparison, we plot the fluorescence decay of the same molecules on a quartz surface at a 10 mW (black) and 40 mW (green) excitation intensity, respectively, in the same figure. The bleaching on the antennas is much slower. (c) A histogram of the fitted decay constants from 24 antennas yields an average life span of 2.1 s, which is a fivefold improvement compared to the same molecules at the same excitation intensity. Adapted with permission from Ref. [1]. Copyright (2013) American Chemical Society.

According to Eq. 9.5, the total number of photons emitted, $\beta\eta k_R/\Gamma$, is independent of the local field enhancement factor f.

The experiment detail is in Ref. [1]. Briefly, after a short time with intensive laser excitation, most of the dyes adsorbed on the antenna's surface are bleached, and we began to observe a single adsorption event: the diffusing fluorophores yield only a constant background, and once a molecule adsorbs onto the antenna, a bright spot appears. The dye eventually bleaches. The adsorbing-bleaching cycle appears as an intermittence ("blinking") pattern. Thus the product of the intensity of the blink with its life span (Fig. 9.16a,b) corresponds to the total number of photons the fluorophore has emitted. We plot the intensity of a fluorophore against the total life span of the fluorescence as a dot in Fig. 9.16c on a log-log plot. The photobleaching limit of the dye—life span multiplied by intensity

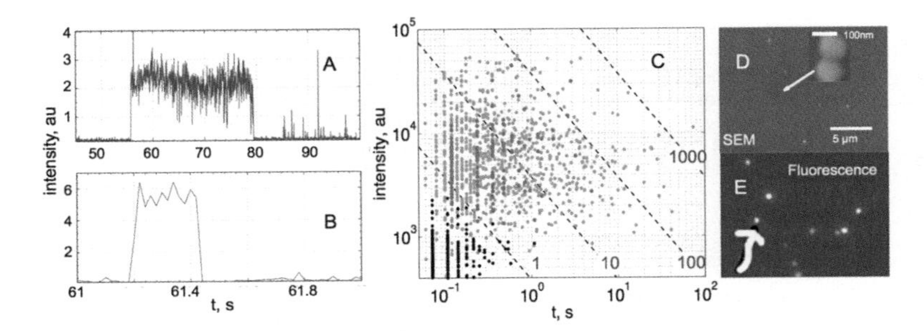

Figure 9.16 (a) The fluorescence intensity trace of a single-molecule adsorption-bleaching event. The rising edge corresponds to the time the molecule adsorbs. The molecule photobleaches after well over 20 s. Another molecule exhibiting stronger fluorescence but a shorter life span is shown in (b). Each single-molecule event is then plotted as a green dot in (c), with the life span and the average fluorescence intensity corresponding to the x and y coordinates, respectively. The power of the excitation laser is 5 mW. The black dots in (c) correspond to the result from a control experiment. A series of black dashed lines are plotted as eye guides, corresponding to 1-, 10-, 100-, and 1000-fold improvement relative to the photobleaching limit. Most black dots, which are from the control experiment, locate below the line marked by 1. In contrast, the green dots, which are from plasmonic structures, locate above the line marked by 1, with one of them even crossing the 1000-fold line. The red dots are the data taken from a dimer structure, whose SEM image and fluorescence images are shown in (d and e). The inset of (d) is a zoomed-in view. Adapted with permission from Ref. [1]. Copyright (2013) American Chemical Society.

as a constant k_R/Γ—is shown as a red dashed line marked as 1 in the figure. The k_R/Γ is determined in ensemble experiments. A data point located to the right or above the photobleaching limit line corresponds to a breaking of the photobleaching limit. From left to right, the series of red lines marked as 1, 10, 100, and 1000 represents the multiples $\beta\eta$ of the photobleaching limit, k_R/Γ. Most of the data points show 10–100-fold improvement, and a few of them even cross the 1000-fold line.

This approach is general and independent on the fluorophore. To confirm that, we repeat an identical single-molecule experiment on a different dye, Chromeo 542, whose emission centers at 561 nm, with a 532 nm laser being used as the excitation source. The result

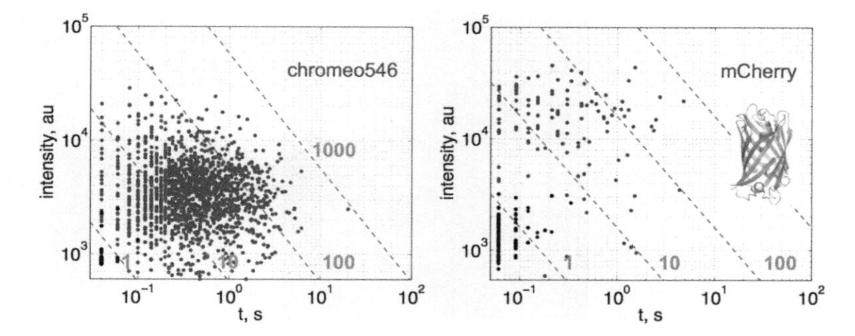

Figure 9.17 Photobleaching suppression of other fluorescence molecules, including Chromeo 546 and fluorescence proteins mChery. Adapted with permission from Ref. [1]. Copyright (2013) American Chemical Society.

is shown in Fig. 9.17; the same antenna structures exhibit similar suppression of the photobleaching.

Our current implementations yields nearly 1000-fold improvement, which can already meet the demand of in vitro applications such as reducing error rates for single-molecule DNA sequencing [10] and improving the detection limit of protein chips [75]. Choosing materials with low ohm loss [76], refining the antenna design [77], improving the fabrication quality, and combining with other commonly used photobleaching suppression methods [15] can improve the suppressing power further. Another promising direction is by taking advantage of the recent progress in the bottom-up approach to synthesize nanoparticles with designed morphology [78] for intracellular and in vivo applications.

9.4 Summary and Prospects

This chapter focuses on the application of plasmonic antennas in overcoming the concentration and resolution barriers in SMS. This is possible, thanks to the progress in (1) nanofabrication that allows fabrication of plasmonic structures with single nanometer-scale features and (2) advanced large-scale electromagnetic field simulation that allows modeling the light–matter interaction in the near-field subwavelength region, both classically and

Figure 9.18 Fluorescence images of antenna arrays fabricated by a hybrid method. The bright spots represent strong enhancement of a single fluorophore adsorbed on the antennas.

semi-quantum-mechanically. Realizing the idea of lab-on-antennas requires fabrication of high-performance antennas on a large scale, which remains challenging, given today's state-of-the-art top-down fabrication capabilities. A promising approach perhaps is by combining top-down with bottom-up fabrication methods. One can use the shape-controlled synthesis method to develop various plasmonic nanocrystals as building blocks and then self-assemble these building blocks into large-scale antenna arrays (Fig. 9.18). Self-assembly provides a bottom-up approach with the potential to accommodate massively parallel, large-scale materials processing. Leveraging the advance of nanoscience, the lab-on-antenna approach will be a general and powerful tool to visualize enzymes working at the single-molecule level with spatial and temporal resolution that was previously unattainable.

References

1. Cang, H., et al. (2014). Giant suppression of photobleaching for single molecule detection via the Purcell effect, *Nano Lett.*, **13**, pp. 5949–5953.
2. Kinkhabwala, A. A., et al. (2012). Fluorescence correlation spectroscopy at high concentrations using gold bowtie nanoantennas, *Chem. Phys.*, **406**, pp. 3–8.

3. Cang, H., et al. (2011). Probing the electromagnetic field of a 15-nanometre hotspot by single molecule imaging, *Nature*, **469**, pp. 385–388.
4. Acuna, G., et al. (2012). Fluorescence enhancement at docking sites of DNA-directed self-assembled nanoantennas, *Science*, **338**, pp. 506–510.
5. Moerner, W. E., et al. (1989). Optical detection and spectroscopy of single molecules in a solid, *Phys. Rev. Lett.*, **62**, p. 2535.
6. Orrit, M., et al. (1990). Single pentacene molecules detected by fluorescence excitation in a p-terphenyl crystal, *Phys. Rev. Lett.*, **65**, p. 2716.
7. Lu, H. P., et al. (1998). Single-molecule enzymatic dynamics, *Science*, **282**, pp. 1877–1882.
8. Zhuang, X., et al. (2000). A single-molecule study of RNA catalysis and folding, *Science*, **288**, pp. 2048–2051.
9. Zhuang, X., et al. (2002). Correlating structural dynamics and function in single ribozyme molecules, *Science*, **296**, pp. 1473–1476.
10. Eid, J., et al. (2009). Real-time DNA sequencing from single polymerase molecules, *Science*, **323**, pp. 133–138.
11. Levene, M. J., et al. (2003). Zero-mode waveguides for single-molecule analysis at high concentrations, *Science*, **299**, pp. 682–686.
12. Flusberg, B. A., et al. (2010). Direct detection of DNA methylation during single-molecule, real-time sequencing, *Nat. Methods*, **7**, pp. 461–465.
13. van den Oever, J. M., et al. (2012). Single molecule sequencing of free DNA from maternal plasma for noninvasive trisomy 21 detection, *Clin. Chem.*, **58**, pp. 699–706.
14. Panchuk-Voloshina, N., et al. (1999). Alexa dyes, a series of new fluorescent dyes that yield exceptionally bright, photostable conjugates, *J. Histochem. Cytochem.*, **47**, pp. 1179–1188.
15. Benesch, R. E., et al. (1953). Enzymatic removal of oxygen for polarography and related methods, *Science*, **118**, pp. 447–449.
16. Rasnik, I., et al. (2006). Nonblinking and long-lasting single-molecule fluorescence imaging, *Nat. Methods*, **3**, 891–893.
17. Nie, S., et al. (1994). Probing individual molecules with confocal fluorescence microscopy, *Science*, **266**, pp. 1018–1021.
18. Axelrod, D. (1981). Cell-substrate contacts illuminated by total internal reflection fluorescence, *J. Cell Biol.*, **89**, pp. 141–145.
19. Vollmer, F., et al. (2008). Whispering-gallery-mode biosensing: label-free detection down to single molecules, *Nat. Methods*, **5**, pp. 591–596.

20. Zhu, J., et al. (2010). On-chip single nanoparticle detection and sizing by mode splitting in an ultrahigh-Q microresonator, *Nat. Photonics*, **4**, pp. 46–49.
21. Hanson, J. A., et al. (2007). Illuminating the mechanistic roles of enzyme conformational dynamics, *Proc. Natl. Acad. Sci. U S A*, **104**, pp. 18055–18060.
22. Yang, H., et al. (2002). Statistical approaches for probing single-molecule dynamics photon-by-photon, *Chem. Phys.*, **284**, pp. 423–437.
23. Montiel, D., et al. (2006). Quantitative characterization of changes in dynamical behavior for single-particle tracking studies, *J. Phys. Chem. B*, **110**, pp. 19763–19770.
24. Bharadwaj, P., et al. (2009). Optical antennas, *Adv. Opt. Photonics*, **1**, pp. 438–483.
25. Garcia-Parajo, M. F. (2008). Optical antennas focus in on biology, *Nat. Photonics*, **2**, pp. 201–203.
26. Kinkhabwala, A., et al. (2009). Large single-molecule fluorescence enhancements produced by a bowtie nanoantenna, *Nat. Photonics*, **3**, pp. 654–657.
27. Novotny, L., et al. (2011). Antennas for light, *Nat. Photonics*, **5**, pp. 83–90.
28. Taminiau, T. H., et al. (2008). Optical antennas direct single-molecule emission, *Nat. Photonics*, **2**, pp. 234–237.
29. Alu, A., et al. (2008). Dynamical theory of artificial optical magnetism produced by rings of plasmonic nanoparticles, *Phys. Rev. B*, **78**, p. 085112.
30. Schuck, P. J., et al. (2005). Improving the mismatch between light and nanoscale objects with gold bowtie nanoantennas, *Phys. Rev. Lett.*, **94**, p. 017402.
31. Vecchi, G., et al. (2009). Shaping the fluorescent emission by lattice resonances in plasmonic crystals of nanoantennas, *Phys. Rev. Lett.*, **102**, p. 146807.
32. Curto, A. G., et al. (2010). Unidirectional emission of a quantum dot coupled to a nanoantenna, *Science*, **329**, pp. 930–933.
33. Muhlschlegel, P., et al. (2005). Resonant optical antennas, *Science*, **308**, pp. 1607–1609.
34. Jeanmaire, D. L., et al. (1977). Surface Raman spectroelectrochemistry: part I. Heterocyclic, aromatic, and aliphatic amines adsorbed on the anodized silver electrode, *J. Electroanal. Chem.*, **84**, pp. 1–20.

35. Fleischman, M., et al. (1974). Raman spectra of pyridine adsorbed at a silver electrode, *Chem. Phys. Lett.*, **26**, pp. 163–166.
36. Albrecht, M. G., et al. (1977). Anomalously intense Raman spectra of pyridine at a silver electrode, *J. Am. Chem. Soc.*, **99**, pp. 5215–5217.
37. Moskovits, M. (1978). Surface roughness and the enhanced intensity of Raman scattering by molecules adsorbed on metals, *J . Chem. Phys.*, **69**, pp. 4159–4161.
38. Creighton, J. A., et al. (1979). Plasma resonance enhancement of Raman scattering by pyridine adsorbed on silver or gold sol particles of size comparable to the excitation wavelength, *J. Chem. Soc.*, **75**, pp. 790–798.
39. Shalaev, V. M., et al. (1988). Fractals: optical susceptibility and giant Raman scattering, *Z. Phys. D*, **10**, pp. 71–79.
40. Stockman, M. I., et al. (2001). Surface plasmon amplification by stimulated emission of radiation: quantum generation of coherent surface plasmons in nanosystems, *Phys. Rev. Lett.*, **87**, p. 167401.
41. Stockman, M. I., et al. (1994). Giant fluctuations of local optical fields in fractal clusters, *Phys. Rev. Lett.*, **72**, p. 2486.
42. Novotny, L., et al. (2006). *Principles of Nano-Optics* (Cambridge University Press, Cambridge).
43. Lu, D., et al. (2014). Enhancing spontaneous emission rates of molecules using nanopatterned multilayer hyperbolic metamaterials, *Nat. Nanotechnol.*, **9**, pp. 48–53.
44. Purcell, E. M. (1946). Spontaneous emission probabilities at radio frequencies, *Phys. Rev.*, **69**, p. 681.
45. Hale, G. D., et al. (2001). Enhancing the active lifetime of luminescent semiconducting polymers via doping with metal nanoshells, *Appl. Phys. Lett.*, **78**, pp. 1502–1504.
46. Malicka, J., et al. (2002). Photostability of Cy3 and Cy5-labeled DNA in the presence of metallic silver particles, *J. Fluoresc.*, **12**, pp. 439–447.
47. Zaiba, S., et al. (2011). Transparent plasmonic nanocontainers protect organic fluorophores against photobleaching, *Nano Lett.*, **11**, pp. 2043–2047.
48. Kena-Cohen, S., et al. (2011). Plasmonic sinks for the selective removal of long-lived states, *ACS Nano*, **5**, pp. 9958–9965.
49. Bharadwaj, P., et al. (2011). Robustness of quantum dot power-law blinking, *Nano Lett.*, **11**, 2137–2141.

50. Jain, P. K., et al. (2012). Near-field manipulation of spectroscopic selection rules on the nanoscale, *Proc. Natl. Acad. Sci. U S A*, **109**, pp. 8016–8019.
51. Uemura, S., et al. (2010). Real-time tRNA transit on single translating ribosomes at codon resolution, *Nature*, **464**, pp. 1012–1017.
52. Anger, P., et al. (2006). Enhancement and quenching of single-molecule fluorescence, *Phys. Rev. Lett.*, **96**, p. 113002.
53. Betzig, E., et al. (1991). Breaking the diffraction barrier-optical microscopy on a nanometric scale, *Science*, **251**, pp. 1468–1470.
54. Betzig, E., et al. (2006). Imaging intracellular fluorescent proteins at nanometer resolution, *Science*, **313**, pp. 1642–1645.
55. Rust, M. J., et al. (2006). Sub-diffraction-limit imaging by stochastic optical reconstruction microscopy (STORM), *Nat. Methods*, **3**, pp. 793–796.
56. Enderlein, J. (2000). A theoretical investigation of single-molecule fluorescence detection on thin metallic layers, *Biophys. J.*, **78**, pp. 2151–2158.
57. Wu, D., et al. (2008). Super-resolution imaging by random adsorbed molecule probes, *Nano Lett.*, **8**, pp. 1159–1162.
58. Camden, J. P., et al. (2008). Probing the structure of single-molecule surface-enhanced Raman scattering hot spots, *J. Am. Chem. Soc.*, **130**, pp. 12616–12617.
59. Lakowicz, J. R. (2006). *Principles of Fluorescence Spectroscopy*, 3rd ed. (Springer, NY, USA).
60. Moerner, W. E., et al. (2003). Methods of single-molecule fluorescence spectroscopy and microscopy, *Rev. Sci. Instrum.*, **74**, pp. 3597–3619.
61. Turro, N. J. (1991). *Modern Molecular Photochemistry* (University Science Books, NY, USA).
62. Vogelsang, J., et al. (2008). A reducing and oxidizing system minimizes photobleaching and blinking of fluorescent dyes, *Angew. Chem. Int. Ed.*, **47**, pp. 5465–5469.
63. Rossetti, R., et al. (1983). Quantum size effects in the redox potentials, resonance Raman spectra, and electronic spectra of CdS crystallites in aqueous solution, *J. Chem. Phys.*, **79**, pp. 1086–1088.
64. Harris, T. D., et al. (2008). Single-molecule DNA sequencing of a viral genome, *Science*, **320**, pp. 106–109.
65. Mukamel, S. (1999). *Principles of Nonlinear Optical Spectroscopy* (Oxford University Press, USA).

66. Yablonovitch, E. (1987). Inhibited spontaneous emission in solid-state physics and electronics, *Phys. Rev. Lett.*, **58**, p. 2059.
67. Kleppner, D. (1981). Inhibited spontaneous emission, *Phys. Rev. Lett.*, **47**, p. 233.
68. Muthu, P., et al. (2007). Decreasing photobleaching by silver island films: application to muscle, *Anal. Biochem.*, **366**, pp. 228–236.
69. Enderlein, J. (2002). Theoretical study of single molecule fluorescence in a metallic nanocavity, *Appl. Phys. Lett.*, **80**, pp. 315–317.
70. Enderlein, J. (1999). Single-molecule fluorescence near a metal layer, *Chem. Phys.*, **247**, 1–9.
71. Hinze, G., et al. (2010). Statistical analysis of time resolved single molecule fluorescence data without time binning, *J. Chem. Phys.*, **132**, p. 044509.
72. Turro, N. J. (1967). Photochemical reactivity, *J. Chem. Educ.*, **44**, p. 536.
73. Johnson, P. B., et al. (1972). Optical constants of the noble metals, *Phys. Rev. B*, **6**, p. 4370.
74. Axelrod, D. (2001). Total internal reflection fluorescence microscopy in cell biology, *Traffic*, **2**, pp. 764–774.
75. Phizicky, E., et al. (2003). Protein analysis on a proteomic scale, *Nature*, **422**, pp. 208–215.
76. Wang, F., et al. (2006). General properties of local plasmons in metal nanostructures, *Phys. Rev. Lett.*, **97**, p. 206806.
77. Barnes, W. L. (2009). Comparing experiment and theory in plasmonics, *J. Opt. A: Pure Appl. Opt.*, **11**, p. 114002.
78. Sun, Y., et al. (2002). Shape-controlled synthesis of gold and silver nanoparticles, *Science*, **298**, pp. 2176–2179.

Chapter 10

Plasmonic Lenses for High-Throughput Nanolithography

Liang Pan
School of Mechanical Engineering, Purdue University, West Lafayette, IN 49707, USA
liangpan@purdue.edu

Concentrating optical energy at the nanoscale has numerous applications in lithography, data storage, biosensing, spectroscopy, molecular trapping, and so forth. However, light diffraction sets a fundamental limit on optical resolution on the order of the wavelength, and it poses a critical challenge to the down-scaling of nanoscale optical focusing. Surface plasmons can circumvent the optical diffraction limit as they have shorter wavelengths. Focusing light with orders-of-magnitude higher efficiency can be realized by focusing surface plasmon polaritons (SPPs) and various plasmonic structures have been proposed and demonstrated in the past. Among them, the plasmonic lens (PL) is a particular type of planar structure capable of generating a nanoscale light spot with high intensity and can be used as a virtual probe for high-speed maskless lithography and many other applications. In this chapter, we will discuss the use of PLs for efficient optical energy focusing. Combining PLs with air-bearing surface (ABS) technology, we demonstrated nanolithography at 22 nm resolution at linear

Plasmonics and Super-Resolution Imaging
Edited by Zhaowei Liu

ISBN 978-981-4669-91-7 (Hardcover), 978-1-315-20653-0 (eBook)
www.panstanford.com

scanning speeds up to 10 m/s. This low-cost scheme has the potential of higher throughput than current photolithography, and it opens a new approach toward next-generation semiconductor manufacturing.

10.1 Introduction to Maskless Nanolithography

Nanolithography is a process of creating a nanoscale pattern, typically from 100 nm down to the molecular and atomic levels. Creating superfine nanoscale patterns with high throughput is essential for high-speed computing, data storage, and broader applications for nanomanufacturing. The semiconductor industry provides the primary demand for economical high-throughput nanolithography approaches and effectively drives the development of nanolithography to keep its trends of growth. As described by the famous Moore's law, the number of transistors that can be placed inexpensively on an integrated circuit doubles approximately every two years in a long-term development trend of computing hardware. To sustain this growth rate, many approaches have been developed to create nanoscale patterns using either top-down or bottom-up schemes. Top-down schemes seek to create and assemble nanoscale patterns using larger devices, while bottom-up schemes seek to assemble smaller entities into larger and more complex ones. The top-down scheme is still dominant in mass production, particularly for the semiconductor industry, but there is a possibility that the bottom-up scheme may take over because of the ever-increasing complexity and cost of the top-down scheme.

Photolithography [1], the process of optically transferring a pattern from a predefined photomask onto a photosensitive resist layer, has been the most important branch of the nanolithography and is exclusively employed in current semiconductor device manufacturing. The smallest achievable feature is primarily determined by the resolution of the optical system on the order of the wavelength limited by the optical diffraction. Therefore, photolithography researchers have been continuously improving pattern density in the past by introducing a shorter working wavelength (from i-line 365 nm, to KrF 248 nm, and then ArF

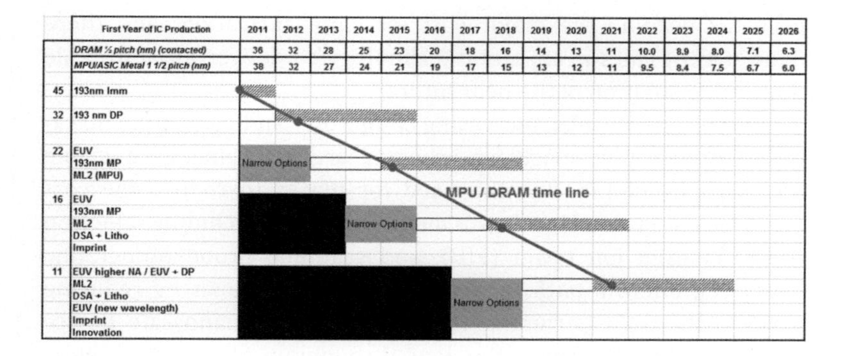

Figure 10.1 ITRS lithography roadmaps updated at year 2012.

193 nm), together with a higher numerical aperture and many other engineering improvements. The state-of-the-art photolithography tools use an ArF light source at a 193 nm vacuum wavelength with water immersion to produce 45 nm optical features. And 22 nm features have been achieved in mass production by patterning a surface multiple times using 193 nm light or using nonoptical methods, such as the self-aligned spacer technique, to further increase the pattern density.

According to the International Technology Roadmap for Semiconductors (ITRS) lithography roadmap (shown in Fig. 10.1), the ever-improving trend of photolithography will inevitably stop due to the fundamental limit of far-field optical diffraction and the infamous problem of absorption of ultrashort-wavelength light. The issue that the current photolithography approach facing is the increasing process complexity and the extremely high cost. The fabrication of high-quality lithography masks has become time consuming and expensive, which hinders device prototyping, where frequent design changes are often needed. The current optical immersion tool cost for dedicated double- and multiple-patterning techniques is exceeding \$50M per tool, and the costs for masks far outweigh those for tools because of the large amount of data required to write these masks and the difficulties in implementing the necessary optical proximity correction. Extreme ultraviolet (EUV) lithography might succeed as the next-generation lithography (NGL) technology at an even higher tool cost. The lithography cost has become the major cost for chip fabrication. And there is an urgent call for an exotic technology to deliver an affordable patterning solution in the near future.

Because of the ever-increasing complexity and cost of the mask-based lithography, maskless schemes are emerging as a viable approach by eliminating the need for masks to reduce the cost and design cycle. Maskless nanolithography, including electron beam, focused ion beam (FIB), and scanning probe lithography (SPL) [2], offers a path to overcome these obstacles by reducing mask costs and shortening the cycle time for nanoscale device validation [3]. However, the low throughput of most maskless methods due to the serial and slow scanning nature remains a bottleneck. Although multiaxial electron beam lithography has been proposed to increase throughput by using multiple beams in a parallel manner, there are difficulties in simultaneously regulating the multiple beam sizes and positions because of the thermal drift and electrical charge Coulomb interactions, which result in significant lens aberration [4]. Zone-plate-array lithography (ZPAL) [5] utilizes a large array of diffractive optical elements or spatial light modulators to improve the throughput, but the ultimate resolution is still restricted to the diffraction limit. SPL, a tip-based low-cost alternative operating in ambient environment, has made a noticeable throughput improvement, as shown in a recent demonstration using 55,000 probes scanning at a speed of 60 μm/s [6]. Its throughput is still 2 to 3 orders of magnitude lower than that required by practical nanofabrication applications. This is because SPL technology relies on the slow scan of the tips at 10–100 nm from the surface, which has limited feedback bandwidth to control the tip–sample distance at higher speed. Another optical maskless approach is to use assisting light beams to control the resist kinetics to achieve subdiffraction features [7–9]. It provides a low-cost alternative; however, the achievable feature size is still greatly affected by the spatial regulation capability of the far-field optics. A major improvement in maskless lithography is thus critical in order to satisfy the demands in mass production for the semiconductor industry.

10.2 Introduction to Plasmonic Nanofocusing Structures

Great efforts have been devoted to achieving subwavelength-scale light spots with high intensity, which is crucial for many applications, including nanolithography, high-density optical and/or magnetic data storage, nanospectroscopy, biosensing, molecular trapping, and so forth. Working at the optical near field of a nanoscale aperture is one of the approaches to overcome the diffraction limit of far-field optical techniques. The subdiffractive size of the focused near-field light spot allows the formation of images with higher resolution and enables direct writing with high-density patterns. The near-field scanning optical microscope (NSOM) probe is one common way to obtain a high confined light spot to study nanoscale optical properties as well as to accomplish optical nanolithography, but this approach still faces key obstacles such as energy throughput and working distance control for high-volume manufacturing [10–12]. The transmission through the aperture of the probe is extremely low for an NSOM's subwavelength-size apertures. The transmission normalized by the aperture area is proportional to $(d/\lambda)^4$ [13], where d is the diameter of the aperture and λ is the wavelength of illumination light in free space. As an example, the optical power transmission of an NSOM probe with a sub-100 nm opening is typically on the order of 10^{-5} to 10^{-7}, resulting in a very low throughput for both imaging and lithography applications.

Plasmonic structures can effectively concentrate optical energy into a nanoscale spot by taking advantage of the short wavelength of surface plasmon polaritons (SPPs). SPPs can be excited by incident light on some metallic interfaces to propagate with a long decay distance, and various structures (shown in Fig. 10.2) have been proposed to focus light to a tiny spot by utilizing SPPs. SPP gratings have been used to excite and construct SPP interference to concentrate optical energy in the near field, such as the bull's-eye aperture (Fig. 10.2e). The bull's-eye aperture is a single nanoscale hole surrounded by circular grooves. The circular grooves can excite SPPs on the metal surface and guide the SPP propagation along the direction normal to the grooves. As a result of interference, the

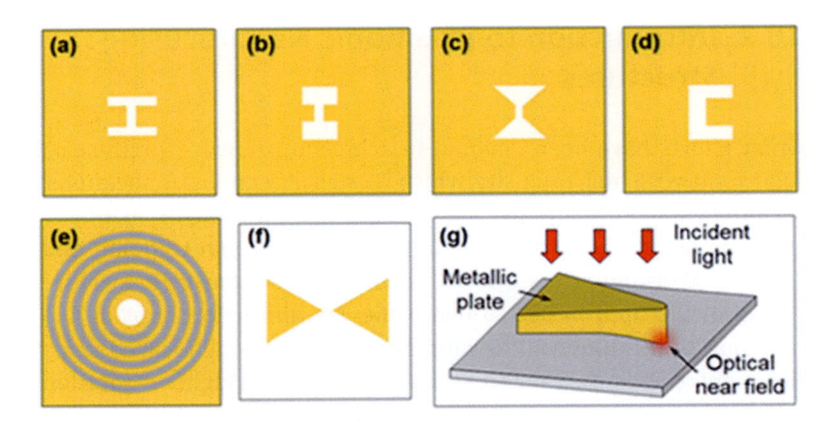

Figure 10.2 Examples of plasmonic resonance structures in the literature.

SPP waves are focused at the center of the circular grooves where the subwavelength hole is located. The concentrated SPP waves can dramatically enhance the evanescent field transmitted through the subwavelength hole [11]. This system was also demonstrated in a recording experiment by Betzig et al. [14]. They used an optical fiber tapered to a 100 nm tip and coated with an aluminum film to deliver light to a recording medium in the presence of an external magnetic field. The bull's-eye structure has stimulated extensive interest in SPPs in the scientific community. However, its application is limited by the trade-off between the brightness of the center spot and the fineness of the spot size.

The second type of plasmonic structure is the SPP resonant aperture. Most common SPP resonant apertures are essentially ridge apertures initially proposed for microwave-frequency applications and later shown to also operate at near-infrared and visible wavelengths [15–17]. As shown in Fig. 10.2a–d, many designs have been studied, including I-shaped, H-shaped, bowtie, and C-shaped apertures. Although the aperture shapes are quite different, they share the same working principal. The long dimension of the rectangular aperture part is on the order of a wavelength to allow light that is polarized perpendicular to this dimension to propagate through the aperture. The ridge in the center of the aperture concentrates the electric field intensity by both the lightning rod effect and a localized SPP resonance. Similarly, other types of SPP apertures have been proposed, such as triangle, modified C-shaped, and L-shaped apertures [18]. These SPP ridge apertures usually have strong wavelength dependency. Similar to the ridge apertures, some SPP nanoantennas are also proposed to focus light in the near field. Among them, bowtie antennas, as shown in Fig. 10.2f, are most widely studied. The antenna can couple the incident light with the electric field polarized along the long axis of the antenna. And the antenna tips also concentrate the electric field intensity by both the lightning pole effect and a localized SPP resonance similar as the ridge apertures. By tuning the SPP-resonant frequency together with the dimensions of the antenna, it is possible to concentrate the surface charges at the points of the tips in order to generate large field intensities. The beaked triangle plate [19] is a modified version of the SPP antenna (Fig. 10.2g), serving as one half of a bowtie antenna. It has the tip of the triangle extending toward the metallic medium to ensure that the optical energy is locally concentrated in the medium. The mirror charges inside the metallic medium, serving as the other half of the bowtie aperture, which greatly enhances the SPP's coupling between the tip and the medium.

10.3 Plasmonic Lenses

The plasmonic lens (PL) [20], as one type of plasmonic nanofocusing structures, is a nanoscale aperture or antenna surrounded by a series of circular coupling slits on a metal film. It can efficiently focus light in its near field, which can be used as a virtual optical probe for maskless high-speed nanolithography.

An example of a PL is illustrated in Fig. 10.3, which is a modified bull's-eye design. In this design, the subwavelength hole is surrounded by circular slits instead of grooves in the bull's-eye structure. This design can achieve better light confinement and higher energy transmission than bull's-eye structures. The field penetration depth is also improved for applications of near-field nanolithography. Although this modification sacrifices the contrast performance of the focus spot by allowing light to be transmitted through the slits, this can be tolerated by using nonlinear resists

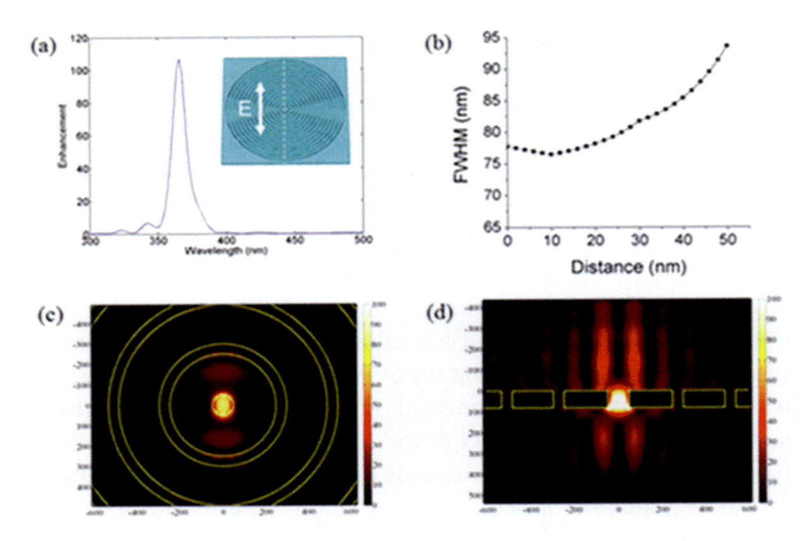

Figure 10.3 Numerical studies of the modified bull's-eye PL. (a) The light intensity profile along the dashed line shown in the inset. (b) The intensity profile plotted as a function of off-plane distance. (c and d) Top and cross-sectional views of the field profile, showing the maximum enhancement about ~100 at a resonant wavelength of 365 nm.

with an exposure threshold. The proposed structure is shown in the inset of Fig. 10.3a, where the light intensity profile along the dashed line is plotted. This numerical simulation to obtain this profile was performed using the commercial FDTD software (CST Microwave Studio) under linearly polarized plane-wave illumination. The polarization direction is along the y axis, as shown in Fig. 10.3a. Under plane-wave illumination the SPPs are excited at both sides of the lens and assist the optical energy to propagate toward the center of the PL [21, 22]. By tuning the period of the rings, the SPPs excited by multiple rings can form a constructive interference. Also the thickness of the metal film is carefully chosen in order to generate constructive interference between the top and bottom surfaces. The focused SPPs launched through the center hole help generate a stronger field profile and improve the field depth. The simulated PL consists of a nanoaperture surrounded by 15 through rings on an aluminum film. The hole diameter, ring periodicity, ring width, and aluminum layer thickness are 100 nm, 250 nm, 50 nm,

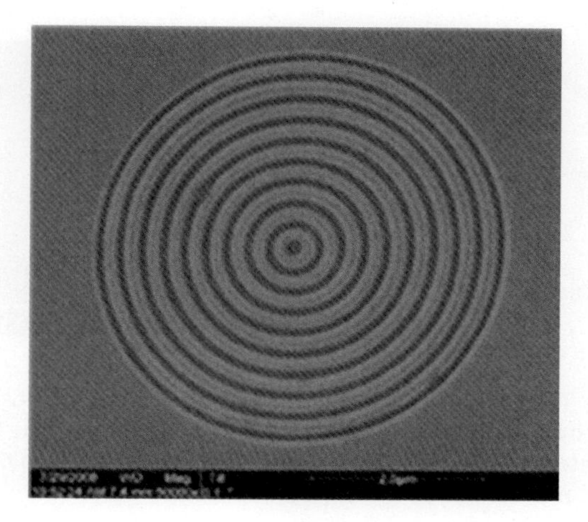

Figure 10.4 An SEM picture of a plasmonic lens consists of concentric slits and a center hole.

and 80 nm, respectively. Figures 10.3c and 10.3d show the maximum enhancement of about ~100 at the designed resonant wavelength of 365 nm. And the intensity profile as a function of the off-plane distance is plotted in Fig. 10.2b. As shown, the full-width at half-maximum (FWHM) of the obtained spot is about 80 nm near the lens surface, and it becomes larger as the distance increases.

A scanning electron microscopy (SEM) picture of the concentric-slit PL is shown in Fig. 10.4. Although this design shows excellent subwavelength-scale optical focusing capability with highly enhanced field intensity, there are still some engineering difficulties to utilize it in high-throughput processes. This design uses on long-range propagating SPPs to concentrate the optical energy at the center hole and has a strong wavelength dependency. It occupies a relatively large area and also limits the choice of metals. There are only a few metals that are both chemically stable and can support long-distance propagating SPPs at the optical frequency range, such as aluminum, silver, copper, and gold. The poor engineering properties of these metals can cause reliability issues during harsh engineering operations.

Designs that can utilize stronger metals as lens materials will allow operations under harsh environments, such as high operating

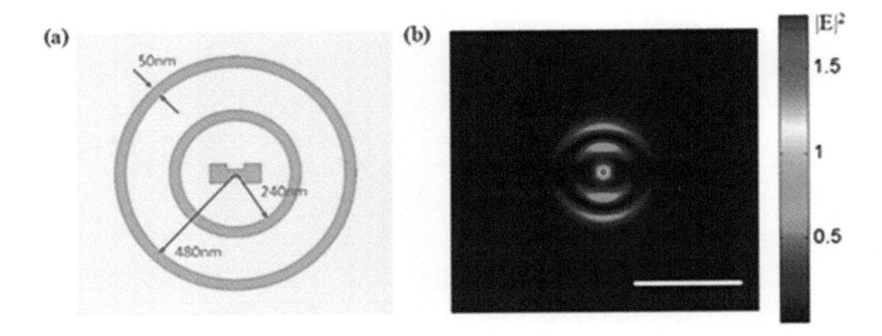

Figure 10.5 Typical H-shaped PLs. (a) The structure of one PL: H-aperture surrounded by two rings. The inset shows the parameters of the H-aperture. (b) E-field intensity distribution at the plane 25 nm away from the PL. Scale bar length: 1 micron.

temperature and high-speed mechanical scanning. Metals such as chromium can support short-range SPPs and have excellent engineering properties. At the working wavelength of 355 nm, the propagation length of the surface plasmons on the Cr surface is several hundred nanometers. To utilize the SPPs on a Cr film to enhance the transmission through the aperture, it is reasonable to design the PL with its diameter around 1 μm, considering the limited propagation length of the surface plasmons on a Cr surface. Many studies [23–25] have been performed recently to obtain a subdiffraction-limited spot with high transmission through a ridge aperture at optical wavelengths. To design a lens using Cr, we replace the circular aperture of the modified bull's-eye PL with a ridge aperture to enhance the intensity of the focus spot by the PL. And only a few rings are needed to achieve reasonable performance.

The PL fabricated on a chromium film can prevent structure damage during high-speed scanning at the optical near field above the substrate. Our typical Cr-based PL has an H-shaped aperture surrounded by two rings, as shown in Fig. 10.5a. Figure 10.5b shows the E-field intensity distribution at a plane 25 nm away from the surface of the H-shaped aperture (calculated by commercial electromagnetic wave software, CST Microwave Studio). The peak intensity of the focus spot is about 4.09 times of the incident

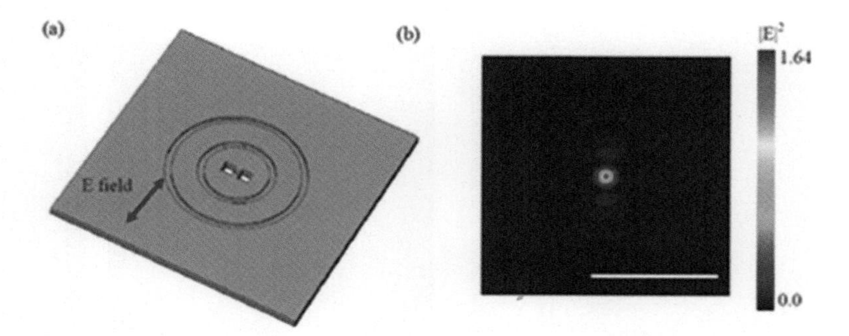

Figure 10.6 (a) An H-shaped aperture surrounded by two partially through grooves. (b) E-field intensity distribution at the plane 25 nm away from the PL surface. Scale bar length: 1 micron.

light. The size of the focus spot, which is defined by the FWHM intensity, is about 80 nm. The half-circular patterns in the intensity distribution are the direct transmission through the two rings surrounding the H-shaped aperture. The local maximum intensity at those patterns is about 1.58, corresponding to a contrast ratio of 2.6. In nanolithography, by properly choosing the resist and exposure condition, their intensity of half-circular patterns can be well under the exposure threshold of the resist.

As shown in Fig. 10.6, by replacing the through rings with shallow grooves to reduce the peak intensity of the half-circular patterns to sub-0.1, the contrast ratio can be greatly enhanced on the order of 20, which is good enough for most lithography applications using linear resists.

The performance of the PL can be further improved by adding a third ring with a radius of 600 nm (see Fig. 10.7). The third ring, placed at a half-period position of the circular grating, acts as a reflector for the outward propagating surface plasmon waves on the Cr film due to destructive interference. Consequently, more light can be concentrated at the focus spot of the PL to generate a stronger peak intensity. And the ring reflector also helps to reduce the crosstalk within an array of closely packed PLs in a parallel patterning process. Figure 10.7b shows the field intensity distribution at the plane 10 nm away from the improved PL. The central focal spot by the PL has a peak intensity 13.1 times the

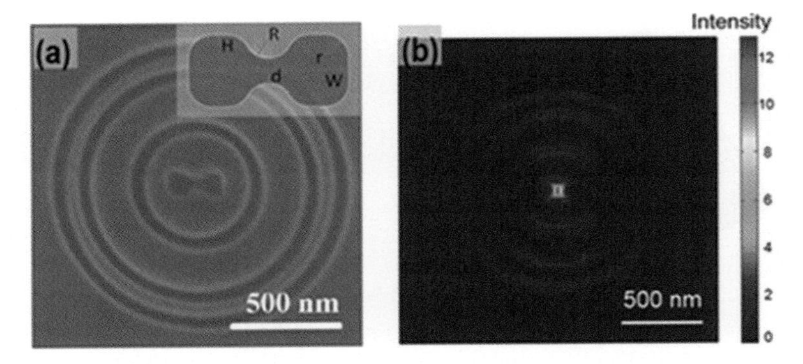

Figure 10.7 (a) SEM pictures of a PL consisting of an H-aperture, a set of ring couplers (two inner rings), and a ring reflector (the outer ring), fabricated on a metallic thin film in a 60 nm thick Cr film. The parameters of the center aperture are shown in the inse , where $W = 240$ nm, $H = 98$ nm, $R = 35$ nm, $r = 40$ nm, and $d = 26$ nm. The radii of the three rings are 240 nm, 480 nm, and 600 nm, respectively. And the width of the rings is 50 nm. (b) The field intensity distribution at the plane 10 nm away from the lens surface normalized to the incident intensity of 355 nm wavelength light. The half-circular-shaped side lopes in the intensity profile are the direct transmissions through the three rings and their intensities are far below the exposure threshold of the resist.

incidence light with a 45 nm FWHM spot size. The ring gratings are etched all the way through the metal film, thereby causing side lobes with a maximum intensity 2.0 times of the incident light, corresponding to a contrast ratio of 6.5 to the focal spot intensity, which is well under the exposure threshold for our current maskless lithography purpose.

As shown in Fig. 10.8, the contrast ratio can be further enhanced to 70 or more by replacing the grating slits with shallow blind grooves. Replacing Cr with other metals, such as aluminum, with better mechanical lens protection can further improve the peak intensity by a few times. Comparing the lens performance between the H-shaped and the optimized modified bull's eye, at the focus spot size of 45 nm defined by the FWHM intensity, we find that the optical intensity enhancement factor of the H-shaped PL is more than 1 order magnitude of the optimized modified bull's-eye design.

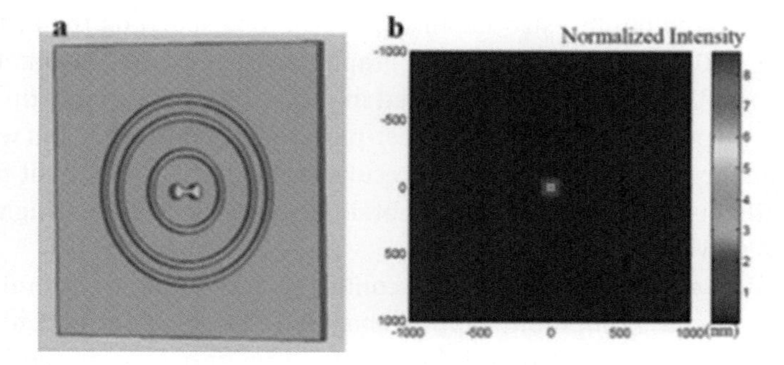

Figure 10.8 (a) Replacing three rings partially through Cr grooves. (b) E-field intensity distribution at the plane 10 nm away from the PL surface.

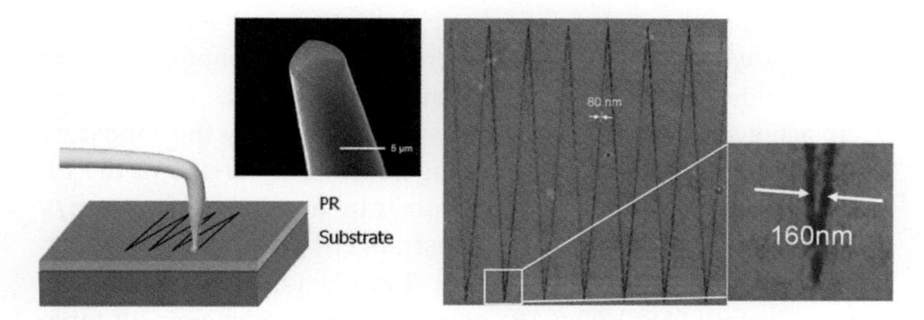

Figure 10.9 (Left) An NSOM probe consists of nanostructured plasmonic structures being fabricated on the end of an optical fiber. (Right) AFM image of a photoresist after near-field scanning exposure using plasmonic structures.

10.4 Scanning Plasmonic Lenses in the Near Field

A PL can focus light to a sub-100 nm spot at the near field with a local intensity orders of magnitude higher than the incident light, but the tightly focused spot only exists at the near field of the lens, normally closer than 100 nm because of the exponential decay of the evanescent field (Fig. 10.3b). Thus, using PLs requires a new mechanism to ensure precise control of the nanoscale gap between the PLs and the photoresist surface.

As shown in Fig. 10.9, lithography using PLs had been demonstrated in an NSOM system [26]. An NSOM can accurately control the PL to the photoresist distance at a scanning speed on the order of 10–100 μm/s. During the lithography process, a laser beam at a wavelength of 365 nm was coupled into the NSOM plasmonic tip to expose a positive photoresist. A best-pattern line width of 80 nm was demonstrated. Similar exposure results through the tip without the plasmonic structures were only obtained by using 10 times' higher input power.

However, it still cannot overcome the limitation of scanning probe–based lithography approaches due to its slow scanning speed. To achieve a practical lithography throughput, it will require scanning 10^6–10^9 NSOM probes in parallel, which is technically challenging.

We introduce here high-throughput plasmonic nanolithography (PNL) to circumvent the critical parallelization and slow scanning challenges which can potentially increase the throughput by orders of magnitude compared to that achieved by an NSOM.

To achieve high-speed scanning while maintaining the nanoscale gap, we designed a novel air-bearing slider to fly arrays of the PL at the height of 20 nm above the substrate at speeds of ~10 m/s (Fig. 10.10). The rotation of the substrate creates an air flow along the bottom surface of the plasmonic flying head, known as the air-bearing surface (ABS). The ABS generates an aerodynamic lift force and it is balanced with the force supplied by the suspension arm to precisely regulate a nanoscale gap between the PL arrays and the rotating substrate, which is covered with the photoresist. With the high bearing stiffness and small actuation mass, this self-adaptive method can provide an effective bandwidth on the order of 100 kHz. The usage of an ABS eliminates the need for a feedback control loop and, therefore, overcomes the major technical barrier for high-speed scanning. The plasmonic flying head is made of a specially designed transparent air-bearing slider with arrays of PLs fabricated on its bottom surface. Employing large arrays of PLs enables parallel writing for high throughput.

In this work, the plasmonic flying heads were fabricated using microfabrication techniques and FIB milling, and they were evaluated using a dynamic flying height tester (DFHT IV, Phase Metrics). The parallelism between the PL array and the substrate needs to be carefully considered in designing an ABS. The ABS for the plasmonic

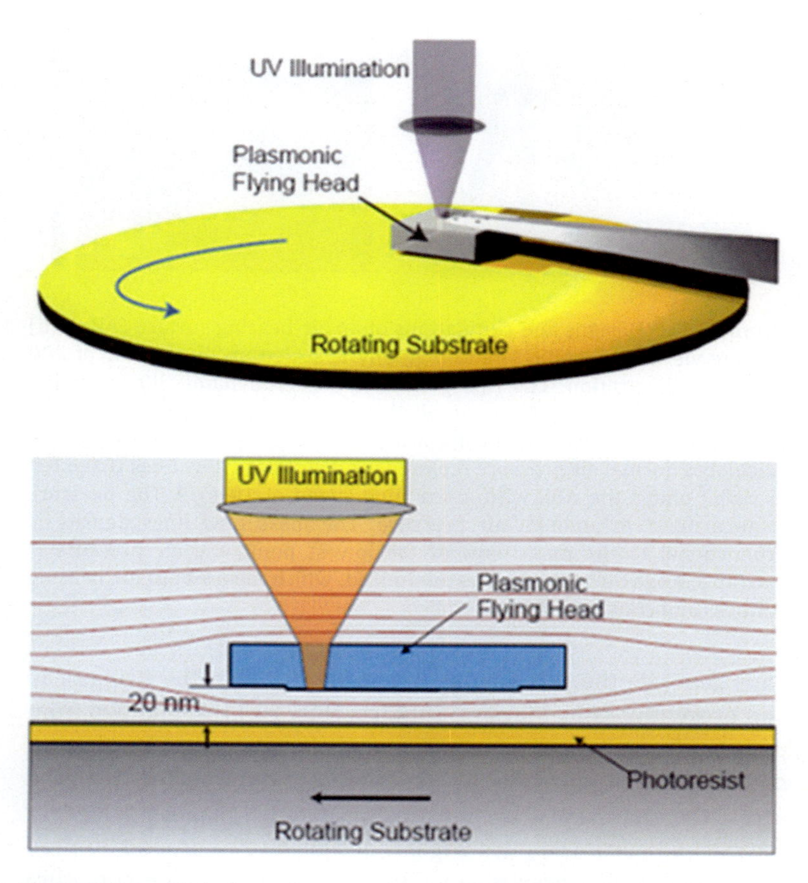

Figure 10.10 High-throughput maskless nanolithography using PL arrays. (Top) Schematic showing the lens array focusing ultraviolet laser pulses onto the rotating substrate to concentrate surface plasmons into sub-100 nm spots. However, sub-100 nm spots are only produced in the near field of the lens, so a process control system is needed to maintain the gap between the lens and the substrate at 20 nm. (Bottom) Cross-sectional schematic of the plasmonic head flying 20 nm above the rotating substrate which is covered with the photoresist.

flying head was designed using in-house-developed air-bearing simulators [27]. The goal was to achieve a consistent flying height at scanning speeds of ~10 m/s. The ABS design consists of a four-pad U-shaped dual rail with a long front bar (Fig. 10.11a). Two large rear

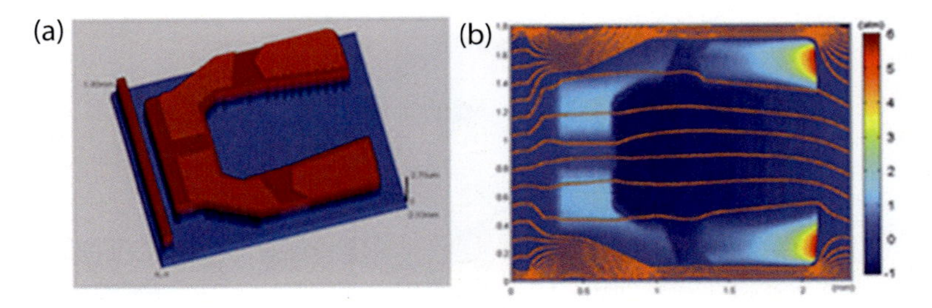

Figure 10.11 Designs and simulations of an air-bearing surface (ABS). (a) Oblique view of the ABS. The topography is scaled up by a factor of 200 for better illustration. The ABS generates an aerodynamic lift force and it is balanced with the force supplied by suspension to precisely retain a nanoscale gap between the PL arrays and the rotating substrate. (b) Calculated normal air pressure (colored) and air–mass flow lines (from left to right) under the ABS with a scanning speed of 10 m/s. The pressure is normalized to ambient air pressure. The mass flow lines density is proportional to the mass flow. At the lowest point, the air pressure is maximized but the mass flow is minimized, which favors both air-bearing stiffness and contamination tolerance.

pads generate the repelling peak pressure to float the flying head and prevent possible physical contacts (Fig. 10.11b). The two front pads produce the steering repelling pressure to increase bearing roll stiffness and minimize the roll angle. A $\sim$2 μrad roll angle is achieved across the rotating disk by adjusting the detailed shape and depth of the rail and pads. The pitch angle is designed to be around 80 μrad rather than being even smaller to compensate for the curvature variations of both the slider and the disk. By throttling injecting air from the leading edge, the long-bar design can significantly reduce the slider's pitch angle and contamination sensitivities and also enhance the slider's damping. The U-shaped dual-rail design efficiently increases the overall subambient ("negative") pressure, which improves both the slider's stability and bearing stiffness. As the disk velocity decreases, both the positive pressure and the negative pressure decrease, which results in lower flying height and bearing stiffness. The effective air-bearing stiffness and damping ratio are about 200,000 N/m and 0.1, respectively. This design provides an effective bandwidth for gap control on the order of 100 kHz. The flying PL array in the optical near field is inspired by the magnetic

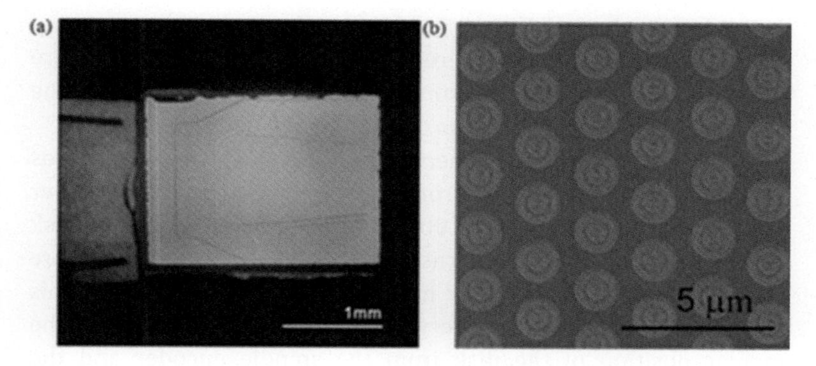

Figure 10.12 Fabricated plasmonic flying head and flying height measurement. (a) Optical micrograph of a plasmonic flying head assembled with suspension. (b) SEM image of an array of plasmonic lenses fabricated on an air-bearing surface (ABS).

recording head in hard disk drives (HDDs). Unlike a conventional HDD ABS which uses only a trailing edge–mounted transducer to serially read and write magnetic bits, we designed a plasmonic head to contain a relatively large area filled by PLs that enable parallel writing and high throughput. Due to the rapid decay of light intensity of the PL, all PLs need to keep the distance to the rotating substrate within 30 nm, which requires the bottom surface to be parallel to the substrate to within a 100 μrad tilt. This stringent parallelism requirement made the design of the plasmonic head different from magnetic head sliders. For example, to fly 1000 lenses within the 30 nm gap tolerance over the usable area of 800 μm × 20 μm on the rear pads with each PL being 4 μm in diameter, the ABS needs to be designed with a less than 100 μrad pitch angle and a 2 μrad roll angle. Also, the ABS needs a larger air-bearing stiffness, higher damping ratio, and better contamination insensitivity than a conventional ABS. In addition, the plasmonic head must be transparent to light.

Figure 10.12a shows an optical microscope image of the fabricated plasmonic flying head where the sapphire ABS coated with a metal film was assembled to the suspension, and an SEM image of a 2D array of PLs fabricated on the ABS (Fig. 10.12b). During operation, the flying height is kept consistently over the disk velocity range of 4 to 12 m/s. Under the above linear velocity range, the experimental measured roll and pitch angle variations are in good agreement with the simulation design, which ensures the entire 1000-lens array is within the designed gap range.

In the lithography experiment, a continuous-wave UV laser was focused down to a several-micrometer spot onto a PL, which further focused the beam to a sub-100 nm spot onto the spinning disk for writing of arbitrary patterns (Fig. 10.13). The laser pulses are controlled by an electro-optic modulator according to the signals from a pattern generator. The writing position is referred to the angular position of the disk from the spindle encoder and the position of a piezostage along the radial direction. We use an inorganic TeO_x-based thermal photoresist [28] deposited on a glass disk by magnetic sputtering. A spindle was used to rotate the disk at 2000 rpm, which is equivalent to the linear speed of 10 m/s at the outer radius. After pattern writing and development in diluted KOH solution, the patterns were examined using an atomic force microscope (AFM). The result demonstrated that we can achieve high-speed patterning with 80 nm line widths at 10 m/s (Fig. 10.13a). Figures 10.13b and 10.13c demonstrate successful patterning of arrays of the acronym "SINAM" with a feature size of 145 nm. The pattern writing in the radial direction involved a coordinate transformation with Cartesian coordinates. The resolution can be improved by careful design using a shorter plasmon wavelength and guiding mechanisms, and theoretical simulation shows it can reach down to 5–10 nm. Due to the fast scanning, a single PL already has higher throughput than most other maskless lithography approaches. The throughput of PNL can be greatly enhanced by employing a larger number of PLs for parallel writing. Let us consider a 1000-lens array occupying the area of 800 μm × 20 μm at the bottom of the ABS, with each PL being 4 μm in diameter. Taking into account the changes of the mean flying height, as well as pitch and roll angles at different linear velocities, our simulation shows that the corresponding flying height variation for all the PLs is in the range from 18 to 24 nm, which is well within the acceptable range of 0–30 nm. Thus, at the scanning speed of 10 m/s, a plasmonic flying head carrying 1000 lenses can write a

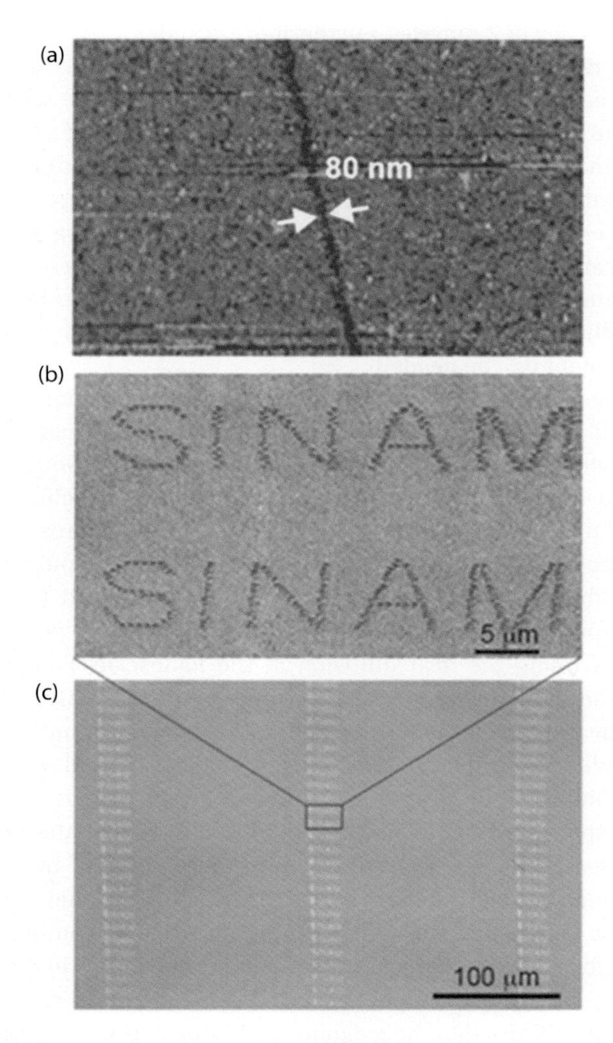

Figure 10.13 Maskless lithography by flying plasmonic lenses at the near field. (a) AFM image of a pattern with a 80 nm line width on the TeO_x-based thermal photoresist. (b) AFM image of arbitrary writing of "SINAM" with a 145 nm line width. (c) Optical micrograph of patterning of large arrays of "SINAM."

12-inch wafer in 2 minutes. Furthermore, a slider a few millimeters in size may take up to 100,000 lenses. Flying PL arrays at the optical near field enables the agile maskless nanoscale fabrication with the potential of a throughput 2 to 5 orders of magnitude higher than conventional maskless techniques. In future industrial implantations of this technology, engineering challenges must be addressed, such as pattern data management, lithography line width control, pattern overlay, and resist defect reduction, which are common for all maskless lithography approaches. Integrated approaches for precision engineering, metrology, and new resist development will be needed.

In the application of PNL, one important consideration is the trade-off between spatial confinement and inherit loss of SPPs. Nanoscale thermal management can help to efficiently squeeze light into the deep subwavelength scale and achieve nanolithography with 22 nm resolution using a 355 nm pulsed laser source.

Instead of a continuous-wave laser we used a pulsed laser with a thermal resist to achieve high resolution and patterning throughput by lowering the required operating laser power level and controlling the heat diffusion at the nanoscale. Figure 10.14 shows the simulated temperature profiles in the thermal resist layer under heating from the optical field of focused plasmons from the lens under two different laser pulses. With 10 ps pulses, we can further improve the feature size down to 22 nm (about half of the optical spot of 45 nm focused by the PL) and reduce the required laser average power from 105 mW to merely 9 mW by utilizing the nonlinear and time-dependent response of the thermal resist [28]. The high-speed plasmonic writing involves the competition of optical absorption at the nanoscale, heat accumulation, and thermal diffusion. The energy deposited into the nanoscale resist volume can rapidly diffuse into the neighboring region within a nanosecond, which enlarges the exposed features, increases the required laser power, and causes pattern distortion. Therefore, the pulsed laser has great advantages over the continuous-wave laser for ensuring good thermal confinement in the resist layer. Application of the pulsed laser also allows the employment of a PL array for parallel patterning.

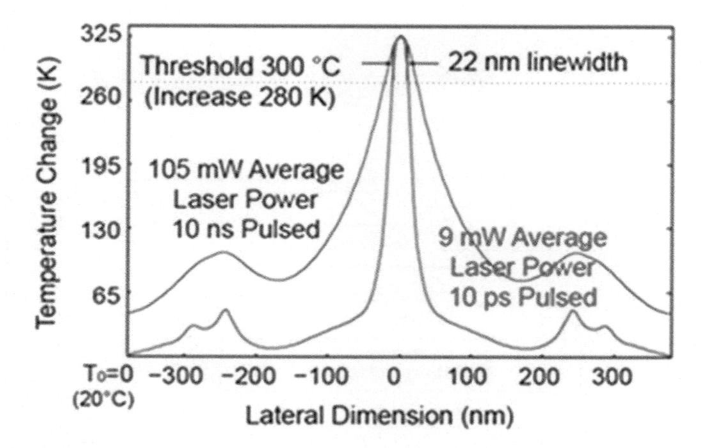

Figure 10.14 Temperature profiles in the thermal-type resist layer under PL heating. Two different time durations of a laser pulse of 10 ps (red) and 10 ns (blue), respectively, have been used in the numerical study. By properly controlling the laser power level and pulse duration, we can further improve the feature size down to 22 nm. The picosecond-pulsed laser has great advantages over the modulated continuous-wave laser in terms of pattern size and contrast.

In the experiment, a 10 mW picosecond pulsed UV laser beam (Spectra-Physics, Vanguard, 355 nm wavelength, 12 ps pulse duration, 80 MHz repetition rate, which is externally doubled to 160 MHz) was used as the exposure light source to manage the critical laser dose and thermal diffusion in order to achieve good pattern contrast, uniformity, and small feature size. The laser pulse train was also probed by an ultrafast photodetector to provide the external clock for the FPGA-based pattern generator for the purpose of improving the pattern stitching accuracy. During the lithographic process, a spindle was used to rotate the substrate with the resist at 2500 rpm, corresponding to disk speeds of 4–14 m/s at different radii. The ABS design used in this experiment has a size of a few millimeters and can carry arrays of PLs (up to 16,000) at sub-10 nm above a resist surface with sub-1 nm variation. The pitch and roll angles of the flying head are kept consistent at 40 μrad and sub-1 μrad, respectively. The array of PLs (SEM image shown in Fig. 10.12b) was later fabricated by FIB milling on a 60 nm

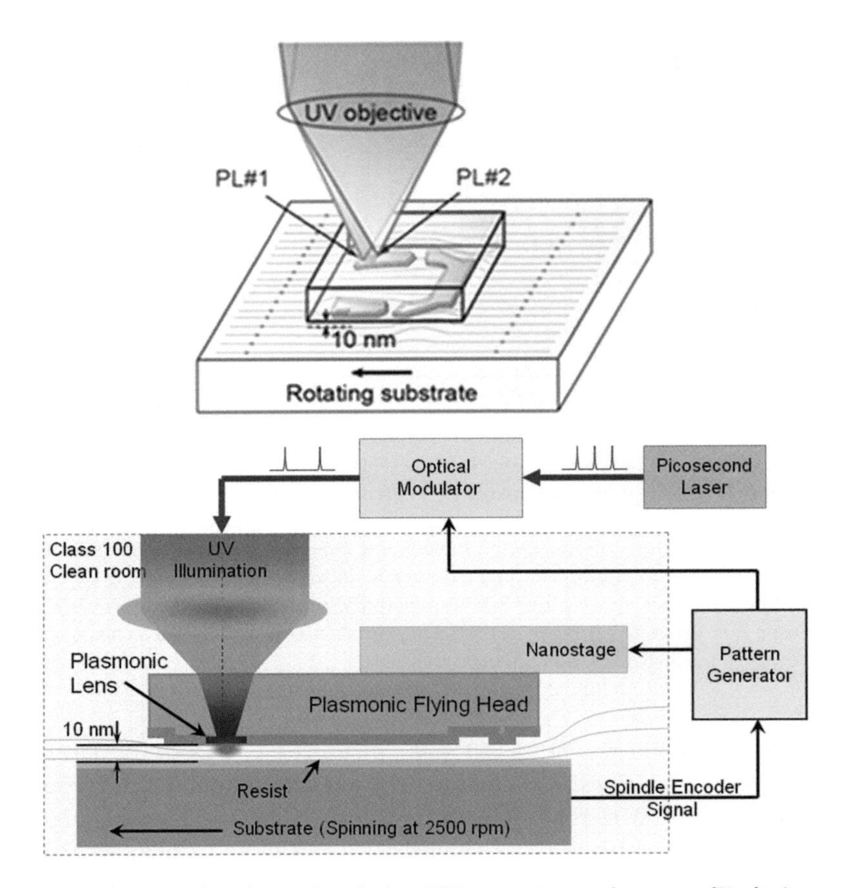

Figure 10.15 A schematic of the PNL experimental setup. (Top) An individual PL is optically addressed using the same UV objective. (Bottom) An advanced air-bearing surface (ABS) technology is used to maintain the gap between the lens and the substrate at 10 nm. A pattern generator is used to pick the laser pulses for exposure through an optical modulator according to the angular position of the substrate from the spindle encoder and the radial position of the flying head from a nanostage.

thick chromium (Cr) thin film coated on the ABS. As schematically shown in Fig. 10.15, each of the PLs can be controlled using an independent laser beam in order to enable high-throughput parallel writing. Through a single UV objective, a few individually modulated laser pulse trains were first focused down to separated spots with a diameter of several micrometers to illuminate the area of the designated PL structure on the ABS (at the bottom of the flying head). Then, each of the PLs further focuses the incident light into a nanoscale spot for patterning the resist layer. The information of the relative position between the flying head and the substrate is provided by the spindle encoder (angular) and a linear nanostage (radial), which feeds to a home-made pattern generator to pick the laser pulse for exposure through an optical modulator. During the test, an interferometry setup and an acoustic emission sensing module were installed to monitor the real-time motion of the flying head during the lithography process. The resist used in our test is $(TeO_2)_x Te_y Pd_z$ ($x \approx 80$ wt%, $y \approx 10$ wt%, $z \approx 10$ wt%), an inorganic thermal type developed on the basis of the Te-TeO_x resist. Pd is added to the Te-TeO_x in order to enhance the exposure uniformities and resist resolution by forming finer crystalline grains during phase transition, and its thermal stability is also beneficially improved. This inorganic resist is employed also because of its good mechanical properties for tribological considerations and good sensitivity for high resolution.

After the PNL experiment, the exposed patterns were developed in diluted KOH solution and examined using an AFM. We experimentally demonstrated high-throughput direct writing at 22 nm half-pitch and the parallel patterning. Figures 10.16a and 10.16b show the AFM image of closely patterned dots with 22 nm half-pitch resolution by PNL with the cross-sectional scan (Fig. 10.16c). The results are in agreement with the experimental conditions (substrate velocity at 7 m/s), and each dot is generated by a single laser pulse. Similar to other maskless approaches, writing at a higher laser power or higher pattern spatial frequency allows the dots to merge into continuous lines with different widths. The result is shown in Fig. 10.16d, where two PLs in the lens array independently write on the thermal resist in parallel. To obtain 50 nm wide solid lines, PL#1 was excited with a laser power twice that used for Fig. 10.16a, and PL#2 simultaneously used a ramping laser power varying from 2 to 4 times. It is shown that the exposure feature size can be controlled by regulating the laser power during the pattern writing. It should also be noted that the pattern definition

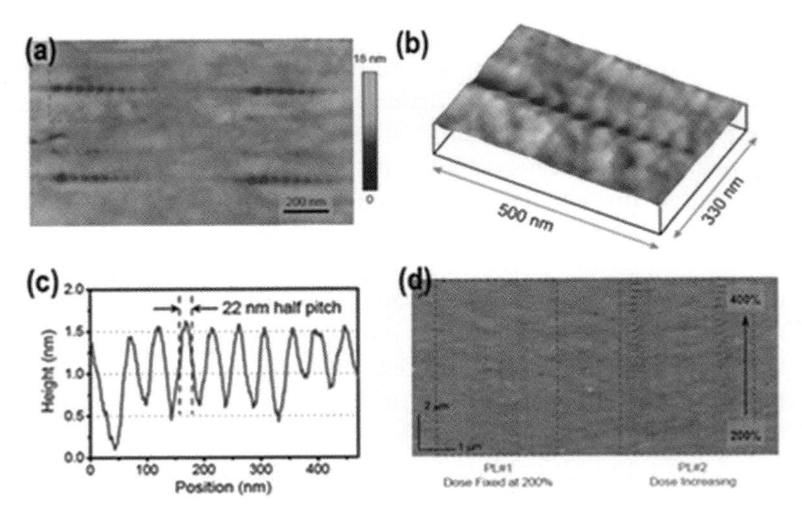

Figure 10.16 AFM image of closely packed dots at 22 nm half-pitch on the thermal resist. (a) AFM image of four trains of dot lines, (b) 3D topography of the boxed dot line in (a), and (c) cross-sectional profile of the dot line in (b). (d) AFM image of a PNL parallel writing result on the thermal resist. Two of the PLs in an array were used to simultaneously write independent patterns, capital letters "PI" and "LI," respectively. PL#1 used a fixed laser power 2 times that used in generating the result in (a), and PL#2 used an increasing power varying from 2 to 4 times. As the laser power level further increases, the side lobe patterns from the PL start to show on the resist layer.

can be greatly improved by the optimization of the resist exposure threshold and postdevelopment conditions.

10.5 Summary

In summary, we have demonstrated high-speed PNL with 22 nm half-pitch resolution. This is achieved by employing plasmon focusing through relatively low-loss propagating surface plasmons focusing and later conversion to localized plasmons. This allows the highly efficient transmission and focusing of the near-field spot, which is the key to improving the throughput, for a given laser power, by increasing the scanning speed, and/or by employing a great number of PLs and flying heads for parallel patterning. The resolution can be further improved to sub-10 nm by utilizing a shorter SPPs wavelength and guiding mechanisms [29]. Also, the lithography throughput can be dramatically enhanced by increasing the scanning speed and employing a greater number of PLs and flying heads for parallel patterning.

Due to the extremely fast scanning, a single PL at a speed of 10 m/s already has a much higher throughput than most other maskless lithography approaches. In principle, this scheme allows a single flying head to carry up to 16,000 PLs, which can pattern a 12-inch wafer in less than 1 minute. Furthermore, a slider a few millimeters in size may take up to 100,000 lenses. This is comparable to conventional production-level photolithography but at a much higher resolution of 22 nm half-pitch size. This new scheme enables low-cost, high-throughput maskless nanoscale fabrication with a throughput a few orders of magnitude higher than conventional maskless approaches.

It opens a new route toward next-generation lithography (NGL) not only for the electronics industry but also for the emerging nanotechnology industry. Besides its application in nanolithography, this technique can also be used for nanoscale metrology and imaging. Furthermore, it has strong potential to facilitate next-generation magnetic data storage, known as heat-assisted magnetic recording (HAMR) and bit-patterned media (BPM), to achieve capacities 2 orders of magnitude higher in the future [30, 31]. In future industrial implementation of this technology, some engineering challenges must be addressed, such as pattern data management, lithography line width control, pattern overlay, and resist defect reduction, which are common for all maskless lithography approaches. Integrated approaches for precision engineering, metrology, and new resist development will be needed. Such a low-cost, high-throughput scheme promises a new route toward next-generation nanomanufacturing.

References

1. Okazaki, S. (1991). Resolution limits of optical lithography, *J. Vac. Sci. Technol. B*, **9**(6), pp. 2829–2833.

2. Piner, R. D., et al. (1999). "Dip-pen" nanolithography, *Science*, **283**(5402), pp. 661–663.
3. Groves, T. R., and Kendall, R. A. (1998). Distributed, multiple variable shaped electron beam column for high throughput maskless lithography, *J. Vac. Sci. Technol. B*, **16**(6), pp. 3168–3173.
4. Pease, R. F., et al. (2000). Prospects for charged particle lithography as a manufacturing technology, *Microelectron. Eng.*, **53**(1–4), pp. 55–60.
5. Menon, R., et al. (2005). Maskless lithography, *Mater. Today*, **8**(2), pp. 26–33.
6. Salaita, K., et al. (2006). Massively parallel dip-pen nanolithography with 55000-pen two-dimensional arrays, *Angew. Chem. Int. Ed.*, **45**(43), pp. 7220–7223.
7. Li, L., et al. (2009). Achieving $\lambda/20$ resolution by one-color initiation and deactivation of polymerization, *Science*, **324**(5929), pp. 910–913.
8. Scott, T. F., et al. (2009). Two-color single-photon photoinitiation and photoinhibition for subdiffraction photolithography, *Science*, **324**(5929), pp. 913–917.
9. Andrew, T. L., Tsai, H.-Y., and Menon, R. (2009). Confining light to deep subwavelength dimensions to enable optical nanopatterning, *Science*, **324**(5929), pp. 917–921.
10. Srisungsitthisunti, P., Ersoy, O. K., and Xu, X. F. (2011). Improving near-field confinement of a bowtie aperture using surface plasmon polaritons, *Appl. Phys. Lett.*, **98**(22), p. 223106.
11. Lezec, H. J., et al. (2002). Beaming light from a subwavelength aperture, *Science*, **297**(5582), pp. 820–822.
12. Liu, Z. W., et al. (2007). Far-field optical hyperlens magnifying sub-diffraction-limited objects, *Science*, **315**(5819), pp. 1686–1686.
13. Valaskovic, G. A., Holton, M., and Morrison, G. H. (1995). Parameter control, characterization, and optimization in the fabrication of optical-fiber near-field probes, *Appl. Opt.*, **34**(7), pp. 1215–1228.
14. Betzig, E., et al. (1992). Near-field magnetooptics and high-density data-storage, *Appl. Phys. Lett.*, **61**(2), pp. 142–144.
15. Jin, E. X., and Xu, X. F. (2006). Enhanced optical near field from a bowtie aperture, *Appl. Phys. Lett.*, **88**(15), p. 153110.
16. Sendur, K., and Challener, W. (2003). Near-field radiation of bow-tie antennas and apertures at optical frequencies, *J. Microsc.-Oxford*, **210**, pp. 279–283.

17. Sundaramurthy, A., et al. (2006). Toward nanometer-scale optical photolithography: utilizing the near-field of bowtie optical nanoantennas, *Nano Lett.*, **6**(3), pp. 355–360.
18. Xu, J. Y., et al. (2005). Design tips of nanoapertures with strong field enhancement and proposal of novel L-shaped aperture, *Opt. Eng.*, **44**(1), p. 018001.
19. Matsumoto, T., et al. (2006). Writing 40 nm marks by using a beaked metallic plate near-field optical probe, *Opt. Lett.*, **31**(2), pp. 259–261.
20. Srituravanich, W., et al. (2008). Flying plasmonic lens in the near field for high-speed nanolithography, *Nat. Nanotechnol.*, **3**(12), pp. 733–737.
21. Liu, Z. W., et al. (2005). Focusing surface plasmons with a plasmonic lens, *Nano Lett.*, **5**(9), pp. 1726–1729.
22. Yin, L. L., et al. (2005). Subwavelength focusing and guiding of surface plasmons, *Nano Lett.*, **5**(7), pp. 1399–1402.
23. Itagi, A. V., et al. (2003). Ridge waveguide as a near-field optical source, *Appl. Phys. Lett.*, **83**(22), pp. 4474–4476.
24. Sendur, K., Peng, C., and Challener, W. (2005). Near-field radiation from a ridge waveguide transducer in the vicinity of a solid immersion lens, *Phys. Rev. Lett.*, **94**(4), p. 043901.
25. Shi, X. L., and Hesselink, L. (2002). Mechanisms for enhancing power throughput from planar nano-apertures for near-field optical data storage, *Jpn. J. Appl. Phys. Part 1*, **41**(3B), pp. 1632–1635.
26. Wang, Y., et al. (2008). Plasmonic nearfield scanning probe with high transmission, *Nano Lett.*, **8**(9), pp. 3041–3045.
27. Juang, J. Y., Bogy, D. B., and Bhatia, C. S. (2007). Design and dynamics of flying height control slider with piezoelectric nanoactuator in hard disk drives, *J. Tribol.*, **129**(1), pp. 161–170.
28. Ito, E., et al. (2005). TeOx-based film for heat-mode inorganic photoresist mastering, *Jpn. J. Appl. Phys. Part 1*, **44**(5B), pp. 3574–3577.
29. Stockman, M. I. (2004). Nanofocusing of optical energy in tapered plasmonic waveguides, *Phys. Rev. Lett.*, **93**(13), p. 137404.
30. Pan, L., and Bogy, D. B. (2009). Heat-assisted magnetic recording, *Nat. Photonics*, **3**(4), pp. 186–187.
31. Kikitsu, A. (2009). Prospects for bit patterned media for high-density magnetic recording, *J. Magn. Magn. Mater.*, **321**(6), pp. 526–530.

Chapter 11

Plasmonic Nanoresonators for Spectral Color Filters and Structural Colored Pigments

Yi-Kuei Wu, Jing Zhou, Kyu-Tae Lee, Ting Xu, Cheng Zhang, and L. Jay Guo

Department of Electrical Engineering and Computer Science, University of Michigan, Ann Arbor, MI 48109, USA

guo@umich.edu

11.1 Introduction and Motivation

Color perception is an important part of our daily life and is normally achieved by incorporating colored pigments or dyes into objects or display devices. These pigments elements produce colors because they absorb a certain band of visible light. In the context of display devices, chemical pigments are widely used as color filters in liquid crystal displays (LCDs). Current color filter technology has room to improve mainly in two aspects: simplifying the manufacturing process and improving brightness [1–3]. In nature, colors can also be produced by light interacting with physical structures, for

Plasmonics and Super-Resolution Imaging
Edited by Zhaowei Liu

ISBN 978-981-4669-91-7 (Hardcover), 978-1-315-20653-0 (eBook)
www.panstanford.com

example, the blue color of the Morpho butterfly is due to the photonic crystal-like volume-diffractive nanostructures on its wings. Inspired by these natural phenomena, structural colors can be produced by exploiting light interaction with photonic crystals [4, 5] and plasmonic nanostructures [2, 3, 6–9]. In the past years many new phenomena have been reported in plasmonic nanostructures and nanocavities and have gathered considerable interest [10–16]. By exploiting plasmonic nanostructures, such as nanohole or nanoslit arrays, efficient conversion between incident and outgoing plane waves and plasmons can be managed at the subwavelength scale and produce color-filtering functions [17]. For example, the resonance effect in a plasmonic nanohole array for filtering color has been reported by sweeping the resonant transmission peaks at the visible spectrum [9]. However, the transmission band of such filters is too broad to meet the bandwidth requirement for display and imaging applications. Other approaches for spectral filtering such as nanoslits combined with periodic grooves [8] or in a metal-insulator-metal (MIM) waveguide [18] also demonstrated color filtering. However, in these structures, two neighboring output slits have to be separated by additional structures or by specific coupling distances; therefore, the device dimension and efficiency are restricted. An even greater limitation is the low efficiency of the plasmonics-based spectrum filters reported previously (typically less than 10%), which cannot satisfy the requirement for practical display applications. Among the above efforts, filters generated by MIM waveguide resonators are of particular interest. MIM waveguide geometries have the ability to support surface plasmon (SP) modes at visible wavelengths and have been widely investigated for various applications, such as guiding waves at the subwavelength scale [19–22], concentrating light to enhance the absorption for photovoltaic applications [23, 24], achieving a near-field plate for super-resolution at optical frequency [25–28], or composing metamaterials for magnetic resonance and negative refraction [25–28]. In addition to enabling efficient subwavelength optical confinement, the top and bottom metal layers of MIM structures can be potentially used as electrodes in an electro-optic system for a compact device size. In this chapter, we introduce a number of structural color designs, and many of them can be traced back to the basic MIM structure. Several important aspects such as color purity, efficiency, and angle dependence will be discussed and correlated to the physical parameters of the structure.

11.2 Transmission Filters Based on MIM Nanoresonators

In this section, we discuss the design of plasmonic MIM nanoresonators capable of spectrum filtering for various colors across the entire visible band. The key concept is to use nanoresonators to realize the photon-plasmon-photon conversion efficiently at specific resonance wavelengths. The new design significantly improved transmission efficiency, pass bandwidth, and compactness. Moreover, the filtered light is polarized because of plasmonics' nature, and therefore it is very attractive for direct integration in LCDs to reduce the number of stacks in the device.

Figure 11.1a shows the schematic diagrams of MIM nanoresonators. The device is designed as a subwavelength periodic MIM stack array on a magnesium fluoride (MgF_2) transparent film with period P (Fig. 11.1a). For each MIM stack, a 100 nm thick zinc selenide (ZnSe) layer is sandwiched by two 40 nm aluminum (Al) layers. The 100 nm thick ZnSe layer ensures the efficient coupling of SP modes at the top and bottom edges of the stack, whereas the two 40 nm thick Al layers support the MIM mode, as well as prohibiting the direct transmission of incident light. Additionally the bottom Al grating is used to couple selectively the incident light into plasmon waveguide modes by diffraction, whereas the top Al grating efficiently reconverts the confined plasmons to propagating waves by scattering and transmits the light to the far field in the forward direction. Here we only consider the normal incidence, and therefore the stack period is related to the plasmon transverse wavevector as $P = 2\pi/kx$ by the $\pm$ first-order diffraction based on the momentum conservation law. It is clear that the SP antisymmetric mode has linear dispersion across the entire visible range. This linear dispersion made it very easy to design filters for any colors across the entire visible spectrum range. The corresponding period

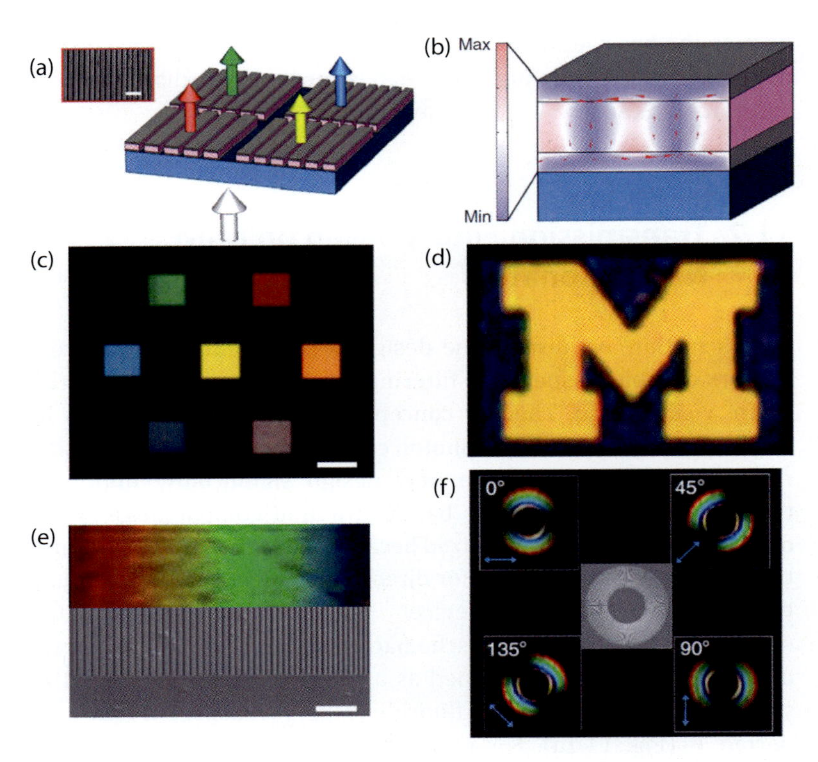

Figure 11.1 (a) Schematic of plasmonic nanoresonators formed by MIM stack arrays. (b) Simulated time average magnetic field intensity (H-field and electric displacement [D-field], arrow) profiles in one MIM stack). (c) Optical microscopic images of seven plasmonic color filters illuminated by white light. Scale bar: 10 μm. (d) Yellow "M" logo in a navy-blue background, (e) gradually changing the periods of the plasmonic nanoresonator array from 200 to 400 nm. (f) $2D$ spoke structure to demonstrate polarization and color effect for RGB.

of the stack is 360, 270, and 230 nm. On the other hand, the transverse electric (TE)-polarized light (the **E**-field is parallel to the Al wire direction) is mostly blocked since this plasmonic mode does not support the excitation of SP modes. This indicates that the proposed transmission color filters can simultaneously function as polarizing optical components, a highly desirable feature for display applications. Figure 11.1b shows the simulated time average magnetic field intensity (H-field and electric displacement [D-field], arrow) profiles in one MIM stack. The transverse magnetic (TM)-polarized incident light has a resonance peak wavelength of 650 nm and the stack period is 360 nm. The magnetic field intensity shows that most of the incident light is coupled into antisymmetric waveguide modes with the maximal intensity near the edges of both top and bottom Al gratings, which supports our design principle. From the electric displacement distribution, we can see that efficient coupling to the SP antisymmetric modes is realized by the strong magnetic resonance response in each sandwiched MIM stack, as in the case of a previous report [26] in which the electric displacement field forms a loop and results in strong magnetic fields opposing that of the incident light inside the dielectric layer.

Figure 11.1c shows the optical microscopy images of the seven 10 μm × 10 μm square-shaped plasmonic color filters illuminated by white light. The filters are fabricated using focused ion beam (FIB) milling of a deposited Al/ZnSe/Al stack on an MgF_2 substrate. The color filters have the stack period changing from 200 to 360 nm, corresponding to the color from violet to red. For TM illumination, stack arrays show the expected filtering behavior with absolute transmission over 50% around the resonant wavelengths, which is several orders of magnitude higher than those of previously reported MIM resonators [18]. This transmission is comparable with the prevailing pigment color filter used in an LCD panel, but the thickness of the plasmonic device is 1 or 2 orders of magnitude thinner than that of the pigment one. The relatively low transmission efficiency for the blue color filter is due to larger material loss of ZnSe in the shorter-wavelength range. The full-width at half-maximum (FWHM) of the resonance in this case is about 100 nm for all three colors. On the other hand, these devices strongly reflect TE-polarized light, as in wire-grid polarizers [29]. This reflected TE feature indicates not only that the multifunctionality of color filters and polarizer could greatly benefit the LCD by eliminating the need of a separate polarizer layer but also that the reflected TE increases light recycling for the LCD panel and reduces the power consumption potentially. In addition, the conductive nature of the Al grating also implies that a separate transparent conductive oxide layer used in the LCD module may not be necessary, as was demonstrated in our previous work on metal-wire-based transparent electrodes [30].

Besides the standard square color filters, one can use different nanoresonator arrays to form arbitrary colored patterns on a micrometer scale. As an example, a yellow character "M" in a navy-blue background is shown in Fig. 11.1d. The pattern size of the "M" logo measures only 20 μm × 12 μm and uses two periods: 310 nm for the yellow letter M and 220 nm for the navy-blue background. It is important to note that the two distinct colors are well preserved even at the sharp corners and boundaries of the two different patterns, which indicates that the color filter scheme can be extended to ultrahigh-resolution color displays. In Fig. 11.1e, by gradually changing the periods of the plasmonic nanoresonator array from 200 to 400 nm, a plasmonic spectroscope for spectral imaging can be produced.

Plasmonic spectroscopes can disperse the whole visible spectrum in distances of just a few micrometers, which are orders of magnitude smaller than the dispersion of the conventional prism-based device. This feature indicates that the color pixels formed by these structures could provide high spatial resolution for application in multiband hyperspectral imaging systems. Our thin-film stack structures can be directly integrated on top of focal plane arrays to implement high-resolution hyperspectral imaging or to create chip-based ultracompact spectrometers. The 2*D* spoke structure shown in Fig. 11.1f consists of 96 slits that form a circular ring. The spacing between neighboring slits changes from 200 nm in the center to 400 nm toward the outer edge of the ring, covering all colors in the visible range as the above linear spectroscope. When illuminated with polarized light and when the polarization is rotated, a clear, dark region appears along the polarization direction. If used with an imaging device, such a structure could provide us with real-time polarimetric information in spectral imaging, or it can be used as a microscale polarization analyzer.

Human eyes typically have a resolution limit of about 80 μm at 350 mm. Therefore, one can use these plasmonic nanoresonators to build colored "superpixels" that are even finer than current retina display pixels. These plasmonics-based color pixels are only several micrometers in a lateral dimension and are much smaller than the resolution limit of the human eyes. The lateral dimension of plasmonic color pixels is 1–2 orders of magnitude smaller than the best high-definition color filters currently available. Furthermore, these plasmonic devices have a longitudinal thickness that is 1–2 orders of magnitude thinner than that of colorant ones and their device multifunctionality (combining electrode, polarizer, and color filter), which are both very attractive for the design of ultrathin panel display devices. As a proof-of-principle, recently we have demonstrated an ITO-free colored LCD structure by using a MIM color filter and grapheme as the electrode to the control the orientation of liquid crystals that can turn on or off the liquid crystal cell [31]. Besides the LCD application, hyperspectral and polarimetric imaging are also promising applications of this plasmonics-based device.

11.3 Metallic Resonant Waveguide Grating Color Filters

The aforementioned works [2, 3] demonstrate that plasmonic color filtering is a promising technology to improve visual display devices. The pigment color filters used in the LCD industry are based on light absorption and the process of making them is costly. Out of tens of components, the manufacturing cost of pigment color filters accounts for 20–30% of the total LCD panel cost due to the complicated manufacturing process. Figure 11.2 shows the process flow for color filters [32], which requires more than 20 process steps. Another issue of current pigment color filters is absorption loss. Currently only 6% light from the backlight unit (BLU) reaches out of the visual display device, and most of the light is absorbed by colorants in the color filters. The structural colors can potentially address these issues by exploiting optical resonances and with fewer fabrication steps involved.

High peak transmission and efficiency of a thin-film color filter is of primary interest. Depending on the specific applications some technologies desire simultaneous polarization of the transmitted light, while others may require narrow spectral peaks for high-

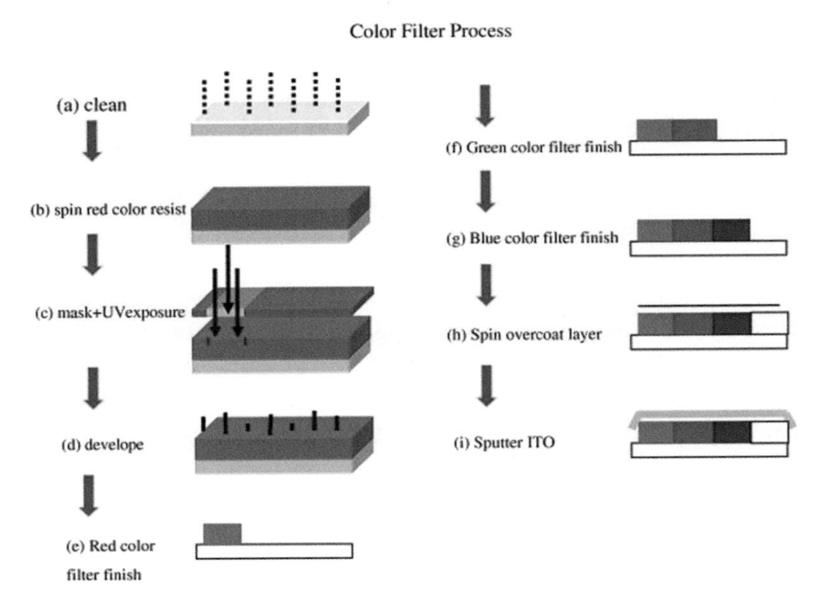

Figure 11.2 Manufacturing process flow for RGB color filter. Reprinted from Ref. [32], Copyright (2008), with permission from Elsevier.

purity colors. The MIM grating structure presented above was fabricated using an FIB. But for large-scale production, we employ nanoimprint lithography to fabricate these structures over large areas. For this purpose we have designed an alternative structure in the form of a metal resonant waveguide grating (MRWG) that is better suited for such a fabrication process.

Here we present a hybrid plasmon/waveguide structure with a single patterned Al layer on top of a dielectric deposited on a glass substrate. This robust, easy-to-fabricate structure can produce various spectra depending on the desired application by altering the period, line width, and thickness of the Al grating or by changing the continuous dielectric layer.

The design is shown in Fig. 11.3a. The metallic grating is placed on a stack of dielectric layers on a glass substrate. The stack of thin films contains a buffer layer and a waveguide layer, the refractive indexes of which are 1.5 (silicon dioxide) and 2.0 (silicon nitride), respectively. The filtered color across the entire visible spectrum

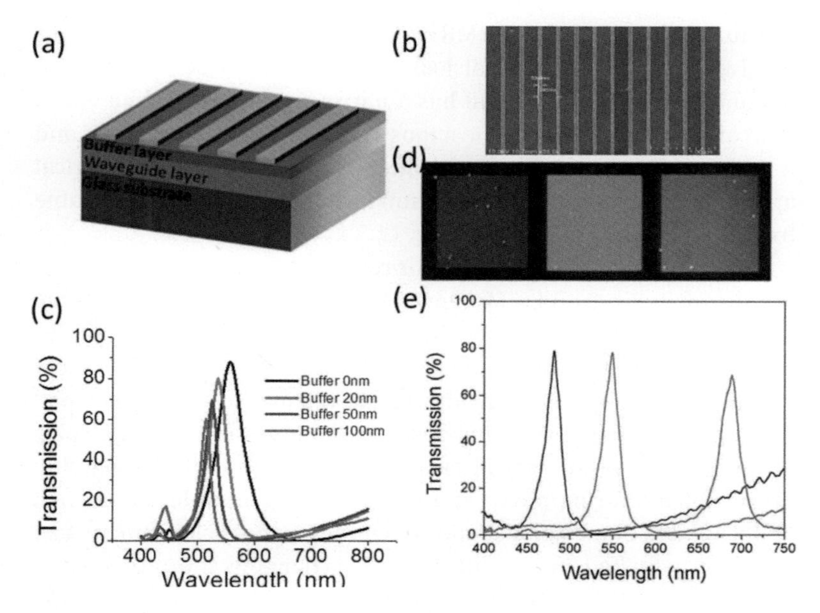

Figure 11.3 (a) Schematic of the metal resonant waveguide grating. (b) The simulated spectrum with the bandwidth between 15 nm and 50 nm by changing buffer layer thickness from 0 to 100 nm. (c) SEM top view of the fabricated device with a period of 300 nm and a gap of 54 nm. (d) Normal incidence transmission images and (e) measured spectra for three square arrays of metal resonant waveguide gratings for blue, green, and red colors. The grating periods are 300, 350, and 450 nm, respectively, with 0.25 duty cycles. The thicknesses of the silica buffer layer and silicon nitride waveguide layer are 50 and 100 nm, respectively.

can be achieved by controlling the period of the grating. Moreover, the line width of the transmission resonance can be adjusted from 10 nm to 50 nm by changing the thickness of the buffer layer. This structure bears similarity to the dielectric resonant waveguide grating (DRWG) consisting of a dielectric grating and a waveguide structure to allow light diffracted by the grating and to couple into the waveguide modes [33, 34]. Because the dielectric medium is lossless and transparent, the transmission spectrum of the DRWG is notch type with an extremely narrow resonance band, typically less than 1 nm. On the other hand, the MRWG structure has the subwavelength grating made of metal rather than of a dielectric medium. Unlike DRWGs, the MRWG structure has a wider resonance bandwidth due to the metal loss, which is about several tens of nanometers. This MRWG still has a narrower resonance line width than other types of plasmonic nanostructure–based color filters and is suitable for the purity of the filtered color required for optical applications. Theoretically, the resonance line width can be tunable from 50 to 15 nm, as shown in Fig. 11.3b.

The measured spectra and color response results for RGB design are shown in Fig. 11.3d,e. Moreover, the resonance bandwidth in the MRWG filter can be tuned by changing the thickness of the buffer layer, which would further increase its practicality [35]. We show in Fig. 11.3e that the bandwidth of the resonance can be as small as 30 nm, indicating that this design guarantees outstanding color purity. Moreover, for real filter applications, the FWHM tunability of the color filtering pass band is a bonus for this device. In real applications, the display panel brightness and color purity can be easily adjusted in this structure by changing the buffer layer thickness to change the FWHM of the pass band. Figure 11.4a shows a design for manufacturable RGB color filters for visual displays. The periods for blue, green, and red are 240, 320, and 420 nm, respectively, where the gaps are 50, 80, and 110 nm, respectively. The calculated spectra for blue, green, and red are shown in Fig. 11.4b.

We demonstrated RGB colors on a 4-inch glass wafer in Fig. 11.4c. The fabrication process of this device is as follows: (1) defining the photolithography pattern with the pitch and duty cycle from the modeling, (2) reactive ion etching (RIE) into the silver film, and (3) striping the photoresist. The color demonstration and simple process for RGB colors show the possibility of simplifying the manufacturing process and reducing its cost. However, this demonstration still has room to improve. Blue and green color filters show good color purity, but the red filter has the blue color component being mixed. The mixed color is proved to result from nonideal trench line width control in the photolithography step.

High resolution, slim dimension, and better power management are the final goals for the development of visual display technologies. Moving toward commercialization, two aspects should be focused on: first, further improvement of their optical performances, that

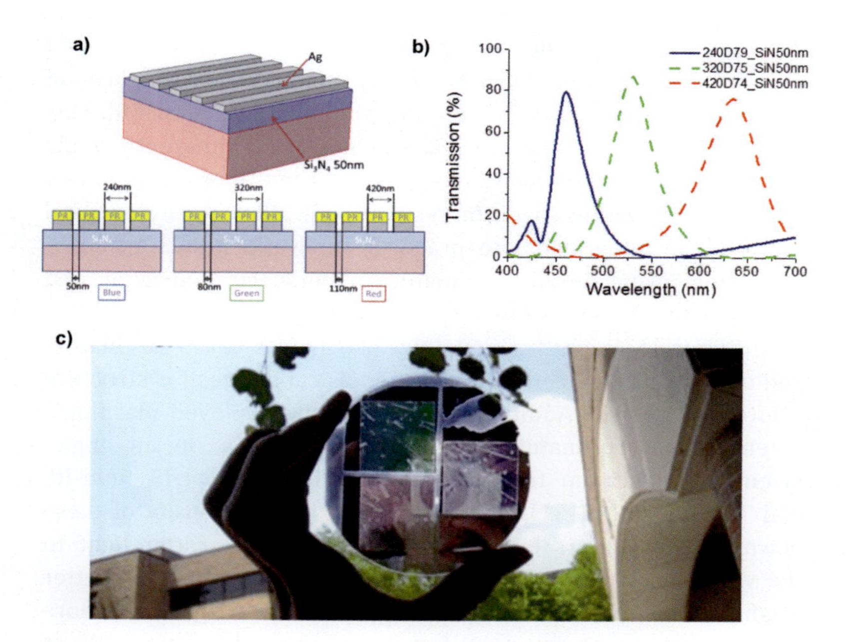

Figure 11.4 (a) The design of manufacturable RGB color filters for visual displays, (b) calculated spectra for RGB colors, and (c) demonstration of RGB colors on a 4-inch glass wafer.

is, transmission efficiency, color purity, and the incident angle independency, and second, development of a more efficient, high-throughput nanofabrication method to realize mass production with low costs. For this purpose, recently developed roll-to-roll-based nanopatterning processes [36, 37] are especially attractive due to their ability to pattern the nanograting structures over large areas with high throughput.

11.4 Angle-Insensitive Plasmonic Spectrum Filtering

The structural colors presented above are attractive to the LCD industry as a potential replacement of the colorant pigments. However, an important issue to address is the color change with different incident angles of light. For the structures described above, this angle dependence is directly related to surface plasmon polariton (SPP) excitation via grating coupling [2, 3, 6, 18, 38]. This has led to high coupling efficiencies [39], but is inherently angle dependent due to momentum matching conditions. Overcoming this angle-dependent spectrum response will allow these structural filters to be integrated into practical applications such as high-resolution visual displays, miniature hyperspectral imaging, and high-sensitivity sensors [40–42].

In contrast to grating coupling, plasmonics-based resonators and antennas have been demonstrated as candidates for structure colors [43, 44–48]. Additionally, horizontally deployed plasmonic antennas and resonators have gained attention on an angle-insensitive spectrum filtering response in the near-IR, mid-IR, and THz bands [49–51]. However, increasing the density of these plasmonic resonators is necessary to effectively scatter light to the viewers' eyes (or detector) at visible wavelengths for better brightness. To increase the scattering efficiency and generate more vivid colors, a vertical plasmonic resonator array has been realized for dense arrangement of plasmonic scatterers.

Next we present a design that exploits the light funneling into nanoslit resonators to generate strong absorption for TM-polarized incident fields [52–54]. Utilizing light funneling, extremely small physical dimensions could have large scattering cross sections and therefore possess plasmonic mode coupling with efficiency comparable to that of grating coupling. We theoretically and experimentally studied angle-robust optical devices with near-perfect absorption, as large as 96%, in the visible spectrum. Moreover, wide color tunability throughout the entire visible spectrum and a pixel size beyond the diffraction limit were demonstrated. We also suggest a design principle for angle-robust reflection by investigating the angular response of the reflection spectra with respect to the periodicity of arrayed 1D structures and further discuss the influence of periodicity and slit opening on field confinement within the nanocavities. More details are available in Ref. [55].

We first discuss metal-slit-based reflective-type colors, with a schematic and a SEM image of the fabricated structure shown in Fig. 11.5(a). To fabricate such a structure, a grating pattern is first imprinted into an MRI-8030 resist using a Nanonex 2000 tool with a soft mold. Use of a soft (or flexible) mold improves the imprinting throughput and quality compared to the case of a hard mold [56]. After de-molding, oxygen plasma is used to remove the residual layer and a lift-off process is carried out to transfer the pattern to Ni. Although the original mold has fixed dimensions, the line width of the Ni grating can be tuned by angled evaporation on the imprinted MRI grating before Ni deposition [57]. Next the grating pattern is etched into the silica substrate using the Ni mask. After removing the Ni mask, Ag is sputter-deposited onto the silica grating conformally, covering the etched silica walls. A fused silica substrate was chosen due to its desirable refractive index in the visible spectrum and Ag was chosen because of its lower optical loss in the visible regime.

The optical measurement is conducted using two systems to obtain the reflection/absorption spectra at normal and angled incidences. Reflection spectra at normal incidence are measured using a Nikon TE300 inverted microscope with a halogen lamp for the light source. Reflection spectra with angled incidence are measured from 45° to 75° (2° increment) with a J. A. Woollam M-2000 ellipsometer. The numerical apertures of the collection optics in both systems are as low as 0.04 to ensure high angular resolution.

As light is incident on the nanogroove array, an induced polarization charge pair accumulates at the top corners of the grooves. This charge pair acts as a dipole and further alters the **E**-field of incident light, redirecting the light into the groove. Figure 11.5b shows this effect with a red-blue surface plot and arrow plot which represent the normalized polarization charge distribution induced by the scattered field [58] and propagation direction of the field's Poynting vector, respectively. It is noticed that the purple arrows near the Ag and silica interface point toward the groove verifying this funneling effect. The intensity distribution of the magnetic field $|\mathbf{H}_y|^2$, under the funneling condition, is depicted in Fig. 11.5c, showing that the light is well confined in the groove at the resonant wavelength.

The above angle-insensitive color filtering results from the fact that light is funneled into metal-insulator-metal Fabry–Perot (MIMFP) cavity modes. By varying the optical path length in the MIMFP cavity through changing the width and depth of the

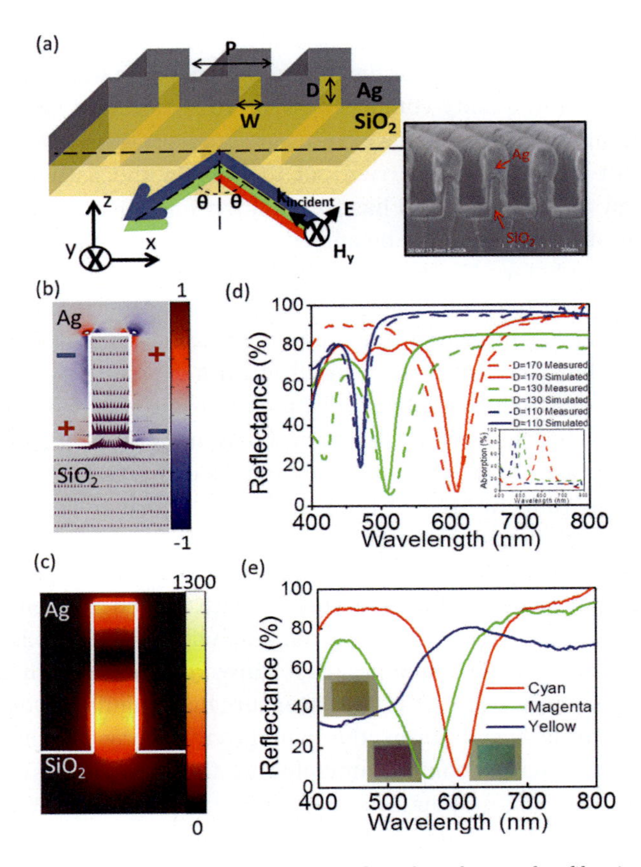

Figure 11.5 Plasmonic-nanocavity-based angle-robust color filtering. (a) A schematic of the proposed structure and corresponding scanning electron microscopy (SEM) image of a fabricated device with width $(W) = 45$ nm, depth $(D) = 160$ nm, and period $(P) = 180$ nm with θ as the incident angle of light. (b) Polarization charge and Poynting vector distribution of light funneled into these nanogrooves, presented with the red–blue surface plot and purple arrows, respectively. (c) Intensity distribution of the magnetic field $\mathbf{H}_y$ at resonance, $P = 180$, $W = 50$, and $D = 170$ nm. (d) Reflection (simulated in solid lines and measured in dash lines) and measured absorption spectra (inset, dashed lines) at $D = 110$, 130, and 170 nm in blue, green, and red curves, given fixed $P = 180$ and $W = 50$ nm at normal incidence. (e) Reflection spectra at fixed $P = 180$ nm and $D =$ 170 nm demonstrating the three basic colors of the CMY color model, cyan (C), magenta (M), and yellow (Y), with varying $W = 40$, 60, and 90 nm at normal incidence.

nanogrooves, the resonance of the structure is able to be tuned across the entire visible spectrum. The MIMFP resonant wavelength, λ, is determined by the effective refractive index, n_{eff}, and the depth, D, in the Fabry–Perot (FP) resonance equation, $(\frac{1}{4} + \frac{1}{2}m)\lambda = n_{\text{eff}}D$, where m is a positive integer and n_{eff} is the effective refractive index of MIM waveguide modes. The effective index is insignificantly dependent on the nanogroove thickness, D, but is highly dependent on acute changes in the width, W. The effective indices, n_{eff}, are calculated for changing values in width, W [52, 59]. For example, when $W = 50$ nm, the effective index dispersion relation of even modes in a MIM waveguide at insulator SiO_2 thickness is found to be $n_{\text{eff}} = 1.85$, 1.90, and 2.02 for wavelengths of $\lambda = 620$, 532, and 460 nm, respectively. When $W = 60$ nm, $n_{\text{eff}} = 1.91$, 1.99, and 2.11, respectively. This design can be integrated into various applications as well as accommodate different manufacturing processes, since there is a wide degree of freedom in modifying the filtered color by changing either D or W. In a set of simulations and experiments, the groove width, W, and period, P, are held constant at 50 nm and 180 nm respectively. The groove depths D corresponding to yellow, cyan, and magenta reflective colors are found to be 110, 130, and 170 nm, respectively. Figure 11.5d presents the simulated and measured reflection spectra (and measured absorption spectra in the inset) of the above three devices with different D at normal incidence and TM-polarized light. These devices are able to trap light as much as 96% at the resonance wavelength and reflect all other wavelengths. The optical propagation loss in silver at shorter wavelengths is nonnegligible, resulting in an 80% absorption peak at the shorter wavelength. This strong absorption at the selected wavelength range can be exploited for high-purity reflective color filtering. In addition to the color tuning based on D change, Fig. 11.5e shows that the three basic colors of the cyan, magenta, and yellow (CMY) color scheme can also be achieved by adjusting the width of the nanogroove, W, from 40 nm to 90 nm, given fixed $P = 180$ nm and $D = 170$ nm. It is noticed that the broad resonance dip on the yellow device results from larger propagation loss of the MIM waveguides at higher frequencies. A better reflection dip for yellow color can be achieved by a D change to 110 nm with $W = 50$ nm, as shown in the blue solid and dotted lines in Fig. 11.5d. This method of holding both the period and depth constant while varying the widths of the groove presents a more viable approach for manufacturing where the etching depth with RIE is held constant and the width of each colored pixel is altered, allowing multiple color pixels to be produced on a single wafer.

To demonstrate the visual performance of these color filters, we have designed and fabricated colored images in the format of the Olympic rings. Figure 11.6a and its insets show SEM images of the devices. The corresponding optical image of these reflective color filters is in Fig. 11.6b. The period, P, and depth, D, of each ring are held constant, while the widths, W, vary from 40 nm to 90 nm to generate the different colors. The angular dependence of these images was unable to be measured because of the small size of the images. The next section presents large-scale devices fabricated through nanoimprint lithography in which we were able to measure the angled reflection spectrum and confirm angle insensitivity. With this technique, cyan, magenta, and yellow, as well as intermediate colors can be achieved. Note that the purple color from the rope held by the gymnast in the top middle ring is produced by two nanogrooves. Moreover, the magenta color from the bow area is produced by several isolated and segmented short lines, as small as 100 nm in length and 60 nm in width, demonstrating ultrahigh color resolution that is beyond the diffraction limit of light. This opens up the possibility of realizing superpixels imaging [5], in which colors are mixed between multiple superpixels before their combined size is comparable to the diffraction limit. In addition, a ring with two colors has been fabricated in order to present an application of the polarization dependence of these filters. Figure 11.6c shows an image with two sets of gratings. As the polarization of incident light is changed, the displayed image is altered. Particularly, the central pattern in Fig. 11.6c is concealed when the incident light polarization changes from TM-polarized light to TE. This polarization dependence can be utilized in applications of cryptography and anticounterfeiting for personal identification cards, credit cards, and currency. Counterfeiters may be able to reproduce the color of the symbol but it would be increasingly difficult for them to make the image additionally dependent on polarization of light, adding another element to the

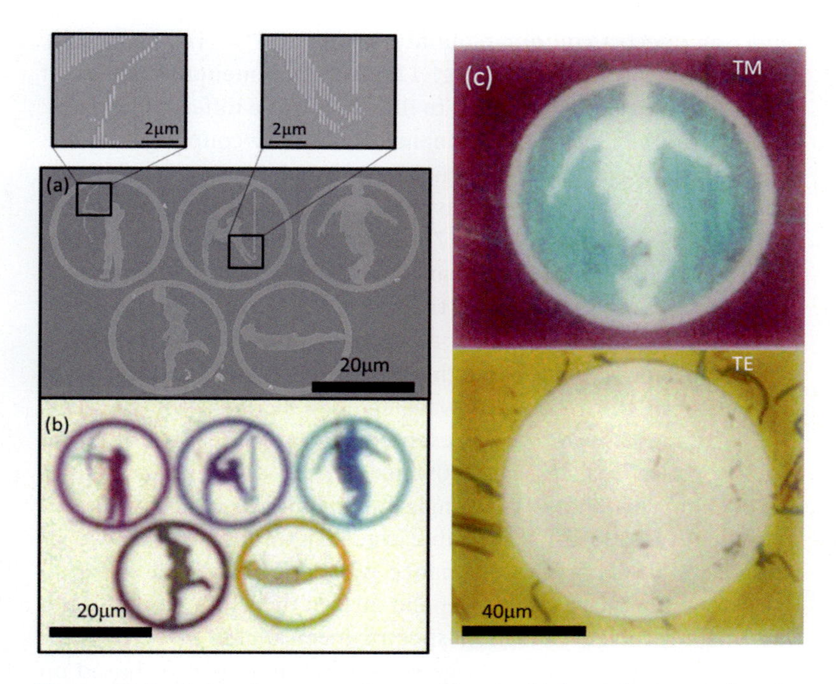

Figure 11.6 Color demonstrations with ultrahigh-resolution and polarization-dependent images. (a) The SEM images and (b) optical image under white light illumination of fabricated colored images in the format of the Olympic rings. The full range of CMY colors is achieved by sweeping W from 40 to 90 nm, with P and D fixed at 180 nm and 170 nm, respectively. (b) Even a single short-segmented nanogroove demonstrates color response. (c) Demonstrates utilizing polarization dependence to actively change the displayed image, with grooves with $W = 40$ for cyan and $W = 60$ nm for magenta under TM illumination.

validity of the identification card. This polarization dependence is also advantageous for implementation in visual display technologies by creating a multifunctional component that can serve as a conductive electrode, polarizer, and color filter simultaneously.

The dependence on optical path length, as opposed to grating coupling into waveguide or plasmonic modes, allows angle-insensitive performance. The effect of periodicity on the angle robustness of these metallic nanogrooves is investigated here. Given $W = 50$ nm and $D = 180$ nm, the simulated angle-resolved reflection spectra contour plots at periods of P = 140, 180, 220, and 260 nm are shown in Fig. 11.7a–d. Incremental changes of 40 nm in the period are chosen to illustrate three different regimes: grating coupling, localized resonance, and cavity coupling between neighboring waveguides, which is explained further in the next section of this chapter. Figures 11.7a and 11.7d indicate higher angle dependence than Fig. 11.7b,c. When the period P = 140 nm and 260 nm, the MIMFP resonance position shifts with increasing incident angle of light, whereas the spectra at the other two periods, 180 and 220 nm, remain at a relatively constant wavelengths over all incident angles. Furthermore, the angular behavior of absorption at the resonance wavelength 630 nm with P = 140, 180, 220, and 260 nm are presented in Fig. 11.7e, showing that the absorption reaches over 90% for a $\pm 90^\circ$ angle range with P = 180 nm. This shows that angle independence is achieved at a periodicity of 180 nm for visible wavelengths of light. Above or below P = 180 nm, the resonance wavelength corresponding to the absorption peak is angle dependent and therefore less efficient. The angle-resolved reflection spectra from angles of 45° to 75° were measured on various large-scale-fabricated devices based on nanoimprint lithography. Two of them are shown in Figs. 11.7a and 11.7b with a period of 180 nm and depths D of 130 and 170 nm, respectively. A device fabricated with a period of 220 nm exhibited angle dependence, further validating this design. The angle-resolved reflection spectra of this device, with P = 220 nm, W = 45 nm, and D = 160 nm, is displayed in Fig. 11.8c. A 25 nm $\Delta\lambda$ is observed per 30° change in the incident illumination angle. This change in reflection dip is not observed at P = 180 nm (Fig. 11.8a,b), showing strong agreement between measured and simulated spectra. Through this analysis we conclude that a range of periods from 160 nm to 200 nm has been found to possess a angle-robust spectrum response.

In Fig. 11.7, two sets of modes can be clearly identified: one is angle independent, that is, associated with the MIMFP mode, and the other has strong angle dependence, which is due to the grating coupling. In this structure the angle dependence has been avoided with a designed periodic structure for visible wavelength resonances employing MIMFP modes. A thorough analysis for the

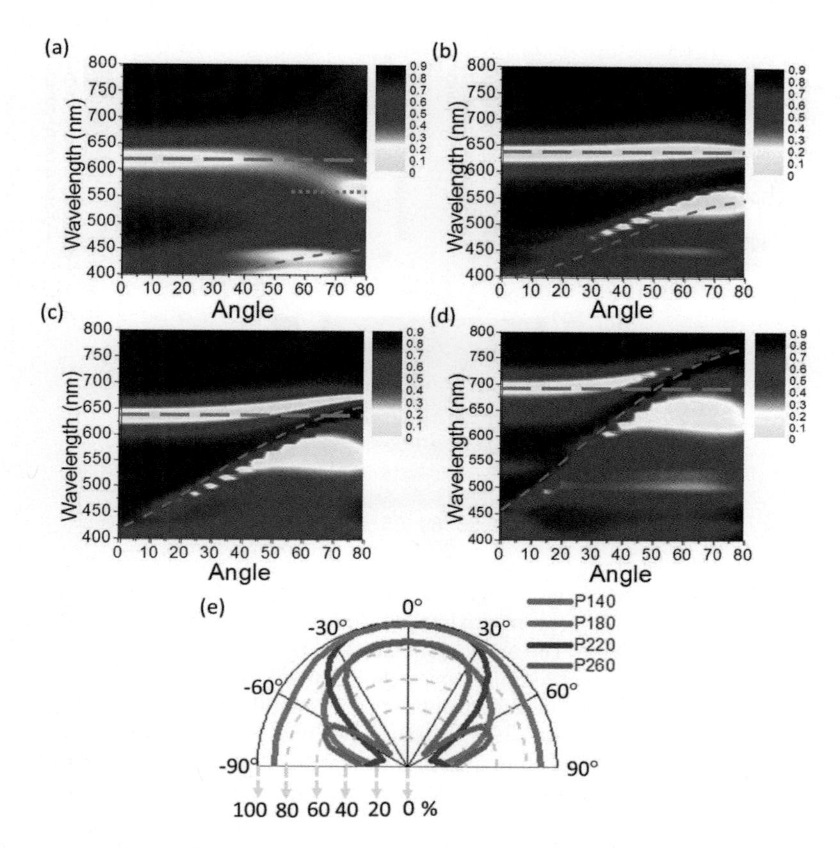

Figure 11.7 Simulated angle-resolved reflection spectra with various periods. The simulated angle-resolved reflection spectra contour plots at periods $P = 140$ (a), $P = 180$ (b), $P = 220$ (c), and $P = 260$ nm (d). The green dashed lines in all four figures indicate the metal-insulator-metal Fabry–Perot (MIMFP) cavity mode, whereas the red dashed lines refer to the grating-assisted surface plasmon (GASP) mode whose dispersion is dependent on the grating period. The crossing between MIMFP and GASP when $P = 220$ and 260 nm indicates coupling between the two modes. The green dashed line in (a) refers to the dispersion of the odd mode defined in (c). (e) The simulated angular absorption maxima in terms of various periods from 140, 180, 220, and 260 nm.

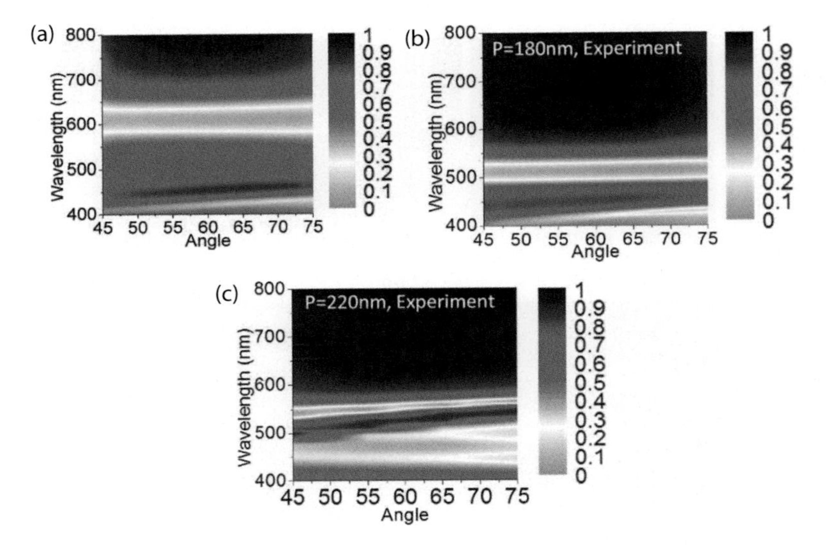

Figure 11.8 Experimental angle-insensitive spectrum filtering. (a) The simulated angular absorption maxima in terms of various periods from 140, 180, 220, and 260 nm. The angle-resolved reflection spectra of this design with sweeping incident illumination angle from 45° to 75° are presented with the following device dimensions: (b) P = 180, W = 50, and D = 130 nm; (c) P = 180, W = 50, and D = 170 nm; and (d) P = 220 nm, W = 50 nm, and D = 160 nm. (b and c) Flat band absorption response indicating angle insensitivity, and (d) 25 nm resonance wavelength shift per 30° change in incident illumination angle representing coupling between MIMFP and GASP modes.

angle dependence of the nanoslit-based color filters will be presented in the next section.

11.5 Wide-Angled Transmission Plasmonic Color Filters

To further apply the aforementioned angle-robust design principles to plasmonic transmission-type spectrum filtering for color filters in LCD panels, the plasmonic structure has to be redesigned. The schematic of Design I is shown in Fig. 11.9a. The design has a very similar schematic with the reflection-type color filter in Fig. 11.5a

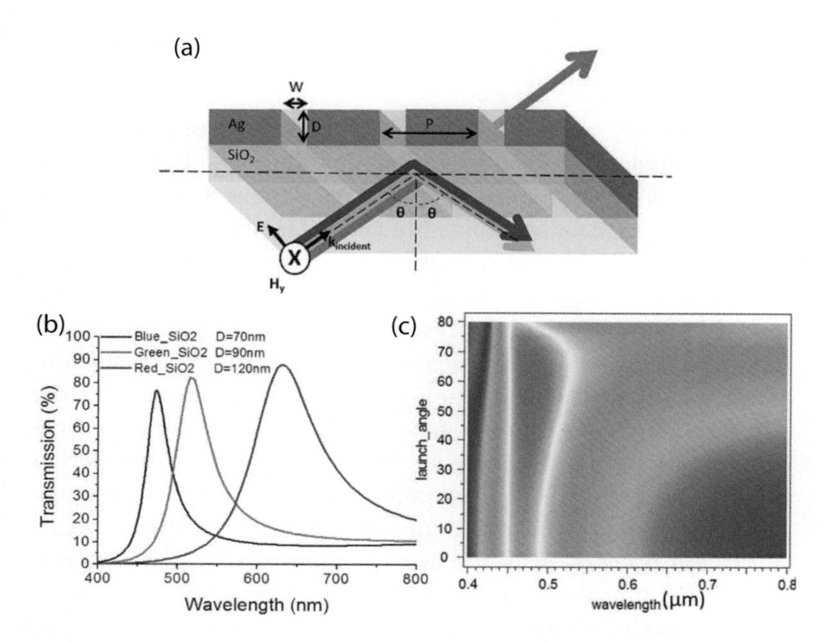

Figure 11.9 (a) Schematic of transmission color filter Design I with the geometrical parameters defined the same as Fig. 11.5a. (b) Spectra for blue, green, and red colors based on Design I with various heights 70–120 nm. (c) Simulated angle-resolved spectrum for the blue color filter, and (d) $\mathbf{H}_y$ field intensity distribution at resonance wavelength = 480 nm.

but with open slit openings. In this figure, the nanostructure is defined by the opening of the nanoslits, W, the height of the slits, D, and period, P. Based on the calculation with a real silver (Ag) refractive index model, the transmission spectra for blue, green, and red can be achieved by varying D from 70 nm to 120 nm, as shown in Fig. 11.9b. This MIMFP resonance condition can be described as $\frac{1}{2}m\lambda = n_{\text{eff}}D$. It is noticed that the missing $\frac{1}{4}\lambda$ is due to the boundary condition of the proposed MIMFP structure with both sides being open ended. Moreover, the angle-resolved spectrum modeling in Fig. 11.9c shows that the resonance holds at 470 nm from a launching angle of 0° to 80°. Therefore, this nanostructure design not only provides a widely tunable resonance through the visible region but also holds the resonance wavelength with a viewing angle up to ±80°.

11.5.1 Color Purity and Suppression of Off-Resonance Transmission

Color tunability and color purity are two important factors of the transmission color filters in real applications. In this section, we would like to introduce another color-tuning mechanism, and further discuss the color tunability and color purity issues for the MIMFP modes in transmission-type structure based on the resonance condition equation in the previous section. This resonance condition indicates that the resonance can be manipulated with varying n_{eff}, which can be achieved by changing W, the width of the nanoslits. The simulation in Fig. 11.10a demonstrates the transmission resonance from blue to red across the entire visible regime by changing W from 60 to 20 nm, given a fixed period of 180 nm. Compared to the color-tuning method by changing D, this method is favorable due to the simple and low-cost fabrication process. Also, the average transmission percentage is above 70%. It is also found that the off-resonance transmission at a longer wavelength for each spectrum increases with W. This effect degrades the color purity and will affect the color filter application and therefore needs to be addressed.

To investigate the relationship between large flat background and slit width W, a simulation is performed and demonstrated in Fig. 11.10b. Given the fixed period P and the resonance wavelength λ, this simulation shows the transmission spectra in various W values swept from 60 nm to 20 nm. As W decreases, the flat background at a longer wavelength is suppressed further. When $W = 20$ nm, this background is totally suppressed. To further understand this background change with W, we apply a MIMFP cavity like the left-handed figure in Fig. 11.10c to be equivalent to a typical dielectric FP model on the right-hand side of the same figure. It is known that FP has transmission T_{FP} determined by reflection coefficient for normal incidence $R = \left(\frac{n_{\text{in}} - n_{\text{out}}}{n_{\text{in}} + n_{\text{out}}}\right)^2$ in the following equation:

$$T_{\text{FP}} = \frac{(1-R)^2}{(1-R)^2 + 4R\left(\sin\left(n_{\text{in}} D/\lambda\right)\right)^2}.$$

According to this equation, the off-resonance T_{FP} will be suppressed when R approaches 1, which indicates n_{in} has to be much larger than

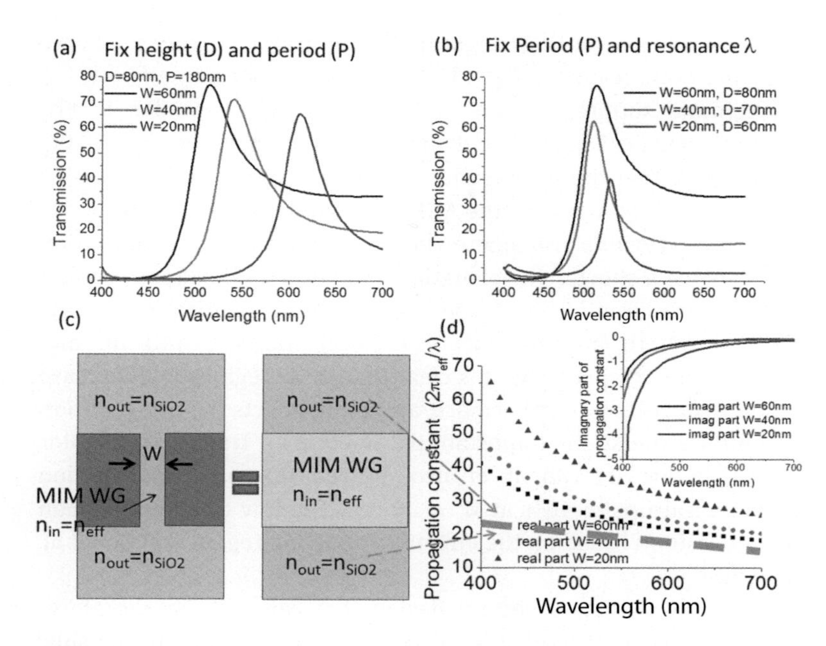

Figure 11.10 (a) Transmission spectra for blue, green, and red colors with various W values from 60 to 20 nm, given P and D fixed, and (b) transmission spectra with various W values from 60 to 20 nm, given resonance wavelength and P fixed. (c) The equivalent Fabry–Perot model for the MIMFP cavity, (d) dispersion curves of the MIM waveguide (WG) at $W =$ 20 (in blue), 40 (in red), and 60 nm (in black), and the green dashed curve indicates the dispersion of n_{out}. The inset shows the imaginary part of the propagation constant, which indicates waveguide loss.

n_{out}, that is, a large impedance mismatch is needed. However, the n_{in} in the case of MIMFP does not always have large values at a longer wavelength due to the waveguide dispersion. The propagation constant dispersion of the MIM waveguide ($2\pi n_{\text{in}}/\lambda$) is calculated in Fig. 11.10d, where the green dashed line represents the optical wave propagation constant ($2\pi n_{\text{SiO2}}/\lambda$) in silica ($n_{\text{SiO2}} = 1.46$). This figure shows that the dispersion curve of the MIM waveguide has a propagation constant value of 18.53, very close to that in silica material, 15.46, at $\lambda = 700$ nm when $W = 60$. On the other hand, this value becomes 26.87 at $\lambda = 700$ nm when $W = 20$ nm. This conclusion indicates that a narrower slit width W helps suppress the background transmission at a longer wavelength, and it agrees well with the simulation results in Fig. 11.10a,b. However, a trade-off between background suppression and color tunability is revealed in Fig. 11.10d, which calculates the dispersion. In the inset of Fig. 11.10d, we observe that the propagation constant dispersion has a visible wavelength region where the MIM waveguide has dramatically increasing waveguide propagation loss (imaginary part of propagation constant) at a shorter wavelength, and the cavity resonance cannot form due to large loss in these regions. These regions in the case of $W = 60$, 40, and 20 nm are 400–420 nm, 400–440 nm, and 400–510 nm, respectively. It is clear that these regions will increase when W decreases. Furthermore, a small W affects the transmission magnitude, one of the important parameters for transmission color filters. In Fig. 11.10b, it is also noticed that the transmission maximum drops to 40% at $W = 20$ nm. The low transmission can be understood based on the effective funneling region that Shi et al. proposed.

We now discuss a second design that can address the aforementioned issue. The idea is to add two thin Ag layers to the slits' cavity in order to increase the impedance of the structure and consequently the reflectivity. As shown in Fig. 11.11a, an additional parameter thickness of Ag mirror DM is defined. Based on the design in Fig. 11.11a, we are able to optimize the parameters of period P, depth D, and slit width W to achieve a set of blue, green, and red color filters which have good color purity and color tunability with fixed $D = 90$ nm and DM = 20 nm, shown in Fig. 11.11b. Therefore, Design II resolves the trade-off issue in Design I.

11.5.2 Effect of Coupling between the Nanoresonator and Grating Resonance

In this section we provide a general discussion of how the slit resonance is affected by the periodic array structure in terms of angle variation of the resonance spectrum and the color purity (i.e., resonance bandwidth). The slit resonance is a result of the standing wave formation due to the two counterpropagating slit waveguide modes, that is, an FP resonance. When the metal layer is perforated by the slit, the transmission shows a peak value at the

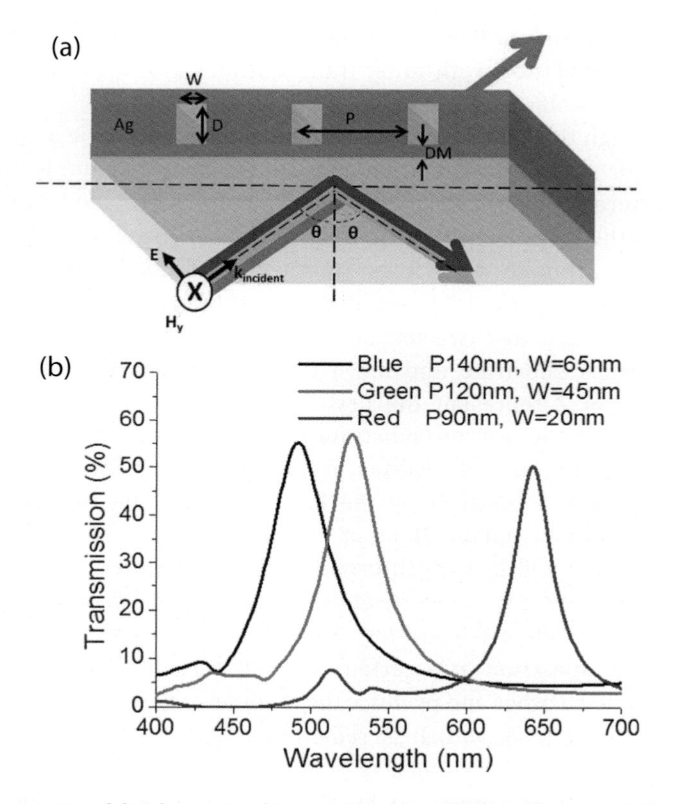

Figure 11.11 (a) Schematic of Design II, where DM is added and defined as the thickness of the top and bottom metal layers. (b) The optimized spectra for blue, green, and red colors, given $D = 90$ nm and DM $= 20$ nm fixed.

resonance and the reflection shows a dip correspondingly. When the slit does not cut through the metal layer, the transmission vanishes, while the reflection is still affected by the resonance in the same way. To enhance the total signal, many slits are needed, for convenience, typically arranged in a periodic array. However, the periodic structure will produce grating resonances (diffractions), which could couple with the slit resonance and influence the spectrum filtering behavior.

To illustrate such a coupling effect on the transmission spectrum, a slit array with a fixed slit width of 110 nm and a varied period was fabricated in a 170 nm thick Al film on a fused silica substrate [60].

Based on an analytical mode derived through mode expansion, a 2D contour plot of the zeroth-order transmission T_0 versus wavelength and period is obtained, as shown in Fig. 11.12a.

Although the analytical model assumes the metal to be a perfect conductor as an approximation, most of the experimental features are captured and a reasonable physical picture is provided. The red zone in the color map corresponds to the resonant transmission peak. The oblique dark lines (marked as dashed white lines) represent the different diffraction orders of the grating resonances. They are divided into two sets due to the different refractive indices of the silica substrate and air. The two lines marked by $N' = 1$ and $N' = 2$ represent the first- and the second-order grating resonance at the grating/substrate interface, while $N = 1$ and $N = 2$ represent the diffractions at the grating/air interface. The slit resonance is marked by the band between the two vertical dash lines. It is independent of the period due to its localized characteristic and it is much broader than the grating resonances due to the low impedance contrast, as will be discussed later in this chapter. When the grating resonance and the slit resonance overlap in spectrum at a certain specific period, the two modes couple to each other and produce the familiar anticrossing behavior. The resonant peak, which is red-shifted with increasing period, is actually the lower branch of the anticrossing. The shift of the resonant peak is confirmed by the measured transmission and reflection spectra in experiment (Fig. 11.12b,c). The higher branch is coupled to a higher-order grating resonance so that it is not collected in the transmission measurement. The interference between a broad resonance (the slit resonance) and a sharp resonance (the grating resonance) gives rise to the typical Fano-type line shape, as confirmed by the asymmetric shape of the resonant peak in the measured transmission or the dip in reflection (Fig. 11.12b,c). At the period of 500 nm, the grating resonance overlaps the central wavelength of the slit resonance so that a strong coupling is formed. Due to the high-Q characteristics of the grating resonance [61, 62], the coupled slit/grating resonance is much sharper than the intrinsic slit resonance. With decreasing period, the grating resonance gradually moves away from the slit resonance toward shorter wavelengths. Consequently, the slit and grating resonance

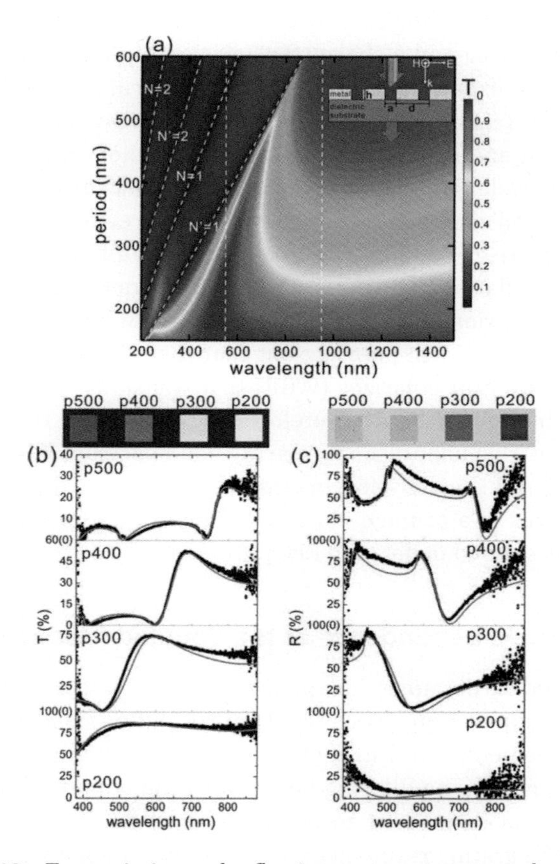

Figure 11.12 Transmission and reflection measurements on the slit arrays with four different periods in comparison with a period-dependent transmission spectrum based on the theoretical model. (a) 2D plot of a period-dependent zeroth-order transmission spectrum based on the theoretical model. The oblique dashed lines mark out the grating resonances of different orders. The band between the two vertical dashed lines marks out the slit resonance. Inset: Sketch of the Al slit array as a spectrum filter. (b and c) The measured transmission and reflection spectra (black dots) together with RCWA simulations (red curves) of the fused silica-supported aluminum slit arrays at four different periods (200 nm, 300 nm, 400 nm, and 500 nm). Slit width (~110 nm) and slit height (~170 nm) remain invariant for different periods. Insets show the transmission and the reflection photographs of the four samples. Each square is 200 μm × 200 μm.

become separated, resulting in a broad transmission peak due to the slit resonance and a sharp transmission dip due to the grating resonance. In experiment, the resonant spectrum is prominent at large periods (e.g., 400 nm and 500 nm), giving rise to the distinctive transmission and complementary reflection colors (Fig. 11.12b,c). At longer wavelength the off-resonance transmissions increase due to the reduced impedance discussed in the previous section. At a small period (e.g., 200 nm), the peak is so broad that the spectrum looks almost flat for the whole measurement range. Accordingly, a white transmission and a dark reflection for TM light are shown in the photos. It is worth pointing out that the latter is the principle behind a wire-grid polarizer (WGP) that transmits broadband TM light but blocks TE light. Therefore, the band-pass transmission behavior of the nanoresonator array for spectrum filtering and the broadband transmission in a WGP can be considered the two limiting cases of a coupled nanoresonator/grating resonance. This aspect was studied in detail in Ref. [60].

11.5.3 Angle Dependence of the Coupled Resonance

As described earlier, angle-independent spectrum filtering is desired for many applications such as high-resolution visual displays, miniature hyperspectral imaging, and high-sensitivity sensors [4–7]. Unfortunately, the filtering behavior of the nanoresonator (slit) array is angle dependent to certain extent, due to the coupling between the grating resonance that is strongly dependent on angle and the slit resonance that is intrinsically angle independent [55]. With an increasing incident angle, each grating resonance splits into two branches due to the phase matching requirement, as shown in Fig. 11.13a,b. The branch shifting to shorter wavelengths represents the forward-diffracted light whose k_x has the same sign as that of the incident light. The branch shifting to longer wavelengths represents the backward-diffracted light whose k_x has the opposite sign to that of the incident light.

One strategy to overcome the angle dependence of the transmission peak is to decouple the slit/grating resonance and let the slit resonance, which is angle independent, dominate the spectrum. By reducing the period, we observed the separation of

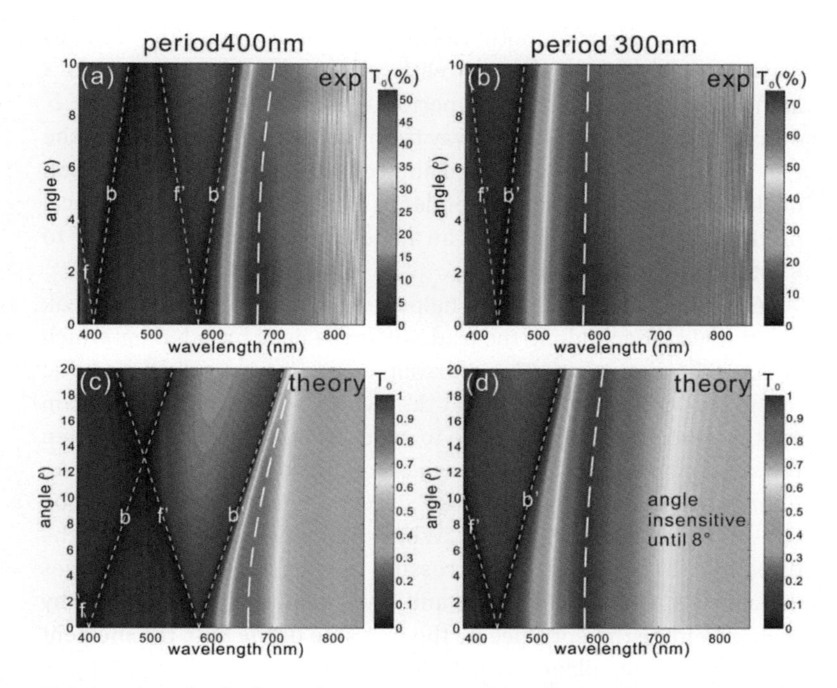

Figure 11.13 Angle-dependent transmission measurement at two periods, together with the theoretical results. (a and b) The measured angle-dependent transmission spectra for fused silica-supported aluminum slit arrays with a period of 400 nm and a period of 300 nm, respectively. The slit width of the structure is ~110 nm and the slit height is ~170 nm. (c and d) The angle-dependent transmission spectra by the theoretical model in an extend angle range. The branches marked out by the thin dashed lines are due to grating resonances. f and b mark the forward and the backward branches, respectively, of the grating resonance at the grating/air interface; f' and b' the mark forward and backward branches, respectively, at the grating/substrate interface. The thick dashed lines mark the transmission peak.

the two resonances and an increased angle-insensitive range of the spectrum. As shown in Figs. 11.13a and 11.13c, at a large period (400 nm), the backward branch from the first-order grating resonance at the grating/substrate interface (marked as *b*′) meets the slit resonance, forming a strong coupling between the two, as shown in an angle-dependent transmission spectrum. Thus, the transmission peak shifts to longer wavelengths with increasing angle, as revealed by both the experimental data (Fig. 11.13a) and the theoretical prediction with an extended angle range (0°–20°) (Fig. 11.13c). When the period decreases to 300 nm, the b' diffraction branch is further away from the slit resonance so that the two resonances are less coupled and consequently the transmission peak becomes less angle dependent (Fig. 11.13b). The theoretical prediction (Fig. 11.13d) shows an angle-insensitive range from 0° to 8° at a period of 300 nm.

Although period reduction helps to make the transmission peak more angle independent, the trade-off is the peak broadening, which affects the color purity. The slit resonance is intrinsically broad due to the lack of high reflectivity at the two interfaces. Another reason for the peak broadening is due to reduced light confinement when the slits become very close to each other as the period decreases. An illuminated metal slit with a deep subwavelength dimension can be considered as a dipole [63–65]. When driven by a TM wave with the direction of the electric field crossing the slit, polarization charges are accumulated at the corners and consequently distort the nearby field. At either the entrance or the exit side of the slits the incident or the outgoing plane waves should couple into or out of the waveguide modes in the slits, which creates an impedance mismatch at the two interfaces. The impedance difference leads to partial light reflection from the two surfaces. Such an impedance mismatch and the constructive field buildup of slit waveguide modes establish an FP-type resonance—this is the nature of the slit resonance discussed above. In the case of an array of slits, the separation between each slit plays an important role in deciding the impedance mismatch and consequently the light confinement. For a 400 nm period, as shown in Fig. 11.14a, p400, the induced dipoles are separated from each other by sufficient distance. Thus, the light field around each dipole is distorted, leading to a large impedance mismatch between the incident plane wave and the slit waveguide mode. Thus, light is strongly reflected at both ends of the slit, which is desirable for forming high-Q FP resonance modes. When the period decreases to 200 nm while the slit width remains 110 nm, the induced dipoles are adjacent to each other end to end, as shown in Fig. 11.14a, p200, so that the accumulated charges tend to cancel out. Thus, the light field is only slightly distorted. Therefore the impedance mismatch to air

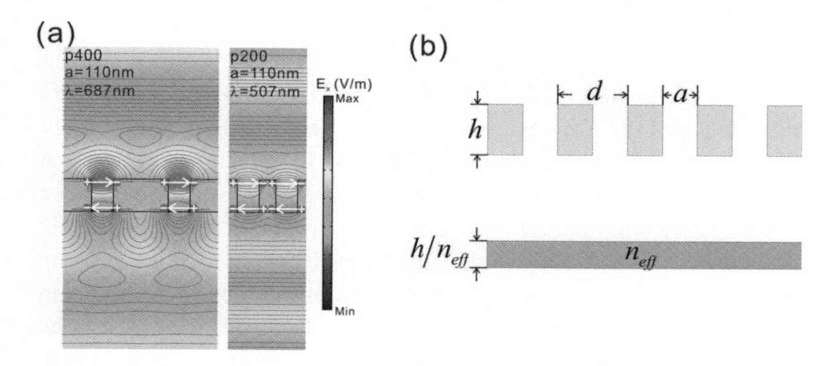

Figure 11.14 Illustration of the slit resonance diminishing with decreasing period. (a) Simulated x component **E**-field distributions and electric field lines of two metal slit arrays at their own resonant wavelengths, that is, the period of 400 nm at its resonant wavelength of 687 nm and the period of 200 nm at its resonant wavelength of 507 nm. As revealed by the calculated electric lines, the light field near the slit openings in the 400 nm period array is distorted more severely than that in the 200 nm period array. Accordingly, the calculated x component of the electric field ($\mathbf{E}_x$) reveals a stronger on-resonance light field inside the slits of the 400 nm period than that of the 200 nm period. The maximum value of the color bar is 2.44×10^5 V/m; the minimum value is -2.90×10^5 V/m. (b) Sketch of the analogy between a metal slit array layer and a dielectric slab.

is significantly reduced, leading to a much weaker FP resonance and significantly broadened transmission peak.

11.5.4 Achieving Angle-Insensitive Spectrum Filter in the Slit Nanoresonator Array Structure

Based on the analysis in the last section, the transmission peak broadening is due to the weak light confinement inside the slit nanoresonator. For a more quantitative description, the metal film with a perforating slit array is mapped into a dielectric slab with an effective refractive index (sketched in Fig. 11.14b) by comparing either the reflection coefficient or the transmission coefficient of a free-standing slotted metal film with that of a dielectric slab [61, 66]. In this sense, the slit resonance corresponds to the FP resonance of an effective dielectric slab. Assuming that the metal is a perfect conductor and the incident free-space wavelength is much larger than the period, the refractive index of the effective medium is approximately expressed as $n_{\rm eff} \approx \frac{d\sqrt{\varepsilon_{\rm s}}}{a}$, where a denotes the slit width, d the period, $\varepsilon_{\rm s}$ the dielectric constant of the medium inside the slit, and λ the free-space wavelength. Now the light confinement or the Q factor of the slit nanoresonator is determined by $n_{\rm eff}$. A large $n_{\rm eff}$ leads to a sharp resonance, while a low $n_{\rm eff}$ leads to a broad FP resonance. Since $n_{\rm eff}$ is proportional to period, the FP resonance (or the slit resonance) diminishes with a decreasing period so that the transmission becomes broadband and flat. From these understandings, two strategies can be used to enhance the light confinement and thus make a sharp resonance even at a small period:

- **Reducing the slit width**: To make the resonance sharp enough even at a small period, we can reduce the slit width a to maintain a high $n_{\rm eff}$ value. This is consistent to the discussion in Section 11.5.1. The physics can be understood as follows: Light confinement inside the slits requires an impedance mismatch at the slit openings due to induced polarizing charges. With slits getting closer to each other, the light-induced polarizing charges cancel out so that the slit resonance gets weakened. To reserve the impedance mismatch, the slits need to be separated at some distance, which can be done by reducing the slit width. At a small period, a much smaller slit width can help confine light inside the slit and produce a sharp resonance. A comparison between two slit arrays with the same period (200 nm) but different slit widths (110 nm and 50 nm) is shown in Fig. 11.15. As revealed by Fig. 11.15a,b, the 50 nm wide slit array shows a stronger field distortion at the slit openings and a higher $\mathbf{E}_x$ field inside the slits. Correspondingly, the intrinsic slit resonance of the 50 nm wide slit array has a much smaller bandwidth than that of the 110 nm wide slit array, as shown by the period-dependent spectra of the slit power enhancement factor, as plotted in Fig. 11.15c,d.

When the period is reduced to $\sim$200 nm (as shown in Fig. 11.15e,f), where the resonance is angle insensitive up to $\sim 60^\circ$ (Fig. 11.15g,h), the 110 nm wide slit array shows a high transmission

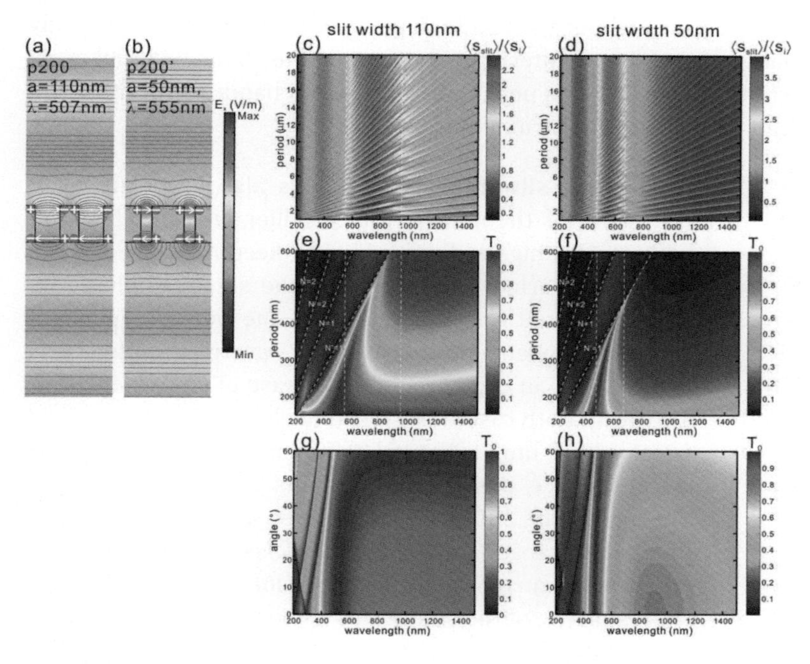

Figure 11.15 The influence of the slit width on bandwidth of the slit resonance. (a and b) Simulated x component $\mathbf{E}_x$-field distributions and electric field lines of two metal slit arrays with different slit widths at their own resonant wavelengths, that is, the slit width of 110 nm at the resonant wavelength of 507 nm and the slit width of 50 nm at 555 nm. The other structural parameters remain the same for these two slit arrays: 170 nm slit height, fused silica substrate, and 200 nm period. As revealed by the calculated electric lines, the light field near the slit opening in the 50 nm slit width array is distorted more severely than that in the 110 nm slit width array. Accordingly, the calculated x component of the electric field reveals a stronger on-resonance light field inside the 50 nm wide slits than that inside the 110 nm wide slits. (c and d) Period-dependent spectra of the slit power enhancement factor of the two 110 nm wide slit arrays and the 50 nm wide slit array over the period range up to 20 μm. (e and f) Period-dependent zeroth-order transmission spectra of the two structures over a small period range from 150 nm to 600 nm. The oblique dashed lines mark out the grating resonances. The band between the two vertical lines marks out the intrinsic slit resonance. (g and h) Angle-dependent zeroth-order transmission spectra of the two structures.

(>0.75) throughout the whole range above 400 nm, which already lacks the band-pass filtering feature, while the 50 nm wide slit array shows a transmission peak with a ~150 nm bandwidth, which still can be used as a spectrum filter.

- **High-index slit filler**: Since n_{eff} is also proportional to $\sqrt{\varepsilon_s}$, which is the index of the slit filler, we can fill the slit with some a high-index medium to keep n_{eff} large enough at a small period. Assuming that the slits are filled with a dielectric with an index of 1.45, the period-dependent and angle-dependent transmission spectra are plotted in Fig. 11.16a,b in comparison to the case of empty slits (Fig. 11.16c,d). Both cases have a slit height of 170 nm and a slit width of 110 nm. In general, the filled slits show a sharper slit resonance than the empty slits at the same period. The period of 250 nm, which gives a quite large angle-insensitive range (up to 50°), is picked out for an angle-dependent transmission spectrum plot (Fig. 11.16b,d). The slit resonance of the empty slits is too weak to be identified from a high-transmission background. However, the filled slits still exhibit a distinct resonance peak in transmission. Although the resonant wavelength of the filled slits is red-shifted due to the increased optical path in a high-index material, the deviation can be compensated by decreasing the slit depth.

We should point out that adding two metal reflectors on both sides of the metal slit array can also increase the transmission resonance Q factor, as was discussed in Section 11.5.

11.6 Ultrathin Metallic Nanoresonators for Angle-Insensitive Reflective Colors

As indicated in the last section, to obtain angle-insensitive structural colors, one should exploit localized resonances rather than relying on grating coupling. Previously we have exploited localized surface

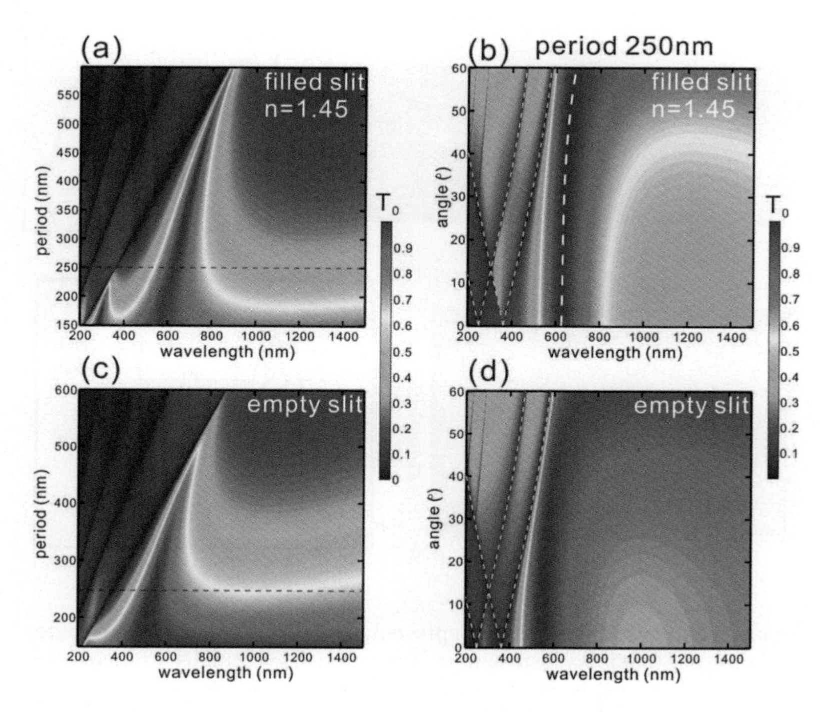

Figure 11.16 Comparison between a filled slit array and an empty slit array. (a) Theoretically calculated period-dependent transmission spectra of the slit array with slit filler whose index is 1.45. The slit array is on a fused silica substrate and has a slit depth of 170 nm and a slit width of 110 nm. (b) Theoretically calculated angle-dependent transmission spectra at the period of 250 nm. (c and d) Period- and angle-dependent transmission spectra of the empty slit array. The black dashed lines in (a) and (c) mark out the period of 250 nm. In the angle-dependent transmission spectra, the thin dashed lines represent the grating resonance branches; the thick dashed lines represent the transmission peak.

plasmon resonance (LSPR) in patterned metallic nanostructures, a topic that has been studied extensively over the past decade. Figure 11.17 shows that light absorption and scattering control over the whole visible band can be easily obtained by tuning the thickness of an array of Au or Ag nanosquares fabricated by using the nanoimprinting process [67]. Clearly distinctive colors can be observed for nanosquares of different Ag thicknesses. In this section,

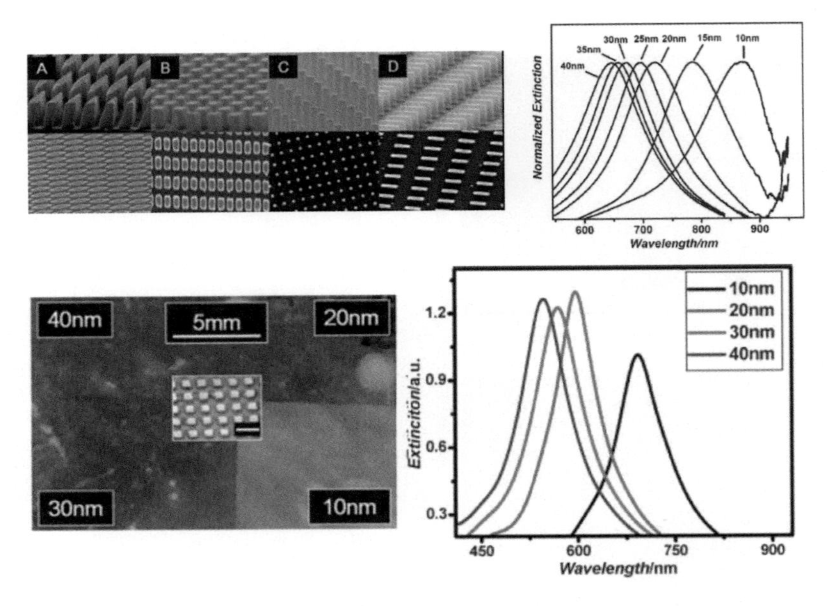

Figure 11.17 SEM images of representative nanoblock molds used to produce plasmonic nanostructures. The resonance shifts as a function of thickness for both Au and Ag. As a result, distinctive colors appear.

we discuss a design that can produce high-purity red, green, and blue reflective colors.

Figure 11.18a shows a schematic of the angle-insensitive reflective RGB color elements, which is composed of ultrathin (<20 nm) metallic nanoresonators on a glass substrate. The silver (Ag) is employed as a metal layer due to the lower absorption loss at visible frequencies. The calculated reflection spectra of the color filters at normal incidence are described in Fig. 11.18b. Altering the thickness of the metal gratings enables the color to be tuned. The thicknesses to create the RGB colors are 6, 10, and 16 nm, respectively. The period (P) and width (W) of the metal gratings are 220 nm and 60 nm, respectively. When the TM-polarized light wave is incident on the structure, the LSPR is excited to produce the reflective colors, which is attributed to the ultrathin thickness of the gratings. As the thickness of the gratings decreases, the effective refractive index of LSPR increases, leading to a red-shifted resonance.

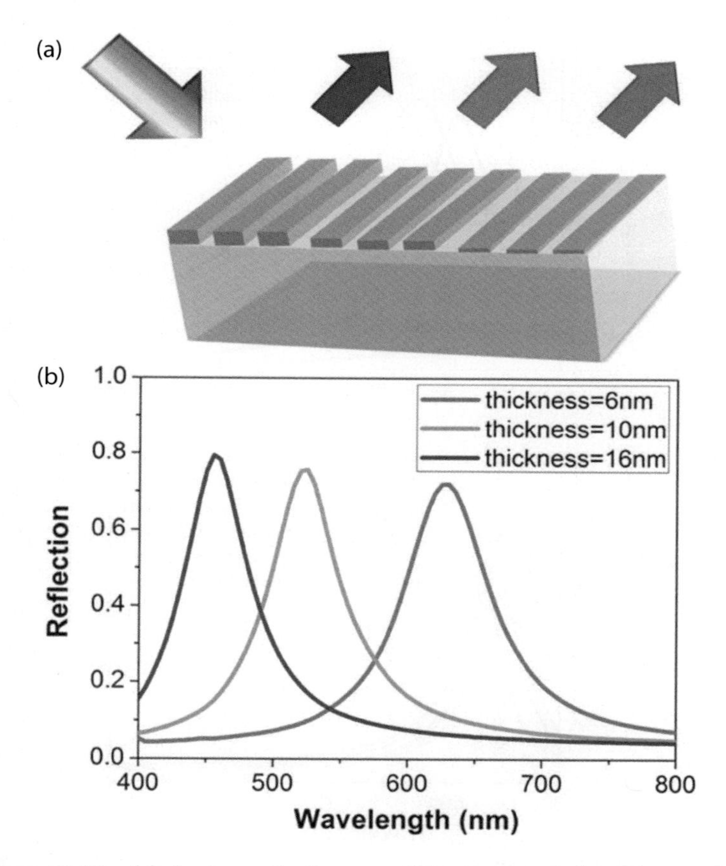

Figure 11.18 (a) A schematic diagram of the proposed ultrathin metallic nanoresonators for reflective red, green, and blue color filters with a wide viewing angle. (b) The calculated reflection spectra of the proposed devices at a normal incident angle.

In Fig. 11.19a, the calculated reflection spectra of the proposed devices for different widths of the metal gratings are depicted. The effective size of the LSPR can be increased by increasing the width of the metal gratings, which will result in a red-shifted resonance. In addition, the profile becomes broad as the width increases, which is attributed to the strong coupling between neighboring modes [60]. The thickness and period of the metal gratings are 16 nm and 220 nm, respectively. By changing the width of the gratings,

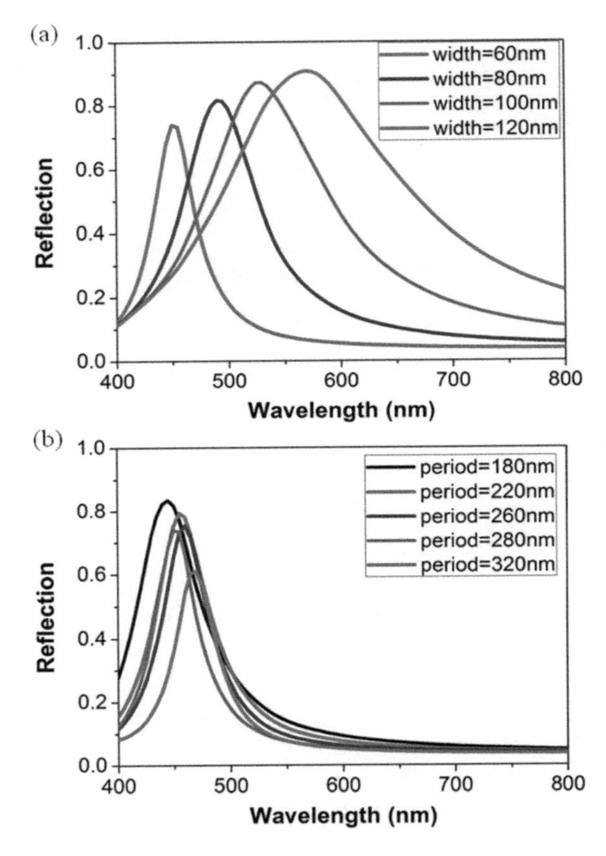

Figure 11.19 The calculated reflection spectra for different widths and periods of the metal gratings in (a) and (b), respectively.

the color can also be varied with sacrificing the purity of the color. Figure 11.19b shows the calculated reflection spectra for different period of the metal gratings. The resonance shift in this case can be explained in a similar manner as discussed in Section 11.5.4. In this design, even though the period is changed, there is a negligible resonance shift due to the relatively weak coupling between the LSPR and the grating resonance. It is therefore expected that the structure shows angle-insensitive performance. The calculated angle-resolved reflection spectra of RGB colors for TM polarization is illustrated in Fig. 11.20a–c. The resonance wavelength does not

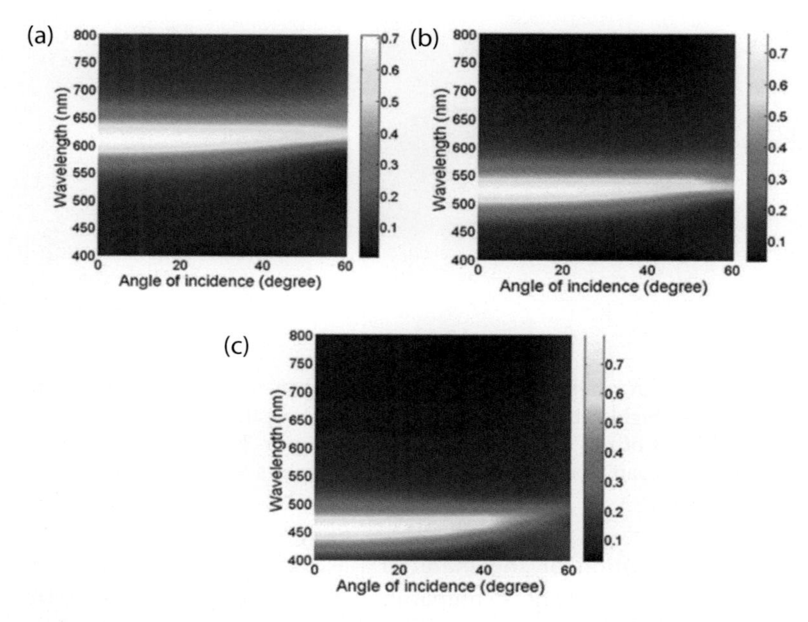

Figure 11.20 Calculated angle-resolved reflection spectra of individual RGB colors under TM-polarized light illumination.

change significantly over a wide angular range up to $\pm 60^\circ$, thus addressing the critical drawback of angle-dependent characteristics of both plasmonic and photonic nanostructure-based color filters [40, 55, 61, 62]. Such structures can be even broken into small pieces or particles, forming nonchemical-based colored pigments that are long lasting without color fading.

11.7 Angle-Insensitive Colors Utilizing Highly Absorbing Materials

Finally we introduce an approach for achieving angle-insensitive color performance that is completely different from the previously discussed concepts. It is based on the recent finding of strong optical interference behaviors in ultrathin, highly absorbing materials deposited on a metal surface. Such resonance is enabled by the

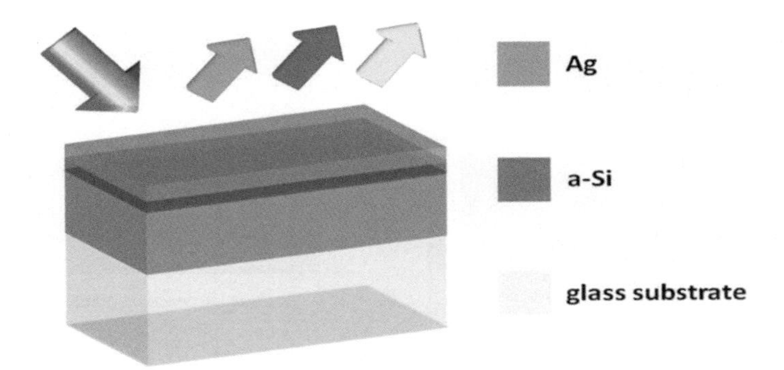

Figure 11.21 A schematic view of the proposed angle-insensitive spectrum filters using highly absorbing materials. The color can be tuned by altering the thickness of the amorphous silicon (a-Si) layer. The thicknesses to be required for cyan, magenta, and yellow (CMY) colors are 34, 20, and 14 nm, respectively.

strong absorption of the material (i.e., an imaginary part of a dielectric constant that is comparable to a real part of the dielectric constant) and the nontrivial reflection phase shift at the interface between absorbing material and metal. The detailed information can be found in Refs. [68, 69]. In this section, the angle-insensitive spectrum filters based on above principle are introduced.

Figure 11.21 exhibits a schematic diagram of the proposed angle-insensitive spectrum filter structure consisting of a highly absorbing material sandwiched by two metals on a glass substrate. Due to the lower absorption loss in the visible ranges, Ag is the best candidate for both top and bottom metals. The top Ag layer is 18 nm that is optically thin, allowing the incident light wave to transmit through it. At the same time, it can increase the reflection that leads to a narrow bandwidth and improve the purity of color. The thickness of the bottom Ag layer is 150 nm in order to block the transmission. As a highly absorbing material, a-Si is utilized due to the presence of a band gap (1.7 eV) in the visible regime. From the numerical simulation, the a-Si thicknesses for creating cyan, magenta, and yellow colors are designed to be 34, 20, and 14 nm, respectively.

In Fig. 11.22a, the calculated reflection spectra at normal incidence are shown. Increasing the thickness of the a-Si layer shifts the resonance toward the longer-wavelength range according

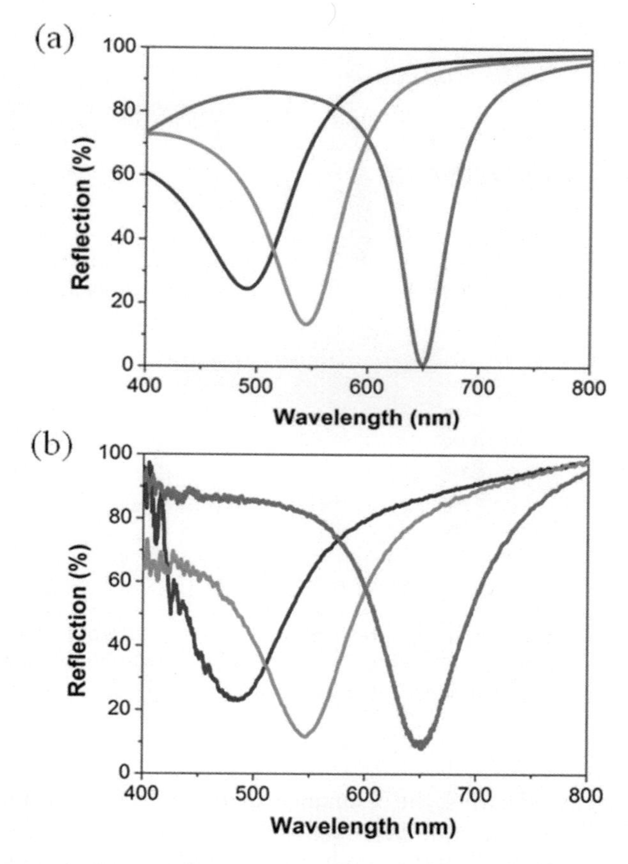

Figure 11.22 The calculated and measured reflection spectra of the proposed angle-insensitive spectrum filters at normal incidence in (a) and (b), respectively. The measured results show the broader profiles than the calculated spectra, which is due to the fabrication imperfections.

to the principle of the FP cavity. The measured reflection profiles depicted in Fig. 11.22b show good agreement with the calculated results. The scattering light loss arising from the nonsmooth surface makes the measured profiles somewhat broader than the calculated data. We note that the sharpest resonance is attained from the case of cyan (i.e., reflection dip is at 650 nm), which is attributed to the insignificant imaginary part of the a-Si material at longer wavelengths.

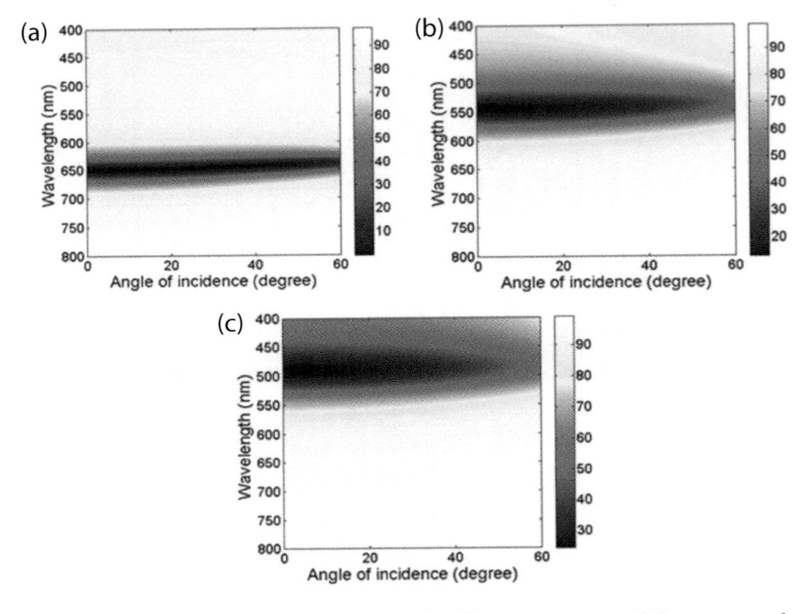

Figure 11.23 Calculated angle-resolved reflection spectra of the proposed devices for individual CMY colors under s-polarized light illumination. The reflection dip (i.e. resonance) is invariant to the angle of incidence up to $\pm 60^{\circ}$.

As discussed earlier, the resonance appears in the ultrathin cavity layer owing to the nontrivial reflection phase shift between the absorbing material and the metal. The ultrathin film leads to a small propagation phase accumulation, which is insignificant as compared to the nontrivial reflection phase change. This implies that the resonance can be insensitive with respect to the angle of incidence. To verify such hypothesis, we perform the simulation of the angle-resolved reflection spectra under the s-polarized light illumination shown in Fig. 11.23a–c, which are a good match with the measured results (Fig. 11.24a–c) obtained by a variable angle spectroscopic ellipsometer from 15° to 60°. As is observed in the figures, each resonance remains similar for the wide incident angle exhibiting flat dispersion property. This incident angle-insensitive characteristic is highly desirable in applications, for example, displays and colored coatings.

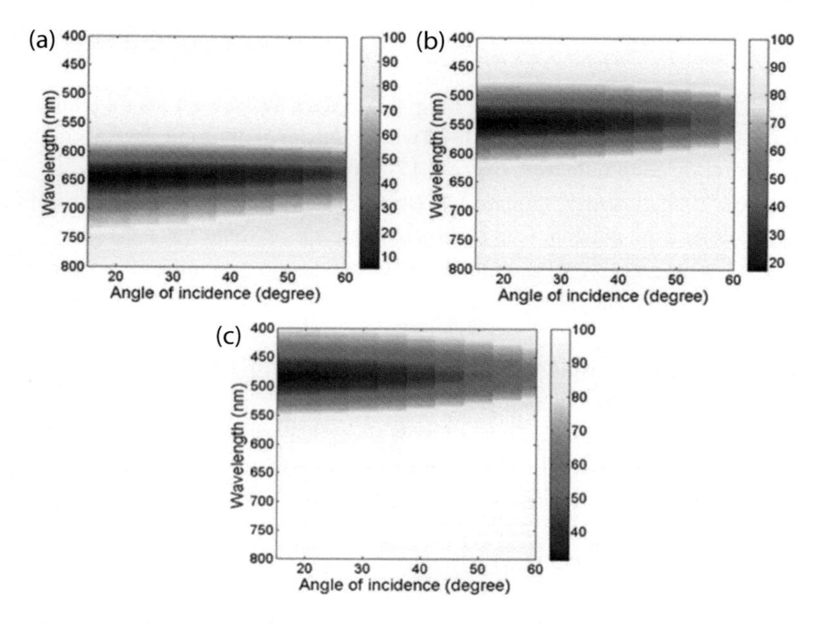

Figure 11.24 Corresponding measured angle-resolved reflection spectra for individual CMY colors.

11.8 Future Outlook

The next step for the research on structural color devices will focus on two aspects: first, continued improvement of their optical performances, including the transmission and reflection efficiencies, the purity of filtered color, and the incident angle independency, and second, the development of a more efficient, high-throughout nanofabrication method to realize mass production with low costs. Newer structures should be pursued that could allow multiple colors to be fabricated in parallel with only one or few different process steps. This will represent a significant reduction in the fabrication complexity and will reduce the manufacturing cost. These structural colored elements could open up a bright future for next-generation high-resolution displays, spectral imaging, and environmentally friendly colored coatings.

References

1. Cho, E., Kim, H., Cheong, B., Oleg, P., Xianyua, W., Sohn, J., Ma, D., Choi, H., Park, N., and Park, Y. (2009). Two-dimensional photonic crystal color filter development, *Opt. Express*, **17**, p. 8621.
2. Xu, T., Shi, H., Wu, Y., Kaplan, A. F., Ok, J. G., and Guo, L. J. (2011). Structural colors: from plasmonic to carbon nanostructures, *Small*, **7**, p. 3128.
3. Xu, T., Wu, Y., Luo, X., and Guo, L. J. (2010). Plasmonic nanoresonators for high-resolution colour filtering and spectral imaging, *Nat. Commun.*, **1**, p. 59.
4. Kinoshita, S., Yoshioka, S., and Miyazaki, J. (2008). Physics of structural colors, *Rep. Prog. Phys.*, **71**, p. 076401.
5. Kim, H., Ge, J., Kim, J., Choi, S., Lee, H., Lee, H., Park, W., Yin, Y., and Kwon, S. (2009). Structural colour printing using a magnetically tunable and lithographically fixable photonic crystal, *Nat. Photonics*, **3**, p. 534.
6. Kaplan, A. F., Xu, T., and Guo, L. J. (2011). High efficiency resonance-based spectrum filters with tunable transmission bandwidth fabricated using nanoimprint lithography, *Appl. Phys. Lett.*, **99**, p. 143111.
7. Yokogawa, S., Burgos, S. P., and Atwater, H. A. (2012). Plasmonic color filters for CMOS image sensor applications, *Nano Lett.*, **12**, p. 4349.
8. Laux, E., Genet, C., Skauli, T., and Ebbesen, T. W. (2008). Plasmonic photon sorters for spectral and polarimetric imaging, *Nat. Photonics*, **2**, p. 161.
9. Lee, H., Yoon, Y., Lee, S., Kim, S., and Lee, K. (2007). Color filter based on a subwavelength patterned metal grating, *Opt. Express*, **15**, p. 15457.
10. Ebbesen, T. W., Lezec, H. J., Ghaemi, H. F., Thio, T., and Wolff, P. A. (1998). Extraordinary optical transmission through sub-wavelength hole arrays, *Nature*, **391**, p. 667.
11. Barnes, W. L., Dereux, A., and Ebbesen, T. W. (2003). Surface plasmon subwavelength optics, *Nature*, **424**, p. 824.
12. Zayats, A. V., Smolyaninov, I. I., and Maradudin, A. A. (2005). Nano-optics of surface plasmon polaritons, *Phys. Rep.*, **408**, p. 131.
13. Ozbay, E. (2006). Plasmonics: merging photonics and electronics at nanoscale dimensions, *Science*, **311**, p. 189.
14. Genet, C., and Ebbesen, T. W. (2007). Light in tiny holes, *Nature*, **445**, p. 39.
15. Atwater, H. A., and Polman, A. (2010). Plasmonics for improved photovoltaic devices, *Nat. Mater.*, **9**, p. 205.

16. Schuller, J. A., Barnard, E. S., Cai, W., Jun, Y. C., White, J. S., and Brongersma, M. L. (2010). Plasmonics for improved photovoltaic devices, *Nat. Mater.*, **9**, p. 193.

17. Garini, Y., Young, I. T., and McNamara, G. (2006). Spectral imaging: principles and applications, *Cytometry Part A*, **69A**, p. 735.

18. Diest, K., Dionne, J. A., Spain, M., and Atwater, H. A. (2009). Tunable color filters based on metal-insulator-metal resonators, *Nano Lett.*, **9**, p. 2579.

19. Zia, R., Selker, M. D., Catrysse, P. B., and Brongersma, M. L. (2004). Geometries and materials for subwavelength surface plasmon modes, *J. Opt. Soc. Am. A*, **21**, p. 2442.

20. Dionne, J. A., Sweatlock, L. A., Atwater, H. A., and Polman, A. (2006). Plasmon slot waveguides: towards chip-scale propagation with subwavelength-scale localization, *Phys. Rev. B*, **73**, p. 035407.

21. Bozhevolnyi, S. I., Volkov, V. S., Devaux, E., Laluet, J. Y., and Ebbesen, T. W. (2006). Channel plasmon subwavelength waveguide components including interferometers and ring resonators, *Nature*, **440**, p. 508.

22. Neutens, P., Van Dorpe, P., De Vlaminck, I., Lagae, L., and Borghs, G. (2009). Electrical detection of confined gap plasmons in metal–insulator–metal waveguides, *Nat. Photonics*, **3**, p. 283.

23. Reilly III, T. H., van de Lagemaat, J., Tenent, R. C., Morfa, A. J., and Rowlen, K. L. (2008). Surface-plasmon enhanced transparent electrodes in organic photovoltaics, *Appl. Phys. Lett.*, **92**, p. 243304.

24. Lindquist, N. C., Luhman, W. A., Oh, S., and Holmes, R. J. (2008). Plasmonic nanocavity arrays for enhanced efficiency in organic photovoltaic cells, *Appl. Phys. Lett.*, **93**, p. 123308.

25. Burgos, S. P., de Waele, R., Polman, A., and Atwater, H. A. (2010). A single-layer wide-angle negative-index metamaterial at visible frequencies, *Nat. Mater.*, **9**, p. 407.

26. Chettiar, U. K., Kildishev, A. V., Klar, T. A., and Shalaev, V. M. (2006). Negative index metamaterial combining magnetic resonators with metal films, *Opt. Express*, **14**, p. 7872.

27. Lezec, H. J., Dionne, J. A., and Atwater, H. A. (2007). Negative refraction at visible frequencies, *Science*, **316**, p. 430.

28. Cai, W., Chettiar, U. K., Yuan, H., de Silva, V. C., Kildishev, A. V., Drachev, V. P., and Shalaev, V. M. (2007). Metamagnetics with rainbow colors, *Opt. Express*, **15**, p. 3333.

29. Wang, J. J., Zhang, W., Deng, X. G., Deng, J. D., Liu, F., Sciortino, P., and Chen, L. (2005). High-performance nanowire- grid polarizers, *Opt. Lett.*, **30**, p. 195.
30. Kang, M., Kim, M., Kim, J., and Guo, L. J. (2008). Organic solar cells using nanoimprinted transparent metal electrodes, *Adv. Mater.*, **20**, p. 4408.
31. Guo, J. B., Huard, C. M., Yang, Y., Shin, Y. J., Lee, K. T., and Guo, L. J. (2014). Compact and ITO-free liquid crystal devices using integrated structural color filters and graphene electrode, *Adv. Opt. Mater.*, **2**, p. 435.
32. Hung, K., Pei, C., Hu, C., and Yang, T. (2008). Transflective TFT-LCDs with adjustable colour gamut, *Displays*, **29**, p. 526.
33. Wang, S. S., and Magnusson, R. (1993). Theory and applications of guided-mode resonance filters, *Appl. Opt.*, **32**, p. 2606.
34. Wang, S. S., Magnusson, R., Bagby, J. S., and Moharam, M. G. (1990). Guided-mode resonances in planar dielectric-layer diffraction gratings, *J. Opt. Soc. Am. A*, **7**, p. 1470.
35. Kaplan, A. F., Xu, T., and Guo, L. J. (2011). High efficiency resonance-based color filters with tunable transmission bandwidth fabricated using nanoimprint lithography, *Appl. Phys. Lett.*, **99**, p. 143111.
36. Ahn, S. H., and Guo, L. J. (2008). High-speed roll-to-roll nanoimprint lithography on flexible plastic substrates, *Adv. Mater.*, **20**, p. 2044.
37. Ok, J. G., Ahn, S. H., Kwak, M. K., and Guo, L. J. (2013). Continuous and high-throughput nanopatterning methodologies based on mechanical deformation, *J. Mater. Chem. C*, **1**, p. 7681.
38. Zhou, W., Gao, H., and Odom, T. W. (2010). Toward broadband plasmonics: tuning dispersion in rhombic plasmonic crystals, *ACS Nano*, **4**, p. 1241.
39. Genet, C., and Ebbesen, T. W. (2007). Light in tiny holes, *Nature*, **445**, p. 39.
40. Lee, K., Chen, P., Wu, S., Huang, J., Yang, S., and Wei, P. (2012). Enhancing surface plasmon detection using template-stripped gold nanoslit arrays on plastic films, *ACS Nano*, **6**, p. 2931.
41. Yao, J., Le, A., Schulmerich, M. V., Maria, J., Lee, T., Gray, S. K., Bhargava, R., Rogers, J. A., and Nuzzo, R. G. (2011). Soft embossing of nanoscale optical and plasmonic structures in glass, *ACS Nano*, **5**, p. 5763.
42. Schmidt, M. A., Lei, D. Y., Wondraczek, L., Nazabal, V., and Maier, S. A. (2012). Hybrid nanoparticle-microcavity-based plasmonic nanosensors with improved detection resolution and extended remote-sensing ability, *Nat. Commun.*, **3**, p. 1108.

43. Kumar, K., Duan, H., Hegde, R. S., Koh, S. C. W., Wei, J. N., and Yang, J. K. W. (2012). Printing color at the optical diffraction limit, *Nat. Nanotechnol.*, **7**, p. 557.

44. Brown, L. V., Sobhani, H., Lassiter, J. B., Nordlander, P., and Halas, N. J. (2010). Heterodimers: plasmonic properties of mismatched nanoparticle pairs, *ACS Nano*, **4**, p. 819.

45. Aksu, S., Yanik, A. A., Adato, R., Artar, A., Huang, M., and Altug, H. (2010). High-throughput nanofabrication of infrared plasmonic nanoantenna arrays for vibrational nanospectroscopy, *Nano Lett.*, **10**, p. 2511.

46. Ross, B. M., Wu, L. Y., and Lee, L. P. (2011). Omnidirectional 3D nanoplasmonic optical antenna array via soft-matter transformation, *Nano Lett.*, **11**, p. 2590.

47. Pasquale, A. J., Reinhard, B. M., and Dal Negro, L. (2011). Engineering photonic–plasmonic coupling in metal nanoparticle necklaces, *ACS Nano*, **5**, p. 6578.

48. Ikeda, K., Takahashi, K., Masuda, T., Kobori, H., Kanehara, M., Teranishi, T., and Uosaki, K. (2012). Structural tuning of optical antenna properties for plasmonic enhancement of photocurrent generation on a molecular monolayer system, *J. Phys. Chem. C*, **116**, p. 20806.

49. Hao, J., Wang, J., Liu, X., Padilla, W. J., Zhou, L., and Qiu, M. (2010). High performance optical absorber based on a plasmonic metamaterial, *Appl. Phys. Lett.*, **96**, p. 251104.

50. Cattoni, A., Ghenuche, P., Haghiri-Gosnet, A., Decanini, D., Chen, J., Pelouard, J., and Collin, S. (2011). $\lambda3/1000$ plasmonic nanocavities for biosensing fabricated by soft UV nanoimprint lithography, *Nano Lett.*, **11**, p. 3557.

51. Le Perchec, J., Desieres, Y., Rochat, N., and de Lamaestre, R. E. (2012). Subwavelength optical absorber with an integrated photon sorter, *Appl. Phys. Lett.*, **100**, p. 113305.

52. Le Perchec, J., Quemerais, P., Barbara, A., and Lopez-Rios, T. (2008). Why metallic surfaces with grooves a few nanometers deep and wide may strongly absorb visible light, *Phys. Rev. Lett.*, **100**, p. 066408.

53. Pardo, F., Bouchon, P., Haidar, R., and Pelouard, J. (2011). Light funneling mechanism explained by magnetoelectric interference, *Phys. Rev. Lett.*, **107**, p. 93902.

54. Polyakov, A., Cabrini, S., Dhuey, S., Harteneck, B., Schuck, P. J., and Padmore, H. A. (2011). Plasmonic light trapping in nanostructured metal surfaces, *Appl. Phys. Lett.*, **98**, p. 203104.

55. Wu, Y. R., Hollowell, A. E., Zhang, C., and Guo, L. J. (2013). Angle-insensitive structural colours based on metallic nanocavities and coloured pixels beyond the diffraction limit, *Sci. Rep.*, **3**, p. 1194.
56. Hollowell, A. E., and Guo, L. J. (2013). Nanowire grid polarizers integrated into flexible, gas permeable, biocompatible materials and contact lenses, *Adv. Opt. Mater.*, **1**, p. 343.
57. Kang, M. G., and Guo, L. (2007). Nanoimprinted semitransparent metal electrodes and their application in organic light-emitting diodes, *Adv. Mater.*, **19**, p. 1391.
58. Marty, R., Baffou, G., Arbouet, A., Girard, C., and Quidant, R. (2010). Charge distribution induced inside complex plasmonic nanoparticles, *Opt. Express*, **18**, p. 3035.
59. Dionne, J. A., Sweatlock, L. A., Atwater, H. A., and Polman, A. (2006). Plasmon slot waveguides: towards chip-scale propagation with subwavelength-scale localization, *Phys. Rev. B*, **73**, p. 035407.
60. Zhou, J., and Guo, L. J. (2014). Transition from a spectrum filter to a polarizer in a metallic nano-slit array, *Sci. Rep.*, **4**, p. 3614.
61. Collin, S., Pardo, F., Teissier, R., and Pelouard, J. (2001). Strong discontinuities in the complex photonic band structure of transmission metallic gratings, *Phys. Rev. B*, **63**, p. 033107.
62. Collin, S., Pardo, F., Teissier, R., and Pelouard, J. (2002). Horizontal and vertical surface resonances in transmission metallic gratings, *J. Opt. A: Pure Appl. Opt.*, **4**, p. S154.
63. Lalanne, P., and Hugonin, J. P. (2006). Interaction between optical nano-objects at metallo-dielectric interfaces, *Nat. Phys.*, **2**, p. 551.
64. Lalanne, P., Hugonin, J. P., Liu, H. T., and Wang, B. (2009). A microscopic view of the electromagnetic properties of sub-λ metallic surfaces, *Surf. Sci. Rep.*, **64**, p. 453.
65. Nikitin, A. Y., Garcia-Vidal, F. J., and Martin-Moreno, L. (2010). Surface electromagnetic field radiated by a subwavelength hole in a metal film, *Phys. Rev. Lett.*, **105**, p. 073902.
66. Shen, J., Catrysse, P., and Fan, S. (2005). Mechanism for designing metallic metamaterials with a high index of refraction, *Phys. Rev. Lett.*, **94**, p. 197401.
67. Lucas, B. D., Kim, J., Chin, C., and Guo, L. J. (2008). Nanoimprint lithography based approach for the fabrication of large-area, uniformly-oriented plasmonic arrays, *Adv. Mater.*, **20**, p. 1129.

68. Kats, M. A., Blanchard, R., Genevet, P., and Capasso, F. (2013). Nanometre optical coatings based on strong interference effects in highly absorbing media, *Nat. Mater.*, **12**, pp. 20–24.
69. Lee, K.-T., Seo, S., Lee, J. Y., and Guo, L. J. (2013). Angle-insensitive reflective color filters using lossy materials, *2013 IEEE Photonics Conference, IPC 2013*, p. 6656349.

Chapter 12

Plasmonic Microscopy for Biomedical Imaging

Chun-Yu Lin, Ruei-Yu He, Yuan-Deng Su, and Shean-Jen Chen
College of Photonics, National Chiao Tung University, Tainan 711, Taiwan
sheanjen@nctu.edu.tw

This chapter describes plasmonic microscopy, which, through surface plasmons (SPs), offers high sensitivity and local electromagnetic (EM) field enhancement to satisfy the requirements of biomedical imaging in real time. For a better understanding, the surface plasmon resonance (SPR) phase and fluorescence-enhanced microscopes based on a prism or objective-coupled attenuated total reflection (ATR) configuration are demonstrated. First of all, an SPR phase imaging system based on modified Mach–Zehnder phase-shifting interferometry (PSI) is adopted to detect DNA hybridization. In addition, a common-path PSI technique via an electro-optic modulator (liquid crystal phase retarder) provides long-term phase imaging stability for real-time kinetic studies in biomolecular interaction analysis (BIA). Furthermore, surface plasmon–enhanced–total internal reflection fluorescence (SPE-TIRF) microscopy provides bright live-cell images via surface plasmon–enhanced fluorescence (SPEF), and has been successfully

Plasmonics and Super-Resolution Imaging
Edited by Zhaowei Liu

ISBN 978-981-4669-91-7 (Hardcover), 978-1-315-20653-0 (eBook)
www.panstanford.com

used for the real-time observation of the thrombomodulin protein of a living cell membrane. In order to exploit SPE-TIRF microscopy, an ultrafast laser is also adopted. With the combination of SP enhancement and two-photon excitation, TIRF microscopy features brighter and more contrasted fluorescence membrane images. Furthermore, a theoretical model based on the Fresnel equation and classical dipole radiation modeling is employed to investigate the SP enhancement and quenching of two-photon excited fluorescence. The local electric field enhancement, fluorescence quantum yield, and fluorescence emission coupling yield via SPs are theoretically analyzed at different dielectric spacer thicknesses between the fluorophore and the metal film. Ultimately, with the combination of the SPR phase and SPE-TIRF techniques, plasmonic microscopy simultaneously enables the sensitive phase and fluorescence-enhanced images of living-cell contacts on the surface of a biosubstrate.

12.1 Introduction

Surface plasmons (SPs) are coherent electron oscillations that exist at the interface between a dielectric medium and a metal. The incidence of light on this interface can cause an excitation of these SPs, and when the momentum of the incident light matches that of the SPs, the so-called surface plasmon resonance (SPR) phenomenon takes place [1]. Since this wave-matching situation is disrupted by even very minute changes in the interface conditions, if the light excitation conditions are constant, then the SPR technique permits any changes in the refractive index or thickness of the medium adjacent to the metal film or in the adsorption layer on the metal surface to be measured precisely. SPR devices are valuable tools for real-time biomolecular interaction analysis (BIA) applications and have the major advantage of not requiring the labeling of analytes. Therefore, SPR biosensors have been applied in a wide range of fields, including molecular recognition, disease immunoassays, etc. [2, 3].

The most conventional SPR biosensing configuration is based on a prism-coupled attenuated total reflection (ATR) method that employs four main detection approaches, namely angular interrogation, wavelength interrogation, intensity measurement, and phase measurement, to investigate any changes in an immobilized layer deposited on a metal surface [3, 4]. It has been shown that modern SPR sensing devices based on the angular interrogation ATR method with an angular resolution of 10^{-4} degrees can achieve a maximum detection limit of approximately 5×10^{-7} RIU (refraction index unit), which corresponds to a 1 pg/mm^2 surface coverage of biomolecules [5]. Although conventional SPR biosensors have high sensitivity compared to other label-free devices, they are nevertheless unable to directly detect interactions involving the biomolecules of low molecular weight ($<$200 Daltons) or biomolecules which are present in low concentrations on the surface of the biosensor. Consequently, various approaches have been developed to enhance the sensitivity and resolution of biosensors by using different SPR modes or advanced detection methods [6–9].

Current research trends concerning advanced SPR biosensors and systems not only include the improvement of their sensitivity, resolution, stability, and speed, but also the development of high-throughput screening capabilities. Achieving this screening capability while simultaneously retaining a detection limit of 1 pg/mm^2 requires the use of SPR microarrays and associated detection systems. In response, a novel SPR phase imaging system was developed based on modified Mach–Zehnder phase-shifting interferometry (PSI) in which the spatial phase variation of the resonantly reflected light can be observed for high-throughput real-time BIA, such as label-free DNA microarray hybridization. It is shown that the detection limit of the developed SPR-PSI imaging system is improved to approximately 1 pg/mm^2 at each individual spot [10]. Nevertheless, in common with other SPR phase imaging systems, the developed system was unable to satisfy the rigorous demands of real-time BIA kinetic studies because it lacks long-term stability [11–13]. Hence, the common-path PSI technique is adopted to develop an SPR phase imaging system with long-term stability and high-resolution capabilities [14]. The proposed system exploits the advantages of the common-path PSI technique to compensate for the phase drift caused by environmental change, mechanical vibration, and light source fluctuation. With the long-term stability issue solved and high sensitivity, SPR phase microscopy can be extended to real-time observation of various cell behaviors such as adhesion, migration, and apoptosis from the contact interactions and distances between cell membranes and biosubstrate [15].

Fluorescence is widely used in biological immunoassays, optical devices, cell and tissue imaging, and medical diagnosis. In total internal reflection fluorescence (TIRF) microscopy, an evanescent field is used to excite and image fluorophores on or very near (i.e., within 100 nm) the liquid/solid interface. Experimental evidence suggests that the molecular interactions that occur on or near the surfaces of cell membranes (typically within 50 nm) have different properties from those which occur in bulk solution [16]. Fluorophores located within or close to the plasma membrane are excited by the shallow evanescent field, resulting in images with very low background fluorescence and no unfocused fluorescence. TIRF microscopy is particularly suited to the investigation of living cells and provides a complementary approach that can be readily combined with other microscopy techniques [17]. Therefore, TIRF microscopy has been widely used in studies of cell–substrate contact regions [18], protein dynamics [19], endocytosis or exocytosis [20], and membrane-associated photosensitizers [21]. Combined with SP enhancement, evanescent field excited SPs on metallic surfaces enhance the local EM field around the fluorophores; as a result, the intensity of the detected fluorescence signal increases, which in turn, increases the brightness and acquisition frame rate of living cell membrane images [22, 23]. Moreover, coupled with an ultrafast laser, SP enhancement can also be employed for enhanced two-photon fluorescence, directional two-photon-induced surface plasmon–coupled emission (SPCE), and local second-harmonic-generation enhancement [24–27]. Here in, two-photon excited surface plasmon–enhanced–total internal reflection fluorescence (SPE-TIRF) microscopy dynamically images the interactions of proteins near the cell membrane. The experimental results demonstrate that the living cell membrane images provided by the two-photon excited SPE-TIRF microscopy not only clearly reveal higher signal-to-noise ratios (SNRs) compared to those of one-photon excited TIRF microscopy but are also approximately 30-fold brighter than those of the two-photon excited TIRF approach with an equivalent image quality [28]. Finally, the combination for SPR phase microscopy and SPEF microscopy enables the simultaneous observation of living-cell contacts on the surface of a biosubstrate [29].

12.2 Metal Film Preparations

A suitable metallic thin film is needed to excite SPs for SPR phase and SPE-TIRF microscopy applications. Gold (Au) and silver (Ag) films are ideal candidates for SP excitation in the visible and near-infrared (NIR) range due to their complex dielectric constants with a large negative real part. For SPR phase imaging, Au was chosen for the metallic thin film since it is rather stable in biochemical procedures and has a shorter SP wave propagation length. A sacrificial layer of 3 nm thick chromium (Cr) was precoated on the cover slip to increase the Au adhesion on the sample substrate; then, a Au film of suitable thickness is deposited to obtain a good sensitivity for the SPR phase imaging. In the SPEF imaging, the local electric field enhancement is an important parameter; hence, an Ag film, which offers superior electric field enhancement, is adopted. When the thickness of the Ag film is about 50 nm, SP excitation based on the ATR method can be optimized to greatly enhance the local electric field. However, collection of the excited fluorescence is significantly reduced because the 50 nm Ag film decreases the SPCE efficiency. To overcome this shortcoming, a thinner Ag film capable of local electric field enhancement and improved fluorescence detection efficiency should be adopted [23]. In addition to the thickness of the metallic film, its surface roughness is also an important parameter for optimizing the SP effects. SPs on rough surfaces lead to phenomena such as broadened SPR curves and decreased SPR sensitivities. The influence of surface roughness can be evaluated via the variation of metal permittivity [1, 30–32]. More specifically, the surface roughness of an Au film deposited on a dielectric substrate by the use of techniques such as sputtering, e-beam evaporation, and chemical vapor deposition, is around 1.0 nm in root mean square (RMS); in contrast, the surface roughness of an Ag film is a few nanometers in RMS, which is unfavorable for the optimization of SP effects [33–37]. In a recent study, an ultrasmooth Ag film was achieved by e-beam evaporation onto Si (100) substrates by the use of a germanium (Ge) nucleation layer [38]. Through this technique, a surface roughness below 1.0 nm in RMS can be achieved. A similar technique by radio-frequency (RF) magnetron sputtering can deposit an Ag film on the Ge nucleation layer placed on a BK7 cover slip. Without the Ge nucleation layer, the average RMS surface roughness of a pure Ag film is around 4.7 nm, with a peak-to-valley height of around 27.0 nm. In comparison, the average RMS surface roughness and the peak-to-valley height of the Ag/Ge film improved to around 1.0 nm and around 8.0 nm, respectively. Therefore, the Ag/Ge film deposited on the cover slip by direct RF sputtering exhibits a smaller RMS surface roughness and a narrower peak-to-valley surface topological height distribution as compared to when the Ag film is directly deposited onto the cover slip [38, 39]. Specifically, artifacts associated with rough surfaces (strong scatterings, broadened SPR curves, and decreased SPR sensitivities) can be reduced [1, 30–32].

12.3 SPR Phase Microscopy

SPR-PSI is a novel technique which combines SPR and modified Mach–Zehnder PSI to measure the spatial phase variation caused by biomolecular interactions upon a sensing chip. The SPR-PSI imaging system offers high-resolution and high-throughput screening capabilities for microarray DNA hybridization without the need for additional labeling and provides valuable real-time quantitative information. The SPR-PSI imaging system has an enhanced detection limit of 2.5×10^{-7} RIU and a long-term phase stability of $\pi/100$ RMS within 30 min [10]. To improve the long-term phase stability, a common-path SPR-PSI imaging system is also proposed. Currently, the common-path SPR-PSI technique by means of a five-step phase reconstruction method with a liquid crystal device (LCD) can greatly extend the long-term stability despite any external disturbances, such as mechanical vibrations, buffer-flow noise, and laser instability, to fulfill the demands of real-time kinetic studies. The SPR-PSI imaging system has achieved the long-term phase stability of $\pi/400$ RMS over four hours [14]. Finally, the developed common-path SPR-PSI imaging system is used to observe cell–biosubstrate contacts. The developed SPR phase microscopy is highly sensitive to cell membrane contact with biosubstrates and also provides the long-term phase stability needed to achieve time-lapse living-cell observation. In this regard, an SPR intensity and phase sensitivity comparison demonstrates that the sensitivity of the phase measurement can be 100-fold greater than that of the intensity measurement. Also, a longer than 2-hour cell apoptosis observation via the SPR phase microscopy is presented [15].

12.3.1 DNA Microarray Sensing

Herein, array biosensing is employed to provide a high-throughput BIA screening capability. To bind the probed DNA onto the thin Au film slide, the slide was first immersed in a 1 mM thiol solution ($HS(CH_2)_{15}COOH$) for 6 hours to form a self-assembled monolayer (SAM) and then placed in an *N*-ethyl-*N*-(3-dimethylaminopropyl) carbodiimide hydrochloride (EDC)/*N*-hydroxysuccinimide (MES) solution (EDC of 100 mg immersed in 40 mM MES of 100 mL) for a further 6 hours for surface activation. After cleaning the slide with deionized (DI) water and alcohol, the probed DNA was mechanically spotted in a matrix arrangement. Finally, a blocking solution (methanol) was applied to modify the functional group of thiol –COOH into $-CH_3$ to prevent the target DNA from being captured on the free thiol/EDC area.

Figure 12.1 presents five interference frames for the 28 thiol/ssDNA spots in a 6×8 mm^2 area scanned by a charge-coupled device (CCD) camera with a resolution of 640×480 pixels. Each spot has a diameter of 0.5 mm and the distance between each probe ssDNA spot is 1 mm. The targeted ssDNA spots can display DNA hybridization if the intense local variations of the interference pattern occur. It should be noted that Frame 5 is created from Frame 1 by consecutively incrementing the phase by $\pi/2$ four times. Therefore, Frame 5 is similar to Frame 1 since the 2π phase shift between them can be neglected. The phase difference is reconstructed using the five-step phase reconstruction algorithm, and then the BIA is analyzed according to the phase difference by means of Fresnel's calculations. Figure 12.2a presents

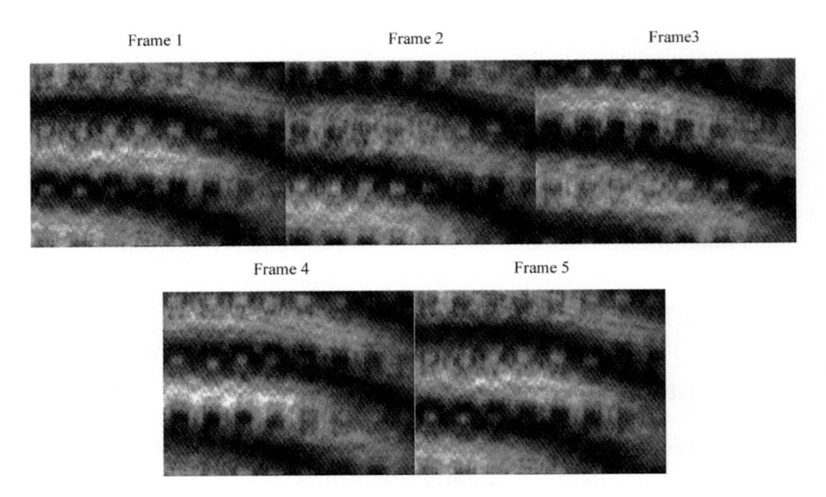

Figure 12.1 Five interference frames for 32 thiol/ssDNA spots recorded by a CCD camera with 640 × 480 pixels.

the reconstructed phase jump associated with six probed ssDNA spots. The phase difference between the areas with and without DNA is seen to be approximately 0.2π. The local resolution is $\pi/100$ with a $100 \times 100\ \mu m^2$ detection area, that is, the size of a targeted DNA spot. Hence, by maintaining a detection resolution of approximately 1 pg/mm^2 surface coverage of biomaterial, the screening area can simultaneously monitor up to 2500 individual spots in a $10 \times 10\ mm^2$ area. As a result, the requirement for a high-throughput screening capability for microarray DNA hybridization diagnostic purposes is achieved. Figure 12.2b shows the reconstructed phase variation in three different scanning lines and clearly distinguishes between the locations of the screening area with and without the probed ssDNA. The results demonstrate that all spots of the SPR DNA microarray can be detected simultaneously in a single experiment and hence meets the requirements of DNA sequence diagnostics. The SPR DNA microarray can be extensively applied to BIAs, such as protein microarrays. Furthermore, the developed SPR phase imaging system and its SPR DNA microarray can be used to observe DNA microarray hybridization in real time, with high sensitivity, and at high-throughput screening rates.

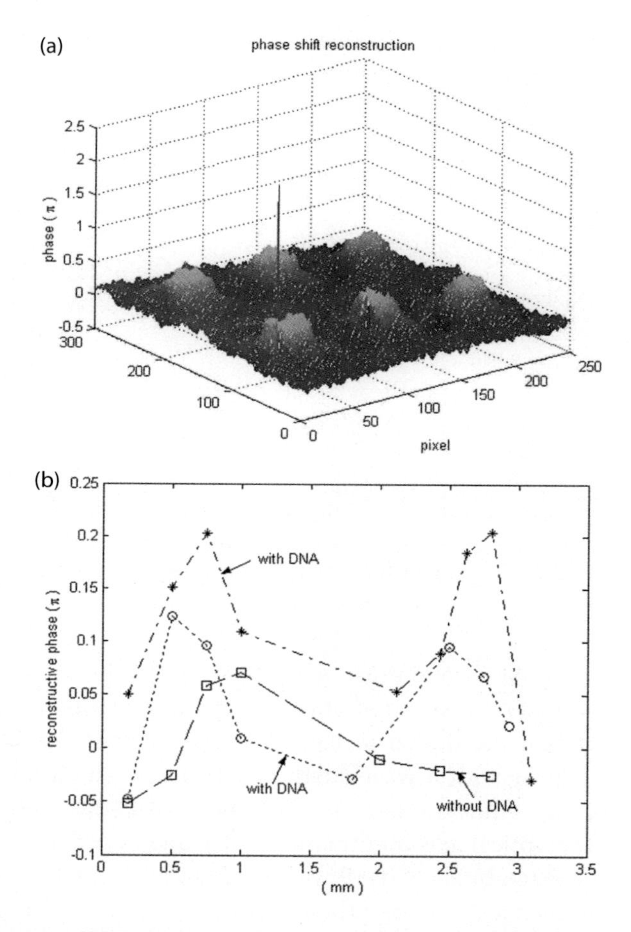

Figure 12.2 SPR-PSI data analysis for six thiol/ssDNA spots. (a) Reconstructed phase shift between sensing area with and without probed ssDNA determined from the five-step phase reconstruction algorithm; (b) reconstructed phase variation in three different scanning lines.

To improve the long-term phase stability for BIA, a common-path PSI technique was developed to compensate for the phase drift caused by environmental change, mechanical vibration, and light source fluctuation. The current experimental setup is illustrated schematically in Fig. 12.3. In this setup, the P-wave and S-wave pass through a common path and experience spatial phase variations that

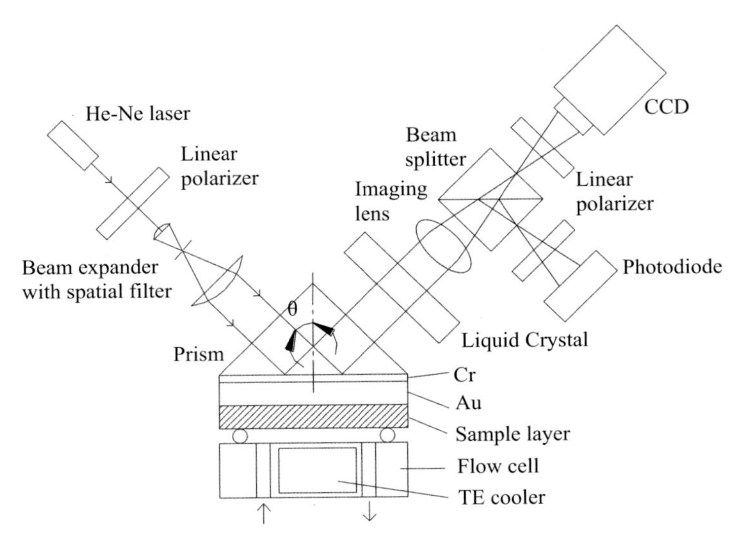

Figure 12.3 System setup of the prism-coupled common-path SPR phase microscope.

would destroy SPR conditions. A five-step PSI technique enables the high-resolution inspection of the spatial distribution of the 2D phase variation. As shown in Fig. 12.3, the incident light source is a He-Ne laser with a wavelength of 632.8 nm. The laser beam passes through a linear polarizer positioned at an adjustable angle between the optical axis and the incident plane in order to adjust the relative magnitudes of the P-wave and S-wave components. The incident beam then passes through a beam expander such that it covers the entire biosample surface. The expanded beam is coupled by a prism to generate SP waves in accordance with the ATR method, allowing a relative phase variation to be introduced between the P-wave and S-wave. The reflected signal beam passes through an electro-optic modulator (liquid crystal phase retarder), which is positioned such that the P-wave falls on the fast axis and the S-wave falls on the slow axis. A sequential phase shift is induced by applying suitable voltages to the modulator's liquid crystal. The biosample information is then mapped by the CCD camera via an imaging lens. Subsequently, the coming beam is split into two separate beams, namely the deflected beam and the straight beam. The former is

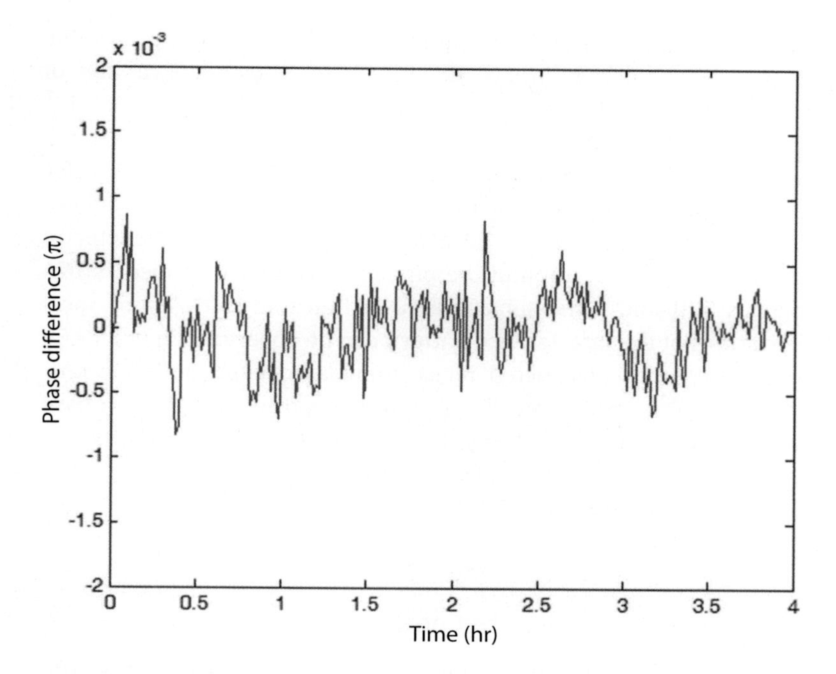

Figure 12.4 Long-term phase stability of $\pi/400$ RMS over four hours.

passed through a polarizer transmitting only the P-wave and is then focused onto a photodiode in order to detect the SPR angle. Meanwhile, the straight beam, containing both P-wave and S-wave components, is adjusted by the orientation of its optical axis for final contrast. Ultimately, five sequential interference patterns are acquired by the CCD camera in order to analyze the phase variation [14].

For testing the system phase stability, the incident angle is adjusted to the SPR angle and nitrogen gas (N_2) is used as the buffer sample with a flow rate of 100 μL/min and a constant temperature of 30°. As shown in Fig. 12.4, the long-term phase stability is found to be $\pi/400$ RMS over four hours. To evaluate the detection limit of the system, the sample was alternated between nitrogen and argon gas every 5 min with a constant flow rate of 100 μL/min and a temperature of 30°. The refraction indices of N_2 and argon (Ar) are known to differ by 1.5×10^{-5}, for which the phase variation is

$\pi/100$. With a short-term (<30 min) phase stability of $\pi/10,000$, the developed common-path SPR-PSI imaging system is capable of resolving a 2D refraction index difference of approximately 2×10^{-7} for sample variation.

Figure 12.5a presents five interference frames of the four 15-mer ssDNA spots captured by a 1.0×1.0 mm^2 area-scan CCD camera with a resolution of 50×50 pixels. Each spot has a diameter of approximately 200 μm and the pitch between each probed ssDNA spot is 500 μm. If the sensing area includes an ssDNA spot, local variations of the interference pattern can be observed. Figure 12.5b presents the reconstructed phase jump associated with the four probed ssDNA spots, while Fig. 12.5c demonstrates the spatial phase variation of a line cut from Fig. 12.5b. The phase difference between the areas with and without DNA is seen to be approximately 0.5π. The local resolution is $\pi/1000$, that is, the size of the probed DNA spot.

12.3.2 Cell–Biosubstrate Contacts

To form a biocompatible monolayer on the surface of the SPR sensing slide, the activated thiol-modified slide was immersed in a bovine serum albumin (BSA) solution with a concentration of 10 mg/mL in phosphate buffered saline (PBS) for 1 hour to form a BSA SAM, and finally rinsed with PBS. Then, B16F10 murine melanoma cells were maintained in Dulbecco's modified Eagle's medium (DMEM) supplemented with 0.292 g/L of L-glutamine, 2% sodium bicarbonate, 1% sodium pyruvate, 1% penicillin, and 10% fetal bovine serum at 37°C under a humid atmosphere and 5% CO_2. For the SPR phase imaging experiments, the cells were cultured on the BSA-coated Au thin film modified with the thiol SAM.

In Fig. 12.6, an oil-immersion objective-based SPR imaging microscope is proposed, and can be used for real-time observation of various cell behaviors such as adhesion, migration, and apoptosis to obtain information of interaction and distance between cell membrane and substrate. The light source is a He-Ne laser (5 mW, Melles Griot) with a wavelength of 632.8 nm. A linear polarizer positioned at an adjustable angle between the optical axes controls the P-wave and S-wave components. The incident beam then passes through

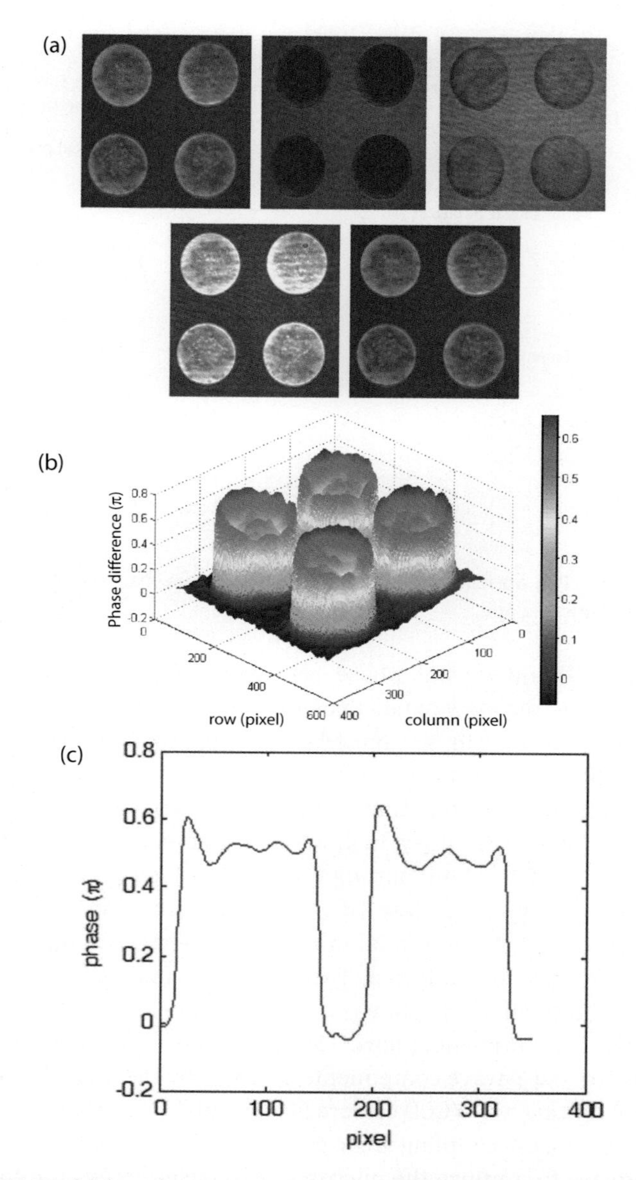

Figure 12.5 (a) Interference patterns at five different phase modulator axis angles in a 50×50 pixel region covering a $1.0 \times 1.0\ \text{mm}^2$ physical area; (b) spatial distribution of phase shift recovered from DNA microarray; and (c) spatial phase variation of a line cut.

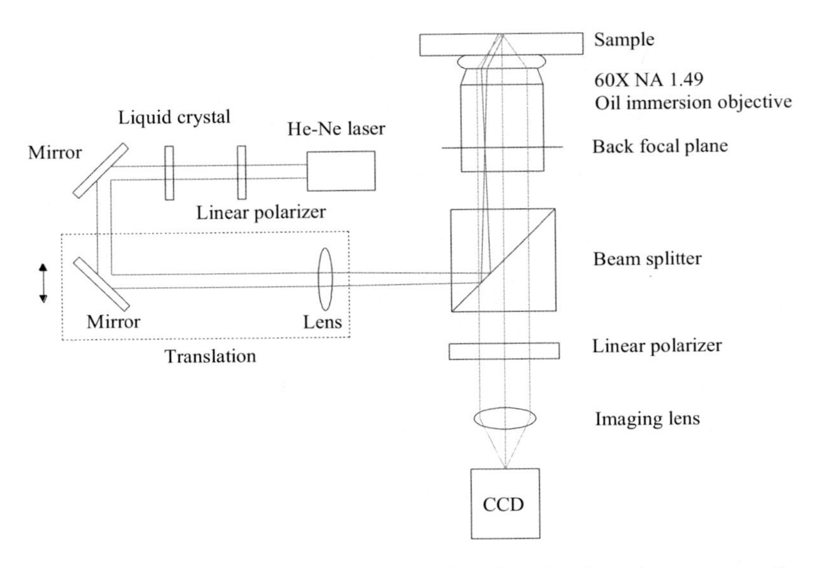

Figure 12.6 The optical configuration of an objective-based common-path SPP phase microscope.

a nematic liquid crystal phase retarder (LVR-200, Meadowlark Optics), while the slow axial is positioned along the S-wave. The S-wave phase delay can be applied by driving appropriate voltages for the PSI. The light passes a beam expander with a spatial filter to form a smooth collimating beam, and is then focused onto the back focal plane (BFP) of a high NA oil-immersion micro-objective (60X, NA = 1.49, Nikon) through a focusing lens (f = 400 mm). A 0.17 mm thin cover slide (BK7) coated with a 46 nm Au thin film is placed on top of the oil-immersion objective, and refraction index matching oil is injected. The SPs are excited at near the SPR angle, and the reflected light passes through a beam splitter, a linear polarizer, and an imaging lens. The interference patterns between the P-wave and S-wave components, induced by the linear polarizer, are then captured by a CCD camera (TM-1320-15CL, PULNiX). In this configuration, the coupling angle can be adjusted from 0° to 79.5°.

Here, we first utilize the microscope to observe the apoptosis of melanoma cells. The cells were placed on the Au-coated and BSA-modified biocompatible substrate. After a 5-hour cell incubation period, the substrate with the cells was observed under the SPR-

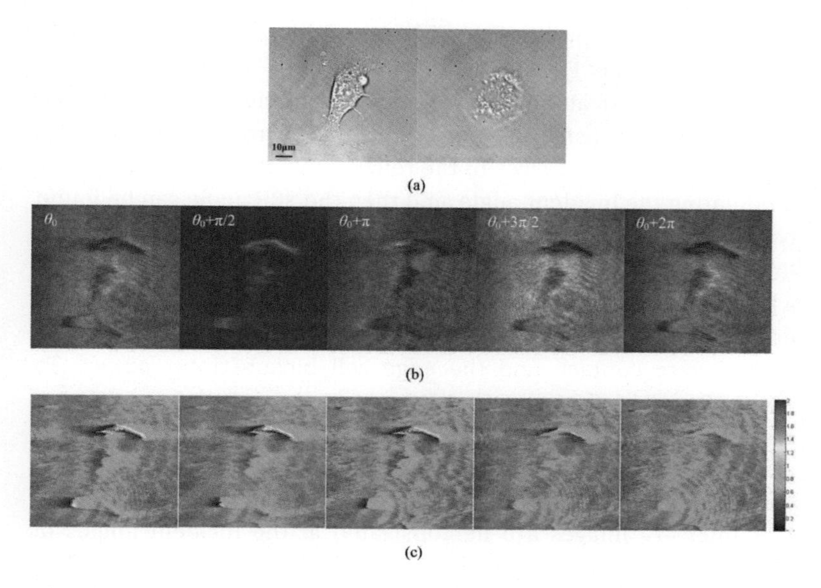

Figure 12.7 Images of the B16F10 cell apoptosis with time-lapse observation. (a) The left picture is the bright-field epi-illuminated image at the 50th minute after the cell was seeded on the biosubstrate, while the right picture is the epi-illuminated image at the 120th minute after the seeding. (b) Five-step interference images by adjusting the S-wave phase delay from 0 to 2π with an initial phase delay θ_0. (c) SP phase images are real-time-measured from the 50th to the 120th minute with a 20 min time interval.

imaging microscope. Figure 12.7 shows the images of the B16F10 cell apoptosis with time-lapse observation. The left-most picture of Fig. 12.7a is an epi-illuminated image of the living cell adhered on the biosubstrate after 50 min of real-time observation. By changing the phase delay of the S-wave via the LCD, the five-step PSI-SPR images of the cell membrane on the biosubstrates were taken, as shown in Fig. 12.7b. The complete SPR-phase image was obtained by the phase-reconstructed algorithm and our developed multichannel phase-wrapping algorithm [40], as shown in the first image of Fig. 12.7c. Also, cell detaching from the biosubstrate was observed after 50 min of seeding. The remaining images of Fig. 12.7c show the complete SPR phase images of the cell apoptosis at the 70th, 90th, 110th, and 130th minute with 20 min time intervals. There still exists slight residual cell–biosubstrate contact, as shown in the last image of Fig. 12.7c. The epi-illuminated image of the cell death is shown in the right picture of Fig. 12.7a.

The cell–substrate contact distance information can be measured by both SPR intensity and phase imaging microscopy. Herein, the intensity and phase curves of the reflected light can be obtained by modulating the incident angle, and then the SPR angle can be found based on the intensity and phase curves with polynomial curve fitting. Figure 12.8a shows the bright-field epi-illuminated image of a living melanoma cell. Three spots, labeled A, B, and C, respectively, denote background locations as follows: no cell contact, inside cell–biosubstrate contact, and the edge cell–biosubstrate contact. To obtain P-wave reflectivity, the linear polarizer, which is positioned before the imaging lens (see Fig. 12.6), is adjusted along the P-wave direction. The incidence angle is scanned from 62° to 78° to estimate the SPR angles for different cell–biosubstrate contact areas. SPR intensity images are demonstrated at the incident angles of 70.75°, 71.45°, 72.19°, 72.96°, 73.75°, 74.59°, 75.48°, and 76.42° in Fig. 12.8b, for which the background images are altered from bright to dark. Also, the intensity of cell location is greater than that of the background at the SPR angle of around 73.75°, as shown in the fifth image of Fig. 12.8b; by contrast, when the incident angle is further increased to approach the SPR angle of the cell location at around 75.48°, the background becomes bright, while the cell location darkens, as shown in the seventh image. Figure 12.8c shows the reflectivity of the three different locations and their corresponding SPR angles. The SPR angle of background (A) is the smallest, while cell locations B and C have greater SPR angles. The greater the SPR angle, the closer the cell is to the biosubstrate.

The SPR phase images can also be obtained by the PSI and phase reconstruction algorithm. The incidence angle is also scanned from 62° to 78° to estimate the SPR angles for different cell–biosubstrate contact areas. Figure 12.9a shows the SPR phase images at the incident angles of 70.75°, 71.45°, 72.19°, 72.96°, 73.75°, 74.59°, 75.48°, and 76.42°. Figure 12.7b shows that the phase jumps of the A, B, and C locations appear at their corresponding SPR angles of around 73.75°, 75.48°, and 74.59°, respectively. Obviously, the phase measurement has a much higher sensitivity than the intensity measurement. The slope of the intensity measurement is about

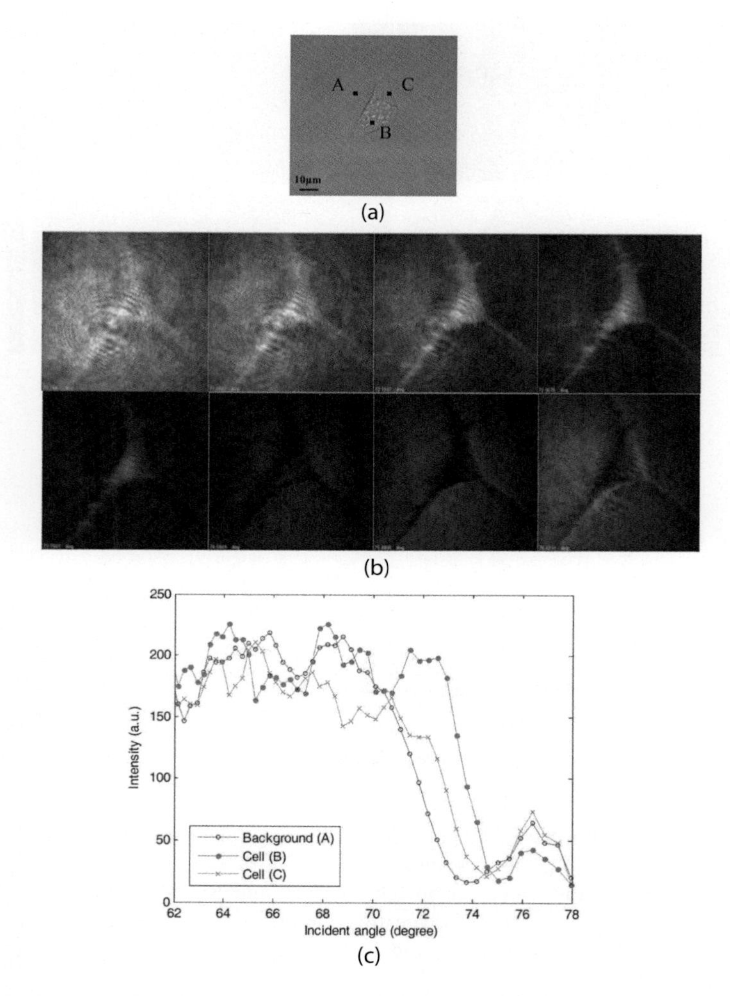

Figure 12.8 SP intensity images with different incident angles: (a) epi-image; (b) SP intensity images at incident angles of 70.75°, 71.45°, 72.19°, 72.96°, 73.75°, 74.59°, 75.48°, and 76.42° from left to right and top to bottom; and (c) SP reflectivity intensity curves of three different locations, A, B, and C, and their corresponding SPR angles at around 73.75°, 75.48°, and 74.59°, respectively.

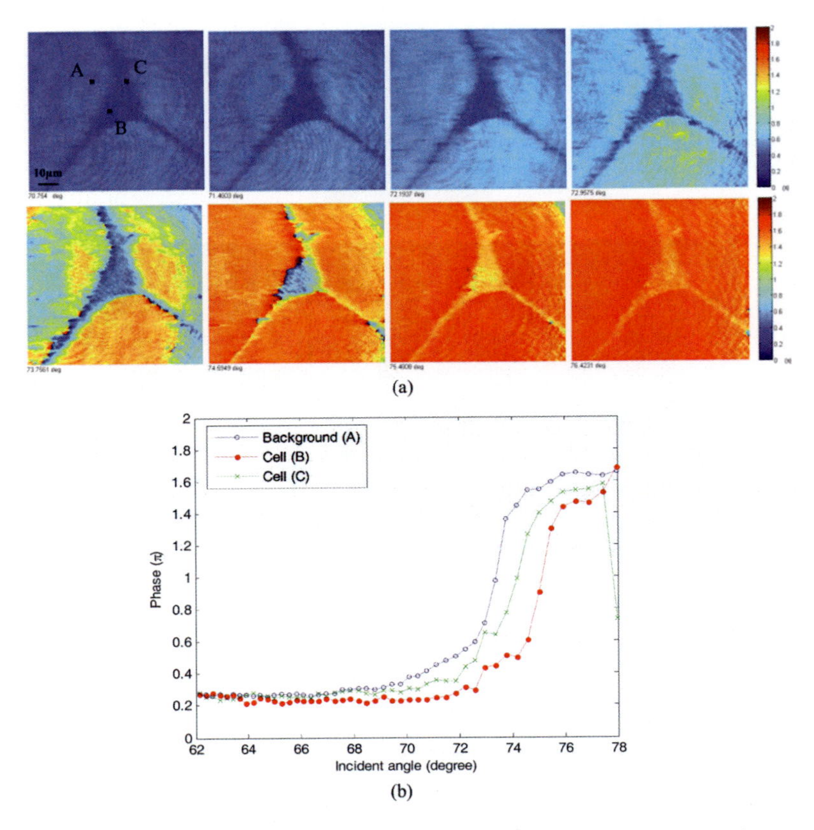

Figure 12.9 SP phase images with different incident angles: (a) SP intensity images at incident angles of 70.75°, 71.45°, 72.19°, 72.96°, 73.75°, 74.59°, 75.48°, and 76.42° from left to right and top to bottom; (b) SP reflectivity intensity curves of three different locations, A, B, and C, and their corresponding SPR angles of around 73.75°, 75.48°, and 74.59°, respectively.

75 a.u. (gray level) per incident angle of degree (a.u./deg) and the slope of the phase measurement can reach 0.938 π/deg. The resolution of the intensity measurement is 13 gray-level units based on the 5% intensity stability in a 256 gray-level CCD camera. The resolution of the phase measurement based on the SPR phase imaging system can be $10^{-3}\pi$ [14]. If the same phase resolution could be achieved in the SPR phase imaging system, the sensitivity of the phase measurement would be about 160 times (0.938

π/deg $10^{-3}\pi$/deg/75 a.u./13 a.u.) higher than that of the intensity measurement, which is similar to the value in SPR biosensing [2, 14]. The phase measurement can achieve a very high sensitivity at a fixed incident angle near the SPR angle, but its linear dynamic range is fairly narrow.

Cell composition is quite complex. It is covered with a plasma membrane and contains cytoplasm and many organelles such as the nucleus, mitochondrion, endoplasmic reticulum, etc. Therefore, the optical property, geometry, and cell structure are difficult to individually analyze with a single imaging technique. Cells can be simplified as a homogeneous and transparent material in the study of cell–biosubstrate contacts. The refractive index (n) and thickness (d) are defined as follows: BK7 cover slip (n = 1.515), Cr (n = $3.135 + 3.31j$, d = 3 nm), Au (n = $0.223 + 3.294j$, d = 46 nm), thiol SAM (n = 1.464, d = 1.49 nm), BSA-SAM (n = 1.4, d = 4.0 nm), PBS solution (n = 1.338), and cell (n = 1.36) at the incident wavelength of 632.8 nm. The multilayer configuration is as follows: BK7/Cr/Au/thiol/BSA/PBS/cell/PBS. The unknown parameters are the cell thickness and cell–substrate distance. Cell thickness is assumed to be greater than 300 nm, so we can simplify the calculation to estimate the cell–substrate distance, which can be inferred by fitting the SPR angle with the phase measurement. Figure 12.10a shows the 3D image of the corresponding SPR angles at different locations. The image is converted by estimating the SPR

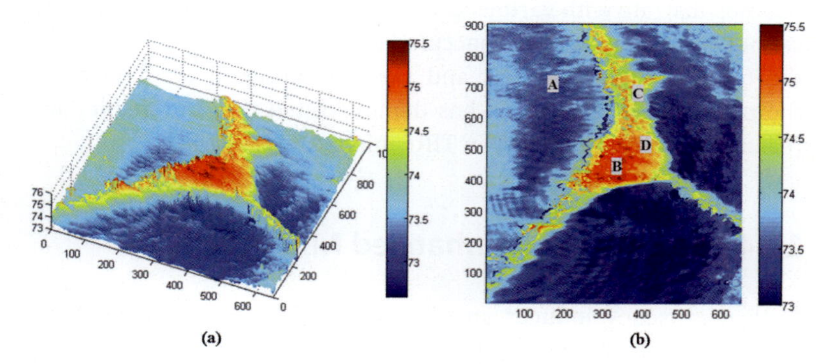

Figure 12.10 (a) 3D image and (b) 2D image of corresponding SPR angles at different locations based on the SP phase measurement. The color scales indicate the SPR angle in degrees.

angles at different locations through the SPR phase images at the incident angles from 70° to 78°. Its 2D image counterpart is shown in Fig. 12.10b.

Four locations are chosen for further discussion, as shown in Fig. 12.10b. First, location A on the background surface is filled by only the PBS solution above the biosubstrate. The simulation result with the configuration of BK7/Cr/Au/thiol/BSA/PBS approaches the corresponding SPR angle of 73.75° from the SPR phase measurement at location A. Locations B and D are situated inside the main cell and contain numerous organelles with typical cell thicknesses of between 10 to 100 μm, which is much thicker than the depth of the SP evanescent field. The upper layer is assumed to be the cell, and hence a configuration of BK7/Cr/Au/thiol/BSA/PBS/cell is adopted. The cell–biosubstrate distances, that is, the PBS gaps, are estimated as 10 nm and 65 nm for B and D, respectively. The biggest SPR angle position locates on B where there is only about a 10 nm space, which means that the cell acts as a focal contact on the biosubstrate. Location C looks like a lamellipodium with an SPR angle of 74.59°. The lamellipodium is very thin compared to the depth of the SP evanescent field, so a configuration of BK7/Cr/Au/thiol/BSA/PBS/cell/PBS is considered reasonable for use. If we assume that the lamellipodium thickness is 40 to 50 nm, then the cell–biosubstrate distance could be estimated as 15 to 30 nm. The experiment results roughly show the cell adhered on the biosubstrate with various contact distances, but the exact values cannot be obtained due to inaccuracies such as the optical property, the topography of the cell, and the measurement error. Despite the inaccuracies, this study has demonstrated that the results are reasonable for the actuality in TIRF microscopy [41–44].

12.4 Fluorescence-Enhanced Microscopy

TIRF microscopy induces an evanescent field from incident light with an incident angle greater than the critical angle to excite fluorescent molecules on or near a surface. TIRF microscopy not only provides enhanced understanding of cellular function but also improved detecting signal SNR in real time. However, the fluorescent emission is needed to be increased when a more dynamic biomolecular image is requested at a frame rate of >100 frames/s. Therefore, the fluorescent signal is enhanced via SPs to match the requirements of more efficiency and quantity. The developed microscope has been successfully used for the real-time observation of SPEF from the thrombomodulin protein of a living cell membrane. Experimental results have demonstrated that SPE-TIRF microscopy can provide brighter living-cell images through SPEF. Further, an ultrafast laser is adopted for SPE-TIRF microscopy. Moreover, a two-photon excited SPE-TIRF microscopy setup not only provides brighter fluorescent images based on the mechanism of local EM field enhancement but also reduces photobleaching. In comparison with one-photon excited TIRF, two-photon excited TIRF can achieve higher SNR cell membrane imaging due its smaller excitation volume and lower scattering. By combining SP enhancement and two-photon excitation TIRF, the microscopy setup has demonstrated its capability for brighter and more contrasted fluorescence membrane images.

Furthermore, the theoretical analysis based on the Fresnel equation and classical dipole radiation modeling is used to verify that two-photon excited fluorescence is enhanced and quenched via SPs. The local electric field enhancement, fluorescence quantum yield, and fluorescence mission coupling yield via SPs are theoretically analyzed at different dielectric spacer thicknesses between the fluorescence dye and the metal film. The theoretical analysis and experimental results both indicate that the two-photon excited SPE-TIRF configuration with a 10 nm SiO_2 spacer can provide an enhanced and less photobleached fluorescent signal via the respective assistance of the enhanced local EM field and quenched fluorescence lifetime.

12.4.1 One-Photon Excited Fluorescence-Enhanced Imaging

Figure 12.11 shows the optical configuration of the proposed prism-coupled SPE one-photon and two-photon TIRF microscope, in which the ATR method is used to excite the SPs in order to enhance the local EM field and to increase the SNR of the

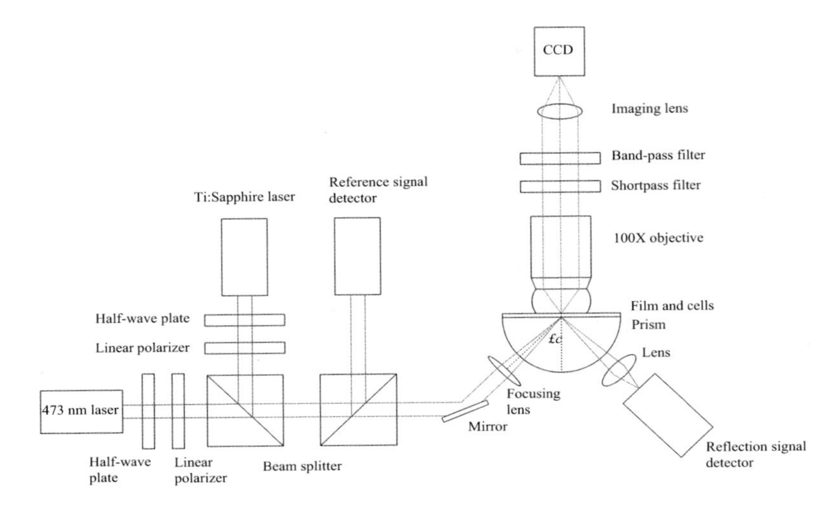

Figure 12.11 Schematic illustration of the experimental configuration employed for live-cell membrane imaging using conventional and SPE-TIRF microscopes with one-photon and two-photon excitations of the same fluorophores.

fluorescence. In this configuration, a thin silver film is deposited via sputtering deposition, followed sequentially by a chemical SAM, and a biomolecular layer, onto an SF-11 slide in accordance with an optimal design that is known to enhance the intensity of the fluorescence and, therefore, increase the acquisition frame rate. To avoid the decay of the local EM field enhancement, an additional chromium layer, that promotes silver adhesion, wasn't deposited on the SF-11 slide. As shown in Fig. 12.11, the beams of a diode-pumped solid-state laser (20 mW, $\lambda = 473$ nm) for one-photon excitation and a 10 W solid-state pumped mode-locked femtosecond Ti:sapphire laser (Tsunami, Spectra-Physics; $\sim$100 fs pulse at 80 MHz, $\lambda =$ 800 nm) for two-photon excitation initially pass through a half-wave plate and a polarizer to adjust their intensity and polarization. Both beams are then focused by a convergence lens ($f = 70$ mm), pass through a coupled SF-11 hemispherical prism, a layer of index-matching oil, and then the SF-11 slide, leading finally to be incident directly upon the interface between the slide and thin silver film. The area of laser illumination is controlled to be approximately 100

$\times$ 100 μm^2, which is comparable to the size of general cells. Using the hemispherical prism, the angle of the incident light can be easily adjusted in order to vary the penetration depth of the evanescent field into the thin silver film (thickness 50.0 $\pm$ 1.0 nm). In the current Kretschmann configuration, the SF-11 hemispherical prism is used to simultaneously increase the wavenumber of the incident light and decrease the SPR angle, that is, the angular position of the dip in the reflectivity spectrum. The fluorescence from the cell membrane excited by the SPs or the evanescent wave is wide-field collected and imaged from the reverse side of the prism by an immersion water objective (100X, NA = 1.0, Olympus) dipped into the flow cell, and is then directed through a short-pass filter (SPF, $\lambda < 680$ nm, Semrock) and a long-pass filter (LPF, $\lambda > 515$ nm, Semrock) into a high-speed frame rate CCD camera (iXon DV885, Andor).

To compare the fluorescence intensity of a conventional TIRF chip with that of the proposed SPE-TIRF chip, cells were added to collagen immobilized by chemical SAMs on a naked SF-11 slide and on a silver thin-film SF-11 slide containing a prewarmed medium, respectively. In fabricating this conventional TIRF chip, the naked SF-11 slide was immersed in 20% (3-aminoproply) triethoxysilane solution in order to form a dense SAM on its surface. To immobilize the protein collagen, covalent activation was conducted by immersing the chip in a solution containing EDC (2 mM) and *N*-hydroxysuccinimide (NHS; 5 mM) for 6 hours. In developing this SPE-TIRF chip, the metal film was immersed in 1 mM 2-aminoethanethiol hydrochloride solution to form a dense SAM on its surface. As the TIR chip, covalent activation was then performed by immersing the chip in a solution containing EDC (2 mM) and NHS (5 mM) for 6 hours to immobilize the protein collagen.

In Figs. 12.12a and 12.12b, the melanoma-GFP-tagged TM cell was cultured on the collagen-coated slide modified with silane and observed by a conventional one-photon TIRF microscope. The exposure times of Figs. 12.12a and 12.12b are, respectively, 0.5 s and 0.04 s. The size of the melanoma-GFP-tagged TM cell is about 30 $\times$ 30 μm^2. Figures 12.12c and 12.12d show a cell cultured on the collagen-coated silver thin film modified with a thiol SAM and observed by the proposed SPE-TIRF microscope. The exposure times of Figs. 12.12c and 12.12d are also 0.5 s and 0.04 s, respectively.

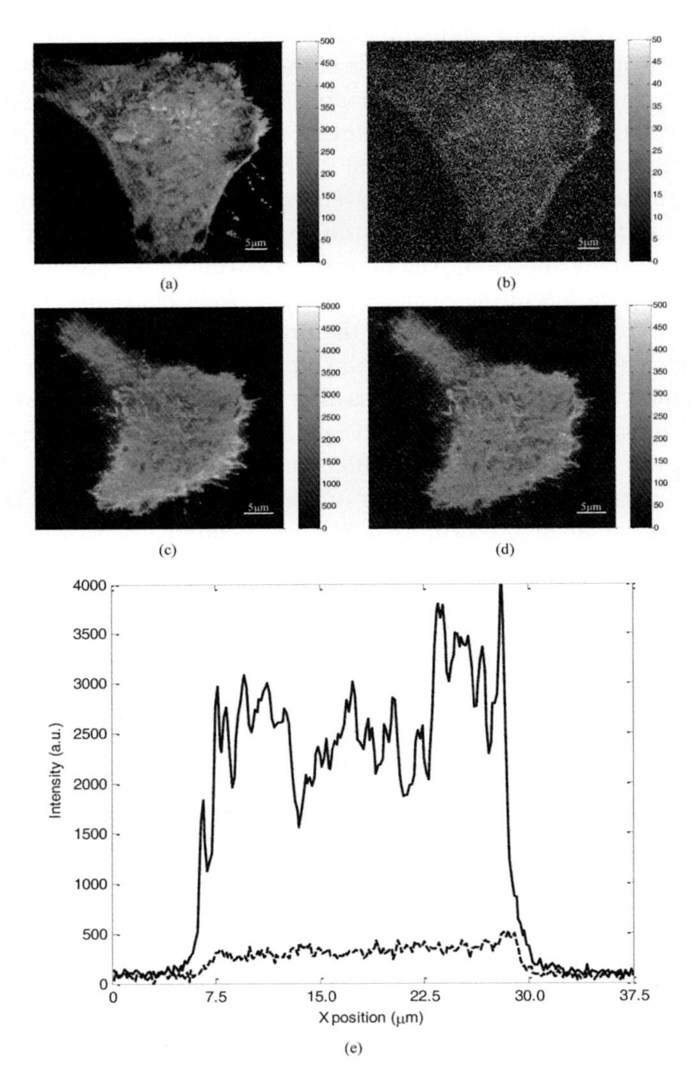

Figure 12.12 TIRF cell membrane images individually exposed for 0.5 s (a) and 0.04 s (b) and SPE-TIRF images individually exposed for 0.5 s (c) and 0.04 s (d). (e) The distribution of fluorescent intensity at various X positions in (a) and (c) with a crosscut line.

In comparison with Fig. 12.12a, the SPEF is observed clearly in Fig. 12.12c. Through exposure times of 0.5 s and 0.04 s, the fluorescence excited by SPs in Fig. 12.12d still can be seen. But the fluorescence excited by TIR in Fig. 12.12b is not as obvious. Also, in comparing Figs. 12.12a and 12.12d, the intensity of TIRF exposed for 0.5 s is almost equal to that of SPEF exposed for only 0.04 s. By using SPs to enhance fluorescence intensity, the imaging frame rate can be improved by roughly 1 order. So, it can handle more kinetic and fast imaging information through using SPs to enhance fluorescence. Figure 12.12e plots the fluorescence intensity versus the *x* position (μm), where the dashed and solid lines respectively represent TIRF and SPE-TIRF, which are from the central crosscut line on the cells in Figs. 12.12a and 12.12c. It also has been demonstrated that the fluorescence intensity can be enhanced by about 1 order from the SPs effect in comparison with that of excitation by the evanescent wave. Because of the variant distance between the cell membrane protein TM and the collagen-coated surface, different SP effects can be applied to interpret the phenomenon of fluorescence emission or quenching. Therefore, the SPE-TIRF microscope via a nanoscalar silver thin film was presented.

According to the stimulated results, we predicted an approximate 20-fold intensity enhancement of fluorescence excited by SPs in comparison with TIRF at the interface between the collagen and buffer. However, the experimental result shows only around 1 order of enhancement of SPEF. In a recent investigation, the fluorescence intensity decreased substantially as the fluorophore approached the metal surface. The decreased fluorescence signal is due to fluorescence quenching via nonradiative resonance energy transfer (RET) to the metal film and dissipates the excitation energy in the metal as heat [45, 46]. As a result of the distance-dependent nature of the RET phenomenon, the SPEF signal exhibited distance-dependent characteristics and the quenching efficiency decreased as an increase of the distance between the metal film and fluorophores. More specifically, the greater the separation of the fluorophores from the metal film, the lower the quenching efficiency [47]. With respect to the separation of metal film and fluorophores, the total thickness of the two layers of collagen and thiol is 6.6 nm, and together constitute a nonabsorbing dielectric layer. This dielectric layer separates the silver thin film and fluorophore-containing cell membrane and serves as a spacer; further, it decreases the phenomenon of fluorescence quenching. So, when cell adheres to the collagen layer, the fluorescence intensity may be decreased by fluorescence quenching.

12.4.2 Two-Photon Excited Fluorescence-Enhanced Imaging

As shown in Fig. 12.11, the optical configuration of the prism-coupled SPE one-photon and two-photon TIRF microscope is used to excite the SPs in order to enhance the local EM field and to increase the SNR of the fluorescence. A diode-pumped solid-state laser (20 mW, $\lambda = 473$ nm) and a 10 W solid-state pumped mode-locked femtosecond Ti:sapphire laser were adopted for one-photon excitation and two-photon excitation, respectively. Figures 12.13a and 12.13b present one-photon and two-photon excited TIRF images of a living COS-7 cell transfected with the eYFP-MEM construct cultured on the collagen-coated TIRF chips. The excitation wavelength of two-photon excited TIRF is roughly twice compared with one-photon excited TIRF; nevertheless, the effective excitation depths of both evanescent fields are comparable due to the two-photon fluorescence is governed by the quadratic EM field intensity. Moreover, two-photon excited TIRF has lower scattering due to its longer excitation wavelength and lower excitation sensitivity to scattering photons due to nonlinear excitation. As a result, the absence of scattered excitation for the two-photon excited TIRF results in a low-background-noise and high-contrast living cell membrane image compared with one-photon excited TIRF. Specifically, the strong fluorescence scattering located at the right-hand side of Fig. 12.13a decreases the SNR of the one-photon excited TIRF image, in which the wave propagates from left to right.

Figures 12.14a and 12.14b show the one-photon and two-photon excited SPE-TIRF images of the living COS-7 cell transfected with the eYFP-MEM construct cultured on the collagen-coated SPE-TIRF chips. The exposure times of the images in Figs. 12.13b and 12.14b are 0.5 s and 0.1 s, the illuminating laser powers are 50 mW and 25 mW, while the corresponding fluorescence intensity scales

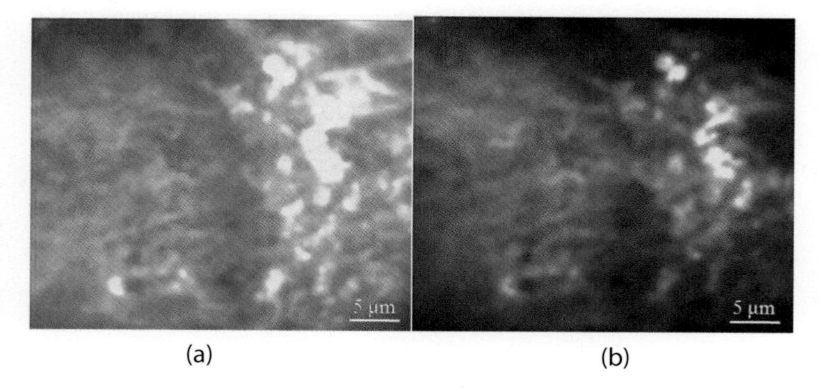

Figure 12.13 Live-cell membrane images utilizing (a) one-photon and (b) two-photon excited TIRF. The evanescent waves propagate from left to right for both figures.

are 950 and 3100 (arbitrary units), respectively. In comparison with images from the two-photon excited TIRF, two-photon excited SPE-TIRF requires about one-fifth of the exposure time and half the illuminating laser power, yet produces a threefold gain in the fluorescence intensity. Consequently, by enhancing the local EM field excited by the ATR method, the SPs introduce a fluorescence gain 30 times higher in the two-photon excited SPE-TIRF image (Fig. 12.14b) than that provided by the two-photon excited TIRF image (Fig. 12.13b). The greater the fluorescence intensity and the higher the SNR value of the image, the shorter the required exposure time and hence the faster the achievable frame rate. In addition to enhancing two-photon fluorescence with two-photon excited SPE-TIRF microscopy, the excitation volume can be efficiently confined by nonlinear two-photon excitation and the SP wave (Fig. 12.14). The cells were observed after 30 min exposure, which revealed that their morphologies and fluorescence intensities changed only slightly with SPE-TIRF microscopy. As such, the cell can be expected to survive under the microscopy. Undoubtedly, two-photon excited SPE-TIRF microscopy provides living cell membrane images with the highest SNR.

The simulation results suggest that the local field intensity of two-photon excited SPE-TIRF offers an approximately 50-fold

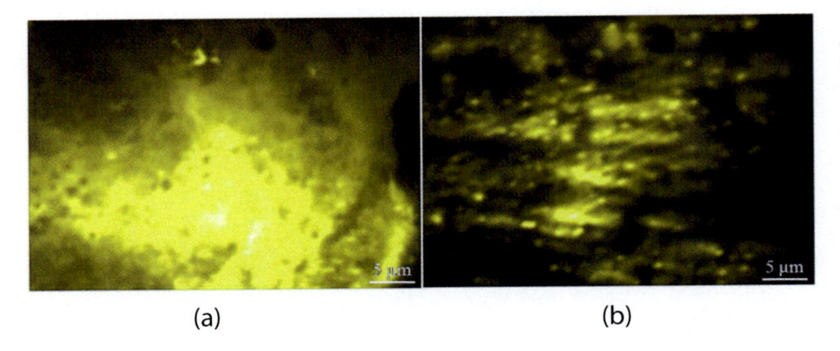

Figure 12.14 (a) One-photon and (b) two-photon excited SPE-TIRF live-cell membrane images. The SP waves propagate from left to right for both figures.

enhancement over two-photon excited TIRF at the interface between the collagen and the buffer. The two-photon fluorescence intensity of fluorophores in free space is theoretically possible according to the quadratic field intensity [25]. However, the experimental results show that the overall fluorescence intensity enhancement on the silver surface is only 30 times in practice. The two-photon fluorescence intensity enhancement is based on at least three factors: (1) the quadratic dependence of the field intensity enhancement via SPs, (2) the emission efficiency of the fluorophores on the metal surface, and (3) the collection efficiency due to SPCE. The emission characterization of the fluorophores on the metal surface can be described as dipole vibrations. Due to the thin-film structure, molecular vibration is affected, resulting in the emission characterization being changed [48]. Furthermore, the SPE-TIRF signal can exhibit distance-dependent characteristics; therefore, the quenching efficiency varies as a function of the distance between the metal film and the fluorophores [49]. Moreover, the detection of a fluorophore is usually limited by its quantum yield and photostability [50]. The two-photon fluorescence intensity (quantum yield) is determined not only by the quadratic dependence of the excitation SP field intensity but also by the quenching influence of the molecular dipole–metal interaction [51]. To compensate for the competing effects, a 6.1 nm thick spacer is placed between the silver layer and the fluorophore-containing cell membrane, and is used to reduce the effect of metallic surface-induced fluorophore quenching at very short distances (approximately 0–5 nm) [52]. However, the key detection issue in fluorescence microscopy is not merely how to increase the quantum yield, but rather how to collect as many photons from the fluorophores as possible before photobleaching occurs. Although the quenching effect reduces the fluorescence emission, it shortens its lifetime to enhance photostability [1]. Therefore, the quantum yield and photostability of fluorophores should be modified and controlled in order to improve the detection limit [1, 50]. This study has demonstrated that proximity to metallic surfaces within 10 nm can increase the two-photon fluorescence intensity more than 30 times. In contrast to the relatively constant radiative decay rate via an increase of incident laser power, the radiative decay rate can be increased by placing the fluorophores at suitable distances from metallic surfaces. An increase in the radiative decay rate results in an increased fluorescence intensity and a reduction in lifetime [49]. Therefore, the fluorescence intensity is enhanced and the fluorophore photostability is improved by the SP excitation.

12.4.3 Fluorescence Enhancement and Quenching

In this section, we investigate theoretically and experimentally that two-photon excited fluorescence is enhanced and quenched via SPs excited by TIR with a silver film. The fluorescence intensity is fundamentally affected by the local EM field enhancement and the quantum yield change according to the surrounding structure and materials. By utilizing the Fresnel equation and classical dipole radiation modeling, the local electric field enhancement, fluorescence quantum yield, and fluorescence emission coupling yield via SPs were theoretically analyzed at different dielectric spacer thicknesses between the fluorophore and the metal film.

In our experimental configuration, the emission wavelength is assumed to be approximately 530 nm for Alq3 [53]. The initial quantum yield Q_0 is 0.3, while the initial fluorescence lifetime is 12.0 ns. The refractive indexes of the SPE-TIRF configuration are 1.5196 (BK7), $0.1294 + j3.1772$ (Ag), 1.4608 (SiO_2), 1.7023 (Alq3), and 1.0 (Air) at an emission wavelength of 530 nm. Figures 12.15a

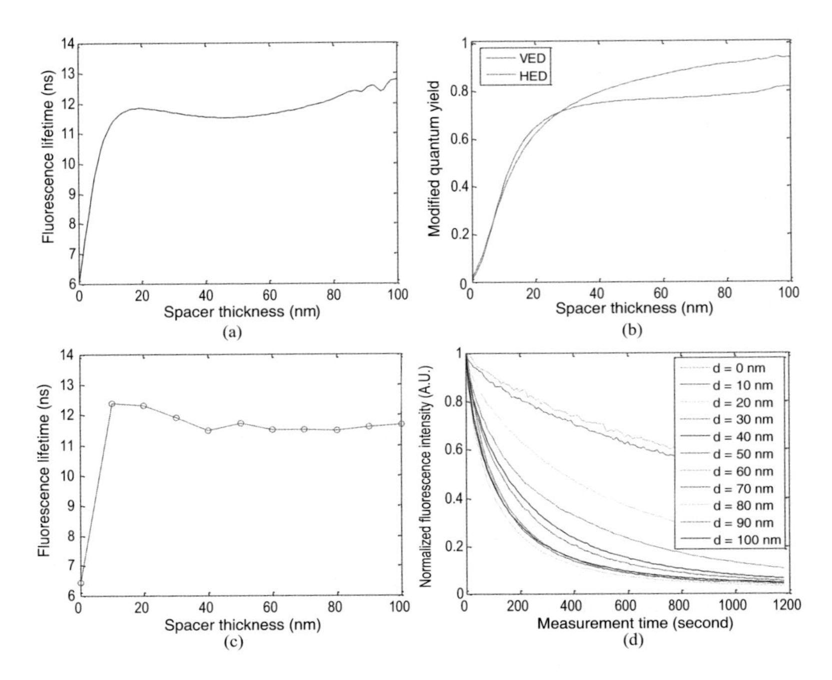

Figure 12.15 Variations of (a) the fluorescence lifetime and (b) modified quantum yield as a function of dielectric spacer thickness from theoretical analysis, and experimental results for the fluorescence (c) lifetime and (d) photostability as a function of spacer thickness.

and 12.15b exhibit the fluorescence lifetime and modified quantum yield with various spacer thicknesses from 0 to 100 nm. The fluorescence lifetime is directly quenched down to less than 9.0 ns via the metal film without the SiO_2 spacer but rises quickly with increasing spacer thickness and approaches 12.0 ns when the spacer thickness is greater than about 10 nm. Similarly, the modified quantum yields rapidly increase when the spacer thickness is less than about 40 nm but rise slowly thereafter. The nonradiated energy is greatly increased via the SPs on the interface and the quenching effect by the silver film when the fluorescent dye is near the silver film. Hence, the fluorescence lifetime and modified quantum yield affected by the quenching effect are decreased. When the spacer thickness is increased, the nonradiated power rapidly lowers and the quenching effect is also reduced. The fluorescence lifetime and modified quantum yield are increased with a thicker spacer thickness. Furthermore, the fluorescence lifetime is estimated at 13.8 ns and the modified quantum yield is set to the standard of 1.0 in the conventional TIRF configuration by using the above theoretical analysis.

Figure 12.15c shows the measured fluorescence lifetime versus different spacer thicknesses in SPE-TIRF excitation, for which the laser power is fixed at 5 mW. The fluorescence lifetime is similar to the theoretical prediction in Fig. 12.15a. The lifetime is extended up to 12.0 ns when the thickness of the spacer is greater than 10 nm. Furthermore, the fluorescence lifetime is also measured from the TIRF excitation. The fluorescence lifetime is 13.9 ns and 13.8 ns for the measurement and the theoretical prediction, respectively. Therefore, the theoretical modeling is appropriate to estimate the actual lifetime in the SPE-TIRF and conventional TIRF configurations. The experimental results and the theoretical simulation both demonstrate that the fluorescence lifetimes and quantum yields are seriously influenced by the metal surface when the fluorescent dye is near the surface. Figure 12.15d shows that the two-photon excited fluorescence intensity decays with time when the laser power is at 15 mW. These experimental results demonstrate that fluorescence intensity decays rapidly when the spacer thickness is increased to greater than 20 nm, and therefore indicates that better photostability can be provided when the spacer thickness is less than 20 nm. After the 10 nm thickness, the photostability quickly decreases as the spacer thickness increases; consequently, the quenching effect dominates the characteristics of the two-photon excited fluorescence lifetime and quantum yield when the spacer thickness is thinner than 10 nm. Although the quenching effect reduces the fluorescence emission, it shortens its lifetime to enhance fluorophore photostability.

For two-photon excited fluorescence, the fluorescence emission intensity is proportional to the modified quantum yield, the emission coupling yield and square of the excitation power and local electric field enhancement factor. The overall fluorescence enhancement factor is defined as the two-photon excited fluorescence intensity in the SPE-TIRF configuration normalized by the intensity in the conventional TIRF configuration. The maximum enhancement factor in the experimental results approached 30-fold at the spacer thickness of 30 nm; however, the fluorescence enhancement reached only 40% of the maximum (70-fold) based on the theoretical simulation. Hence, some other effects that influence the two-photon excitation of the fluorescent dye in the SPE-TIRF configuration are apparently not considered in the theoretical analysis. We believe that the two-photon absorption cross section of the fluorescent dye differs depending on configuration. Based on this assumption, the two-photon absorption cross section in the conventional TIRF configuration is about 2.5 times (70-fold/30-fold) higher than that in the SPE-TIRF configuration. The maximum enhancement factor of 30-fold is attained with a 30 nm spacer, but better fluorophore photostability can be achieved when the spacer thickness is decreased to less than 20 nm (Fig. 12.15d). As such, a compromise between the fluorescence enhancement and the fluorophore photostability must be made. We found that the SPE-TIRF configuration with an SiO_2 spacer of 10 nm not only clearly reveals an enhanced fluorescent signal compared to that of conventional TIRF but also significantly improves the fluorescence photostability compared to that of a less quenching setup with a thicker spacer.

12.5 Combination of SPR Phase and Fluorescence-Enhanced Imaging

SPR phase microscopy with a common-path PSI technique can provide high-sensitivity phase information with long-term stability. Simultaneously, fluorescence microscopy with the enhancement of a local EM field can supply bright fluorescent images. Herein, the combination for wide-field SPR phase microscopy and SPE-TIRF microscopy to simultaneously image living-cell contacts on the surface of a biosubstrates is shown in Fig. 12.16 [29]. The SP phase and fluorescence images were mainly obtained by utilizing the diode-pumped solid-state laser light source (20 mW, $\lambda = 473$ nm) to excite SPs on the thin silver film. A 632.8 nm He-Ne laser which can be switched by a flipper mirror was incorporated. The laser beam

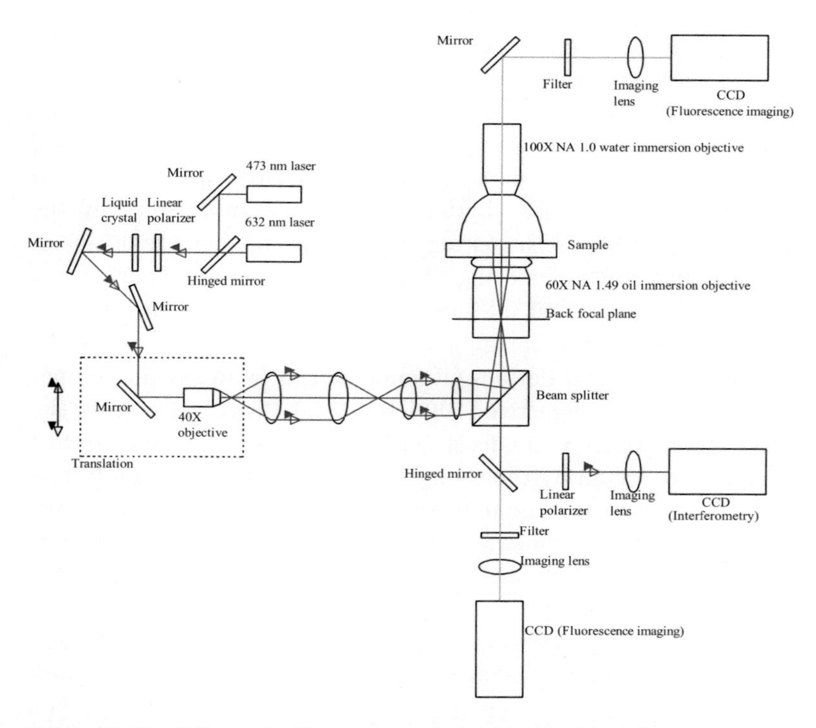

Figure 12.16 Schematic illustration of experimental configuration employed for simultaneous live cell–substrate contact imaging using a combination of a wide-field oil-immersion objective based on SP phase microscopy and SP-enhanced fluorescence microscopy.

passes through two linear polarizers to control the polarization and intensity. The liquid crystal adjusts the phase delay between the P-wave and S-wave for the PSI, as aforementioned. Then, the light is focused onto the BFP of a high-NA oil-immersion objective (60X, NA = 1.49, Nikon) through another objective (40X, Nikon) and a relay lens pair; finally, the light emerges from the objective as a parallel beam. The incident angle is controlled by means of adjusting the focusing spot position on the BFP through a linear translation stage. A 0.17 mm cover slide (BK7) coated with a metal film via an RF sputtering deposition process is coupled into the oil-immersion objective by adding index-matching oil. The setup's range of available excitation angles is up to 79.5°. The SPs are excited and the reflection from the metal film enters the objective, and then passes through a beam splitter, a linear polarizer, and an imaging lens; finally, the SPs are imaged onto a regular CCD camera. The linear polarizer is adjusted at a suitable angle between its optical axis and the incident plane in order to ensure the P-wave and S-wave interference with better contrast. In comparison with the 50 nm thin silver film, a coating of 30 nm is an adequate thickness for simultaneously achieving SP phase imaging and fluorescence imaging with the 1.49 oil-immersion NA objective. Concurrently, the SP fluorescence imaging of living cells is also able to be achieved by using a bottom oil-immersion objective and an upper water-immersion objective (100X, NA = 1.0, Olympus) as well. The eGFP fluorescence excited via SPs from the cell-substrate contact region is collected by the oil-immersion objective and the water-immersion objective, and individually imaged onto two high-speed-frame-rate EMCCD cameras (Luca and iXon DV885, Andor) after passing through band-pass filters (BPF, $\lambda = 500$–535 nm, Semrock) and imaging lenses.

Recent investigations have supported the contention that thrombomodulin, an integral membrane glycoprotein, plays an important role in extravascular activities [54] and thrombomodulin-mediated cell adhesion [55]. In Figs. 12.17a and 12.17b, the SPE-TIRF imaging of a living melanoma-GFP-tagged thrombomodulin cell

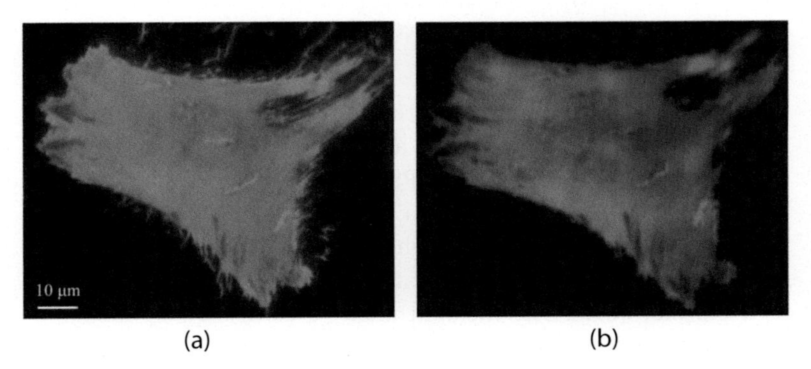

Figure 12.17 Live melanoma-GFP-tagged thrombomodulin SPE-TIRF images utilizing (a) the 1.49 NA oil-immersion objective and (b) the 1.0 NA water-immersion objective, both with an exposure time of 0.5 s at an incident angle of 77.5°.

can be realized at the incident angle of 77.5° associated with the maximum fluorescence enhancement by making use of the bottom oil-immersion objective through the thinner silver film as well as the upper water-immersion objective [29]. The fluorescence signal collected from the bottom objective is quenched when a thicker sliver film is adopted. A five times brighter fluorescent live-cell image via SPs can be observed by the bottom oil-immersion objective based on the appropriate microscopy setup. Compared to the observation of melanoma-GFP-tagged thrombomodulin cells through the culture medium with the water-immersion objective, better collection efficiency and spatial resolution of SPE-TIRF images can be obtained via the oil-immersion objective with a higher NA value, as shown in Figs. 12.17a and 12.17b. The decrease in silver film thickness solves the two problems regarding imaging in fluid via the 1.49 NA oil-immersion objective and improvement in the lateral resolution. Moreover, a simultaneous observation of phase image can be made with this setup.

A candidate tumor suppressor WW domain-containing oxi-doreductase, known as marine WOX1 or WWOX, and monkey kidney COS7 fibroblasts are used to demonstrate the imaging capabilities of the developed microscopy [56]. WOX1 proteins play an important role in the regulation of a wide variety of cellular functions, such as protein degradation, transcription, and RNA splicing. Figures 12.18a, 12.18b, and 12.18c demonstrate the simultaneous fluorescence and phase imaging of a single living COS7 fibroblast transiently transfected with the eGFP-WOX1 construct cultured on the collagen-coated SPE-TIRF chip based on the developed SPE-TIRF and SPR phase microscopy. In contrast to the flat adhesion of the melanoma-GFP-tagged thrombomodulin cell illustrated in Figs. 12.17a and 12.17b, the expression of WOX1 proteins, which possesses two N-terminal WW domains (containing conserved tryptophan residues), a nuclear localization sequence, and a C-terminal short-chain alcohol dehydrogenase/reductase domain (which contains a mitochondria-targeting sequence) tagged with eGFP, is imaged with the punctuated adhesion of the COS7 fibroblast near its plasma membrane area, as shown Fig. 12.18a. Also, it presents the localization to the perinuclear region and mitochondria of the WOX1-containing clusters. Figures 12.18b and

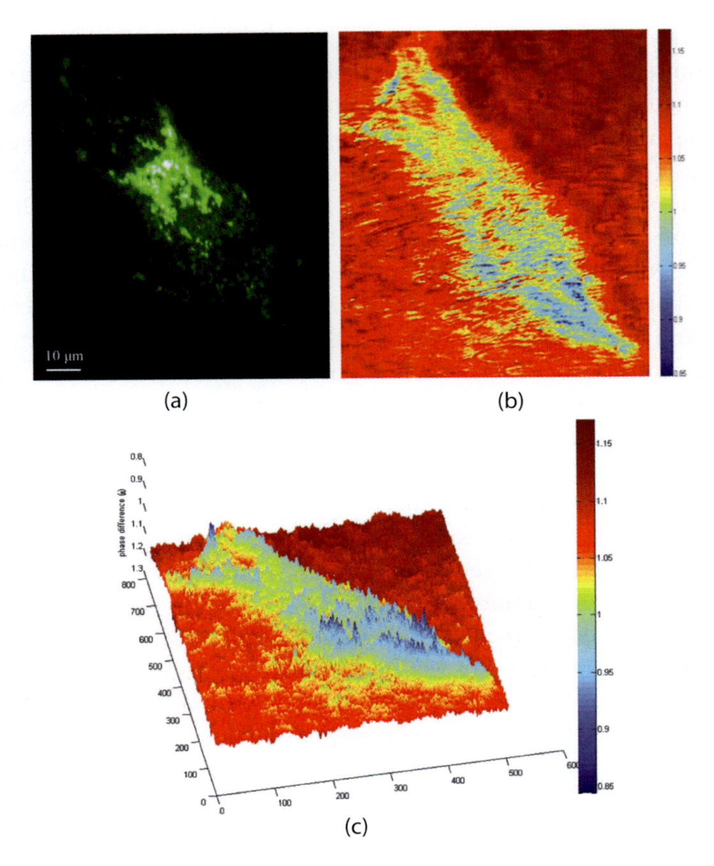

Figure 12.18 Simultaneous (a) SPE-TIRF image, (b) 2D SP phase image, and (c) its 3D phase image counterpart of a living COS7 fibroblast transfected with an eGFP-WOX1 construct. The color bars indicate the phase difference in π.

12.18c show the simultaneously acquired 2D and 3D SPR phase images of the same cell, the phase difference of which between the areas with and without the cell is seen to be approximately 0.35 π. The regions with large phase differences indicate the localities that are directly in contact with the biosurface and the areas of greater density, which previous research has explained reflect where the contraction of microfilaments in the cytoskeleton have sustained the adhesion of the cell [57]. Contrary to the dense perinuclear localization of the WOX1-containing clusters, the SPR phase image corresponding to the cell's footprint exhibited a greater phase difference resulting from shorter cell–substrate distance and greater density in the cell cytoplasm close to the membrane. Overall, this chapter demonstrated SPR phase microscopy, SPE-TIRF microscopy, and the combination microscopy. The cell–substrate interaction can be qualitatively illustrated through the SPE-TIRF and SPR phase images, as well as via comparisons of the images.

References

1. Raether, H. (1988). *Surface Plasmons on Smooth and Rough Surfaces and on Gratings* (Springer-Verlag, Berlin, Heidelberg).
2. Homola, J., and Yee, S. S. (1999). Surface plasmon resonance sensors: review, *Sens. Actuators B*, **54**, pp. 3–15.
3. Ligler, F. S., and Taitt, C. A. R., ed. (2002). *Optical Biosensors: Present and Future* (Elsevier, NY, USA).
4. Chien, F.-C., and Chen, S.-J. (2004). A sensitivity comparison of optical biosensors based on four different surface plasmon resonance modes, *Biosens. Bioelectron.*, **20**, pp. 633–642.
5. Stenberg, E., Persson, B., Roos, H., and Urbaniczky, C. (1991). Quantitative determination of surface concentration of proteins with surface plasmon resonance using radiolabeled protein, *J. Colloid Interface Sci.*, **143**, pp. 513–526.
6. Nikitin, P. I., Grigorenko, A. N., Beloglazov, A. A., Valeiko, M. V., Savchuk, A. I., Savchuk, O. A., Steiner, G., Kuhne, C., Hucbner, A., and Salzer, R. (2000). Surface plasmon resonance interferometry for microarray biosensing, *Sens. Actuators A*, **85**, pp. 189–193.
7. Nenninger, G. G., Homola, J., Yee, S. S., and Tobiska, P. (2001). Long-range surface plasmons for high-resolution surface plasmon resonance sensors, *Sens. Actuators B*, **74**, pp. 145–151.
8. Salamon, Z., and Tollin, G. (2001). Optical anisotropy in lipid bilayer membranes: coupled plasmon-waveguide resonance measurements of molecular orientation, polarizability, and shape, *Biophys. J.*, **80**, pp. 1557–1567.
9. Chen, S.-J., Chien, F. C., Lin, C. Y., and Lee, K. C. (2004). Enhancement of the resolution of surface plasmon resonance biosensors by control

of the size and distribution of nanoparticles, *Opt. Lett.*, **29**, pp. 1390–1392.

10. Su, Y.-D., Hsiu, F.-M., Tsou, C.-Y., Chen, Y.-K., and Chen, S.-J. (2005). Surface plasmon resonance phase-shift interferometry: real-time DNA microarray hybridization analysis, *J. Biomed. Opt.*, **10**, p. 034005.
11. Kabashin, A.V., and Nikitin, P. I. (1998). Surface plasmon resonance interferometer for bio- and chemical-sensors, *Opt. Commun.*, **150**, pp. 5–8.
12. Nikitin, P. I., Grigorenko, A. N., Beloglazov, A. A., Valeiko, M. V., Savchuk, A. I., Savchuk, O. A., Steiner, G., Kuhne, C., Huebner, A., and Salzer, R. (2000). Surface plasmon resonance interferometry for micro-array biosensing, *Sens. Actuators A*, **85**, pp. 189–193.
13. Notcovich, A. G., Zhuk, V., and Lipson, S. G. (2000). Surface plasmon resonance phase imaging, *Appl. Phys. Lett.*, **76**, pp. 1665–1667.
14. Su, Y.-D., Chen, S.-J., and Yeh, T.-L. (2005). Common-path phase-shift interferometry surface plasmon resonance imaging system, *Opt. Lett.*, **30**, pp. 1488–1490.
15. Su, Y.-D., Chiu, K.-C., Chang, N.-S., Wu, H.-L., and Chen, S.-J. (2010). Study of cell-biosubstrate contacts via surface plasmonpolariton phase microscopy, *Opt. Express*, **18**, pp. 20125–20135.
16. Axelrod, D. (2001). Total internal reflection fluorescence microscopy in cell biology, *Traffic*, **2**, pp. 764–774.
17. Toomre, D., and Manstein, D. J. (2001). Lighting up the cell surface with evanescent wave microscopy, *Trends Cell Biol.*, **11**, pp. 298–303.
18. Truskey, G. A., Burmeister, J. S., Grapa, E., and Reichert, W. M. (1992). Total internal reflection fluorescence microscopy (TIRFM) II. Topographical mapping of relative cell/substratum separation distances, *J. Cell Sci.*, **103**, pp. 491–499.
19. Sund, S. E., and Axelrod, D. (2000). Actin dynamics at the living cell submembrane imaged by total internal reflection fluorescence photobleaching, *Biophys. J.*, **79**, pp. 1655–1669.
20. Betz, W. J., Mao, F., and Smith, C. B. (1996). Imaging exocytosis and endocytosis, *Curr. Opin. Neurobiol.*, **6**, pp. 365–371.
21. Sailer, R., Strauss, W. S., Emmert, H., Stock, K., Steiner, R., and Schneckenburger, H. (2000). Plasma membrane associated location of sulfonated meso-tetraphenylporphyrins of different hydrophilicity probed by total internal reflection fluorescence spectroscopy, *Photochem. Photobiol.*, **71**, pp. 460–465.

22. Fort, E., and Gresillon, S. (2008). Surface enhanced fluorescence, *J. Phys. D: Appl. Phys.*, **41**, pp. 1–31.

23. Lin, C.-Y., Chiu, K.-C., Chang, C.-Y., Chang, S.-H., Guo, T.-F., and Chen, S.-J. (2010). Surface plasmon-enhanced and quenched two-photon excited fluorescence, *Opt. Express*, **18**, pp. 12807–12817.

24. He, R. Y., Chang, G. L., Wu, H. L., Lin, C. H., Chiu, K. C., Su, Y. D., and Chen, S.-J. (2006). Enhanced live cell membrane imaging using surface plasmon-enhanced total internal reflection fluorescence microscopy, *Opt. Express*, **14**, pp. 9307–9316.

25. Kano, H., and Kawata, S. (1996). Two-photon-excited fluorescence enhanced by a surface plasmon, *Opt. Lett.*, **21**, pp. 1848–1850.

26. Wenseleers, W., Stellacci, F., Friedrichsen, T. M., Mangel, T., Bauer, C. A., Pond, S. J. K., Marder, S. R., and Perry, J. W. (2002). Five orders-of-magnitude enhancement of two-photon absorption for dyes on silver nanoparticle fractal clusters, *J. Phys. Chem. B*, **106**, pp. 6853–6863.

27. Gryczynski, I., Malicka, J., Lakowicz, J. R., Goldys, E. M., Calander, N., and Gryczynski, Z. (2005). Directional two-photon induced surface plasmon-coupled emission, *Thin Solid Films*, **491**, pp. 173–176.

28. Anceau, C., Brasselet, S., Zyss, J., and Gadenne, P. (2003). Local second-harmonic generation enhancement on gold nanostructures probed by two-photon microscopy, *Opt. Lett.*, **28**, pp. 713–715.

29. He, R.-Y., Lin, C.-Y., Su, Y.-D., Chiu, K.-C., Chang, N.-S., Wu, H.-L., and Chen, S.-J. (2010). Imaging live cell membranes via surface plasmon-enhanced fluorescence and phase microscopy, *Opt. Express*, **18**, pp. 3649–3659.

30. He, R.-Y., Su, Y.-D., Cho, K.-C., Lin, C.-Y., Chang, N.-S., Chang, C.-H., and Chen, S.-J. (2009). Surface plasmon-enhanced two-photon fluorescence microscopy for live cell membrane imaging, *Opt. Express*, **17**, pp. 5987–5997.

31. Liu, H., Wang, B., Leong, E. S. P., Yang, P., Zong, Y., Si, G., Teng, J., and Maier, S. A. (2010). Enhanced surface plasmon resonance on a smooth silver film with a seed growth layer, *ACS Nano*, **4**, pp. 3139–3146.

32. Kolomenski, A., Kolomenskii, A., Noel, J., Peng, S., and Schuessler, H. (2009). Propagation length of surface plasmons in a metal film with roughness, *Appl. Opt.*, **48**, pp. 5683–5691.

33. Fontana, E., and Pantell, R. H. (1988). Characterization of multilayer rough surfaces by use of surface-plasmon spectroscopy, *Phys. Rev. B*, **37**, pp. 3164–3182.

34. Kapaklis, V., Poulopoulos, P., Karoutsos, V., Manouras, T., and Politis, C. (2006). Growth of thin Ag films produced by radio frequency magnetron sputtering, *Thin Solid Films*, **510**, pp. 138–142.
35. Yin, L., Vlasov, V. K. V., Pearson, J., Hiller, J. M., Hua, J., Welp, U., Brown, D. E., and Kimball, C. W. (2005). Subwavelength focusing and guiding of surface plasmons, *Nano Lett.*, **5**, pp. 1399–1402.
36. Yuan, H.-K., Chettiar, U. K., Cai, W., Kildishev, A. V., Boltasseva, A., Drachev, V. P., and Shalaev, V. M. (2007). A negative permeability material at red light, *Opt. Express*, **15**, pp. 1076–1083.
37. Chi, Y., Lay, E., Chou, T.-Y., Song, Y.-H., and Carty, A. J. (2005). Deposition of silver thin films using the pyrazolate complex [Ag(3,5-(CF3)2C3HN2)]3, *Chem. Vap. Deposition*, **11**, pp. 206–212.
38. Logeeswaran, V. J., Kobayashi, N. P., Islam, M. S., Wu, W., Chaturvedi, P., Fang, N. X., Wang, S. Y., and Williams, R. S. (2009). Ultrasmooth silver thin films deposited with a germanium nucleation layer, *Nano Lett.*, **9**, pp. 178–182.
39. Chiu, K.-C., Lin, C.-Y., Dong, C. Y., and Chen, S.-J. (2011). Optimizing silver film for surface plasmon-coupled emission induced two-photon excited fluorescence imaging, *Opt. Express*, **19**, pp. 5386–5396.
40. Chyou, J.-J., Chen, S.-J., and Chen, Y.-K. (2004). Two-dimensional phase unwrapping using a multichannel least-mean-square algorithm, *Appl. Opt.*, **43**, pp. 5655–5661.
41. Betz, W. J., Mao, F., and Smith, C. B. (1996). Imaging exocytosis and endocytosis, *Curr. Opin. Neurobiol.*, **6**, pp. 365–371.
42. Sund, S. E., and Axelrod, D. (2000). Actin dynamics at the living cell submembrane imaged by total internal reflection fluorescence photobleaching, *Biophys. J.*, **79**, pp. 1655–1669.
43. Reichert, W. M., and Truskey, G. A. (1990). Total internal reflection fluorescence (TIRF) microscopy. I. Modelling cell contact region fluorescence, *J. Cell Sci.*, **96**, pp. 219–230.
44. Schmoranzer, J., Goulian, M., Axelrod, D., and Simon, S. M. (2000). Imaging constitutive exocytosis with total internal reflection fluorescence microscopy, *J. Cell Biol.*, **149**, pp. 23–32.
45. Knobloch, H., Brunner, H., Leitner, A., Aussenegg, F., and Knolla, W. (1993). Probing the evanescent field of propagating plasmon surface polaritons by fluorescence and Raman spectroscopies, *J. Chem. Phys.*, **98**, pp. 10093–10095.

46. Anger, P., Bharadwaj, P., and Novotny, L. (2006). Enhancement and quenching of single-molecule fluorescence, *Phys. Rev. Lett.*, **96**, p. 113002.

47. Ekgasit, S., Yu, F., and Knoll, W. (2005). Displacement of molecules near a metal surface as seen by an SPR-SPFS biosensor, *Langmuir*, **21**, pp. 4077–4082.

48. Sullivan, K. G., and Hall, D. G. (1997). Enhancement and inhibition of electromagnetic radiation in plane-layered media. II. Enhanced fluorescence in optical waveguide sensors, *J. Opt. Soc. Am. B*, **14**, pp. 1160–1166.

49. Aslan, K., Gryczynski, I., Malicka, J., Matveeva, E., Lakowicz, J. R., and Geddes, C. D. (2005). Metal-enhanced fluorescence: an emerging tool in biotechnology, *Curr. Opin. Biotechnol.*, **16**, pp. 55–62.

50. Huang, Z., and Thompson, N. L. (1993). Theory for two-photon excitation in pattern photobleaching with evanescent illumination, *Biophys. Chem.*, **47**, pp. 241–249.

51. Stranik, O., McEvoy, H. M., McDonagh, C., and MacCraith, B. D. (2005). Plasmonic enhancement of fluorescence for sensor applications, *Sens. Actuators B*, **107**, pp. 148–153.

52. Liebermann, T., and Knoll, W. (2000). Surface-plasmon field-enhanced fluorescence spectroscopy, *Colloids Surf. A*, **171**, pp. 115–130.

53. Kishore, V. V. N. R., Narasimhan, K. L., and Periasamy, N. (2003). On the radiative lifetime, quantum yield and fluorescence decay of Alq in thin films, *Phys. Chem. Chem. Phys.*, **5**, pp. 1386–1391.

54. Boffa, M. C., Burke, B., and Haudenschild, C. C. (1987). Preservation of thrombomodulin antigen on vascular and extravascular surfaces, *J. Histochem. Cytochem.*, **35**, pp. 1267–1276.

55. Huang, H. C., Shi, G. Y., Jiang, S. J., Shi, C. S., Wu, C. M., Yang, H. Y., and Wu, H. L. (2003). Thrombomodulin-mediated cell adhesion, *J. Biol. Chem.*, **278**, pp. 46750–46759.

56. Chang, N. S., Hsu, L. J., Lin, Y. S., Lai, F. J., and Sheu, H. M. (2006). WW domain-containing oxidoreductase: a candidate tumor suppressor, *Trends Mol. Med.*, **13**, pp. 12–22.

57. Lee, K. H., Su, Y.-D., Chen, S.-J., Tseng, F. G., and Lee, G. B. (2007). Microfluidic systems integrated with two-dimensional surface plasmon resonance phase imaging systems for microarray immunoassay, *Biosens. Bioelectron.*, **23**, pp. 466–472.

Index